CdTe and CdZnTe Materials

Kris Iniewski
Editor

CdTe and CdZnTe Materials

Material Properties and Applications

Springer

Editor
Kris Iniewski
Redlen Technologies
Port Moody, BC, Canada

ISBN 978-3-031-64520-4 ISBN 978-3-031-64521-1 (eBook)
https://doi.org/10.1007/978-3-031-64521-1

This Springer imprint is published by the registered company Springer Nature Switzerland AG
The registered company address is: Gewerbestrasse 11, 6330 Cham, Switzerland

Contents

About the Editor

Krzysztof (Kris) Iniewski is a director of detector architecture and applications at Redlen Technologies Inc., a detector company based in British Columbia, Canada, owned by Canon Inc in Japan. During his tenure at Redlen he has managed the development of highly integrated Cadmium Zinc Telluride (CZT) detector products in medical imaging and security applications. Prior to Redlen, Kris held various management and academic positions at PMC-Sierra, University of Alberta, Simon Fraser University (SFU), University of British Columbia (UBC), and University of Toronto. Dr. Iniewski has published over 150+ research papers in international journals and conferences. He holds 25+ international patents granted in the USA, Canada, France, Germany, and Japan. He wrote and edited 75+ books for Wiley, Cambridge University Press, McGraw-Hill, CRC Press, and Springer. He is a frequent invited speaker and has consulted multiple organizations internationally.

Introduction to CdTe/CZT Physics and Applications

Conny Hansson and Kris Iniewski

1 Introduction

There have been a large number of technological advances in cadmium telluride (CdTe) and cadmium zinc telluride (CdZnTe or CZT) materials recently, and a new generation of advanced applications using these materials are poised to enter the commercial market. The main application area in the field of X-ray detection is provided by direct-conversion-detector-based CdTe and CdZnTe sensors. Direct conversion quantum counting detectors improve performance in several diverse ways. First, the electronic noise in quantum counting can be lowered because of the thresholding effect. Second, quantum counting can provide better energy weighting. Rather than increasing the weights of higher-energy photons because of the integrating effect, it is possible to increase the weights of lower-energy photons with better contrast. Third, the fill factor of these detectors can be higher than in energy-integrating detectors as antireflection septa are eliminated.

Photon-counting-detector-based X-ray computed tomography (CT) has the potential to not only improve current medical images but also enable new clinical applications. High-impedance (high-Z) materials like CdTe and CZT offer the best possibility for the implementation of direct conversion detectors conversion detectors and are the subject of this book. We discuss material options and their characteristics, detector operation physics and technology and readout integrated circuits required to detect signals processes by CdTe and CdZnTe sensors.

CdTe and CdZnTe materials are extensively used in medical imaging, industrial testing, and security applications, with medical imaging being the largest application. X-ray computed tomography (CT) is not the only application of CdTe and

C. Hansson
SLAC National Accelerator Laboratory, Menlo Park, CA, USA

K. Iniewski (✉)
Redlen Technologies, Saanichton, BC, Canada

K. Iniewski (ed.), *CdTe and CdZnTe Materials*,
https://doi.org/10.1007/978-3-031-64521-1_1

CdZnTe technologies as they are also used in other branches of medicine, for example, in nuclear medicine. Nuclear medicine tests are different from most imaging techniques in that the main function of the diagnostic test is to show the physiological function of the system being investigated, unlike traditional anatomical imaging such as magnetic resonance imaging (MRI) or CT. Nuclear medicine imaging studies are typically more organ or tissue specific (e.g., lung scan, heart scan, bone scan, brain scan, etc.) than those on traditional radiology imaging, which instead focus on a specific section of the body (e.g., chest X-ray, abdomen/pelvis CT scan, head CT scan, etc.).

This chapter provides an overview of the CdTe/CZT material properties, device physics, and readout electronics. Many of the topics described here are discussed in more depth in the book chapters that follow and have been covered in recent book publications [1–10].

CdTe and CdZnTe (CZT) are not well-known compound semiconductor materials used in X-ray and gamma-ray detection applications. The purpose of the chapter is to introduce these interesting materials. We will start this chapter with a review of the basic material properties, followed by a discussion on radiation detection principles of operation. The chapter will end with a description of radiation detector readout electronics and application examples.

2 CdTe and CZT Material Properties

2.1 *Crystal Structure*

Both CdTe and CdZnTe (CZT) crystals have a zinc-blende structure, which is comprised of two interpenetrating face-centered cubic (FCC) lattices of cadmium (Cd) and tellurium (Te) atoms (when zinc is used in CZT, zinc atoms substitute cadmium's positions). One of these is displaced one-quarter of the way along the main diagonal of the other, as shown in Fig. 1. This displacement of the sublattice results in an asymmetrical structure of the crystal.

Most crystal properties (mechanical, chemical, magnetic, and electrical) are orientation dependent. For CZT, crystal growth is faster in the <111> direction as compared to other orientations. Cleavage is faster in <110> as compared to the other ones. Etching is faster in <100> as compared to others. These properties depend on both the density of atoms and atomic bond types.

Besides the quality of the bulk crystal, the properties of the surfaces and metal/semiconductor interfaces are often dominant factors influencing detector performance. Various surface treatments, such as sputtering, annealing, and mechanical and chemical processing, change surface morphology and induce various defects, affecting in turn the electronic structure and hence transport properties. There is a wealth of information available in the literature on both the theoretical and experimental aspects of the importance of crystal surfaces and their influence [1, 4].

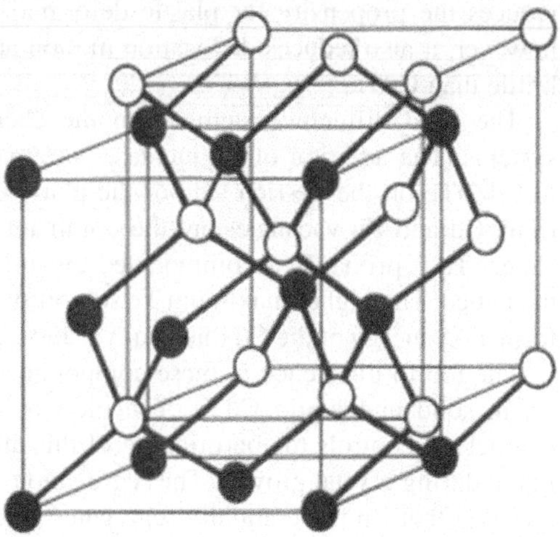

Fig. 1 CdTe/CdZnTe crystal structure. The white circles indicate Te atoms, while the black circles indicate Cd atoms. Zn atoms when present substitute Cd atoms

The first difference between CZT and CdTe is obviously zinc (Zn in CdZnTe). Historically, CdTe was grown first since growing a binary (two components) semiconductor compound is easier than growing a ternary (three components) one. The addition of the third component provides extra benefits, which are explained below. Today, there are attempts to add a fourth component, i.e., selenium, for material growth (CdZnTeSe) or to replace zinc with magnesium (CdMnTe), both highly experimental and far from maturity, so we will not be covering those early research attempts here.

The introduction of zinc to the CdTe lattice makes important changes to the material's properties. CdZnTe is an alloy of CdTe and ZnTe, with typically about 10% Zn alloyed in the ternary compound for detector applications. Alloying with Zn causes several changes to the properties of CdTe, as described below. There are other applications of CdZnTe that use a lower amount of zinc (4%) to prepare mechanical substrates for infrared (IR) detector film growth; this is unrelated to X-ray detection and will not be discussed here as it is a niche application of CZT.

By adding Zn to CdTe, the band gap is increased. The wider band gap enables a higher maximum resistivity of the ternary compound. For CdZnTe with 10% Zn, the band gap increases from 1.5 eV to 1.6 eV, and the maximum achievable resistivity increases by a factor of three, typically from 2×10^{10} Ohm-cm for CdTe to 6×10^{10} Ohm-cm for CdZnTe. It is to be noted, however, that these are only the maximum resistivity limits allowed by the band gap of the material, values that are not always achieved in practice because of incomplete electrical compensation.

The second difference between CdTe and CZT comes from the different chemical properties of CdTe and ZnTe. ZnTe has a lower iconicity and a higher binding energy than CdTe, and the bond length is shorter in ZnTe. Thus, the CdTe lattice is strengthened by the incorporation of Zn, leading to an increase in the shear modulus and solution hardening of the ternary compound. The solution hardening of CdZnTe reduces the propensity for plastic deformation and the formation of dislocations; however, it also reduces dislocation motion and makes the ternary compound more brittle than CdTe.

The third difference stems from the chemical potentials of the two material systems. The addition of Zn increases the maximum deviation from stoichiometry in CdZnTe on the Te-rich side of the phase diagram. Because, in thermal equilibrium, Cd and Zn vacancies are the dominant native defects in CdTe and CdZnTe, excess Te is primarily accommodated by an increase in the number of vacancies in the lattice. The higher maximum Te solubility in CdZnTe therefore indicates that the formation energy of the Cd (and Zn) vacancy is reduced in CdZnTe relative to CdTe.

The fourth difference in these compounds stems from the different solubility of Zn in solid and liquid CdTe. The ratio of the solubility, called the segregation coefficient, controls the partitioning of this alloying element between the solid and liquid during crystal growth. The segregation coefficient of Zn is about 1.3 in CdTe, meaning that Zn preferentially segregates at the solid-liquid interface into the solid rather than the liquid. This leads to an axial Zn concentration distribution in CdZnTe produced by melt growth techniques. For a 10% Zn-doped CdZnTe, the actual Zn concentration starts at about 14% in the first-to-freeze section of the ingot and decreases to about 6% in the last-to-freeze section. With an appropriate design of the solvent and feed material, constant Zn concentrations can be achieved with solution growth techniques such as the traveling heater method (THM).

2.2 Impurities and Defects

As in all semiconductor materials, there are two types of impurities in CZT: intentional, both extrinsic and intrinsic, and unintentional. The intentional extrinsic process, called doping, is achieved by introducing impurities during the fabrication process. This can be accomplished during crystal growth or after, at the wafer fabrication stage, by means of ion implantation or diffusion. In the doping process, impurities may take either interstitial or substitutional sites by displacing the host atom from its lattice position.

Intentional intrinsic impurities happen during crystal growth due to deviation from stoichiometric composition and/or from equilibrium processing conditions. During this process, the host atoms or impurities occupy an interstitial or another substitutional site, resulting in the formation of lattice defects such as vacancies (cadmium vacancy (V_{Cd}), zinc vacancy (V_{Zn}), and tellurium (V_{Te})), interstitials (Te or Cd atoms in the spaces between lattice positions), antistites (for example, Te at the Cd site), and their complexes. Some of these defects create donor traps while the other ones acceptor traps. The resultant lattice and impurity defects give rise to localized energy levels within the band gap. These defects are classified as shallow or deep level, depending on their energy position in the band gap. These levels are also sometimes referred to as trap levels and/or recombination centers.

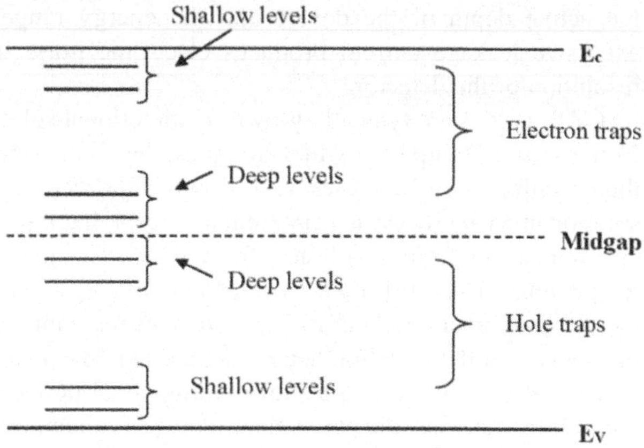

Fig. 2 Schematic representation of traps within the bandgap of CZT

Finally, unintentional impurities are introduced during the crystal growth process or subsequent surface contaminations. Examples include copper (Cu), chromium (Cr), nickel (Ni), iron (Fe), chlorine (Cl), and aluminum (Al). These elements create additional acceptor or donor traps that might impact device performance.

There are two types of traps: electron and hole traps. An electron trap is a defect in which the electron capture rate (R_{cn}) is much larger than the hole capture rate (R_{cp}), and vice versa for the hole trap. An electron trap is a majority carrier trap in an n-type and a minority carrier trap in a p-type semiconductor. Also, it is important to know the difference between donor and acceptor levels. Donor levels or states are those that become positively charged; in other words, they donate an electron when ionized. Acceptor states become negatively charged when ionized. The levels of these traps can be classified as shallow or deep, majority or minority, electron or hole, and donor or acceptor. Figure 2 illustrates the concept. E_c is the conduction band, and E_v is the valence band.

In CZT, due to its iconicity, crystals tend to have a much higher density of traps or defects (electron and hole traps) as compared to traditional semiconductors like silicon (Si) or germanium (Ge). Due to the presence of various trapping levels, there is a wide variation in electron and hole lifetimes reported in the literature. The manufacturing control of growth processes is essential.

2.3 *Crystal Growth*

Semiconductor radiation detectors require high-quality, semi-insulating single crystals for satisfactory performance. High resistivity is required in order to attain sufficient carrier depletion of the semiconductor crystal and to realize active detector thickness in the few millimeter to a few centimeter range and to maintain a high electric field across the detector with a low leakage current. Too low depletion limits the active depth of the device and the energy range of the detector, while an excessive leakage current produces electronic noise that deteriorates the energy resolution of the detector.

CZT crystals are typically grown by directional solidification from their melts or from solution. In melt growth techniques, the CZT is melted at a few degrees above their melting points and then very slowly solidified in a temperature gradient. The solution growth process is very similar, except that the melt is enriched in one of the constituents to form a solution from which the material crystallizes at a lower temperature. The most often used techniques for the growth of CZT single crystals are the Bridgman, gradient freeze, and electrodynamic gradient-freeze melt-growth processes and the traveling heater method (THM), a solution growth technique.

Crystal growth aims to achieve atomic-level perfection of the crystal lattice by controlling macroscopic parameters. This is a fundamentally very challenging problem: the thermodynamics of the material system dictates the formation of point defects during crystallization. The unavoidable formation of defects during solidification and the ensuing defect interactions during cooling lead to a rich and complex defect structure in the crystals. In addition to point defects, the formation of extended structural defects is also unavoidable because the physical system lowers its total energy under the fields and forces imposed during crystal growth. These external factors are the driving forces of extended defect generation. Gaining an increased control over the atomic perfection of the crystals and the suppression of crystal defects are the continuing, principal challenges for CZT crystal growth technologies. Through 30 years of research and development (R&D) and years of process development, the industry has mastered the THM process up to the point that people have become successful at controlling the multitude of microscopic interfacial phenomena with far-field macroscopic parameters to minimize the formation of crystal defects at the growth interface and in the solidified crystal.

Such a process control is paramount due to detector applications requiring high-quality, single crystals as defects in the crystals can cause carrier trapping, recombination, and the distortion of the internal electric field, which all contribute to the deterioration of detector performance. Large-angle grain boundaries are typically strong carrier traps in CZT. To avoid the detrimental effects of grain boundaries on detector performance, single crystals are used to be mined from the typically polycrystalline ingots grown using the Bridgman technique. The single-crystal yield from this mining process was the largest cost driver of CZT detectors and has delayed the introduction of CZT technology to universal use in X-ray detection for several years.

Growing as perfect single-crystal ingots as possible or achieving the highest single-crystal yield from polycrystalline ingots is the primary goal and challenge of CZT crystal growth technologies. To minimize carrier trapping, the extracted single crystals themselves have to be as perfect as possible, with a low concentration of point defects and their clusters, as well as a low density of extended defects, such as dislocations, low-angle grain boundaries (subgrain boundaries), twins, and second-phase precipitates and inclusions. These achievements were possible with the traveler heat method (THM), which has been created and perfected over the last 20 years of technology development.

3 CdTe/CZT Radiation Detectors

3.1 Basic Principles of Operation

The operation of any radiation detector depends on how the radiation interacts with the sensor material of the detector. The interaction needs to cause a measurable change in the physical properties of the sensor, typically a change in the number of free carriers available to conduct current. Electronic circuits then measure and further process this change in sensor property to extract information about the detected photon. The subject of this chapter pertains to the detection of X-rays and γ-rays using semiconductor sensors and the associated front-end electronics. More specifically, a largely qualitative treatise on the physics and principles of hybrid and monolithic pixel detectors is given.

The basic steps in detecting an X-ray photon by a semiconductor sensor can be described as follows: an X-ray impinges on a semiconductor crystal and interacts with the atoms of that crystal, creating a charge cloud consisting of electrons and holes. The electron-hole clouds are accelerated in opposite directions due to the electric field applied across the sensor via the electrodes. The movement of these charge carriers is sensed at the electrode, creating a current in the amplifier that is measured and evaluated per what property is being determined. This basic measurement process and the associated detection chain are illustrated in Fig. 3.

Depending on the sensor and the electronics design, it is possible to extrapolate different properties related to the measured photons. The most common properties of interest are the position of interaction, photon energy, and the time of arrival of the photon. The position is required to get an image of the photon source. As such, detailed information regarding the photon interaction point in the sensor is required. This is achieved by segmenting, i.e., pixelating, the sensor. Energy determination is required to measure the source's spectrum, i.e., the energy deposited by each interacting photon. This requires us to ensure that the full photon energy is deposited and subsequently measured accurately. Finally, the photon arrival time can be used to perform timing measurements and evaluate how long the photon has traveled with

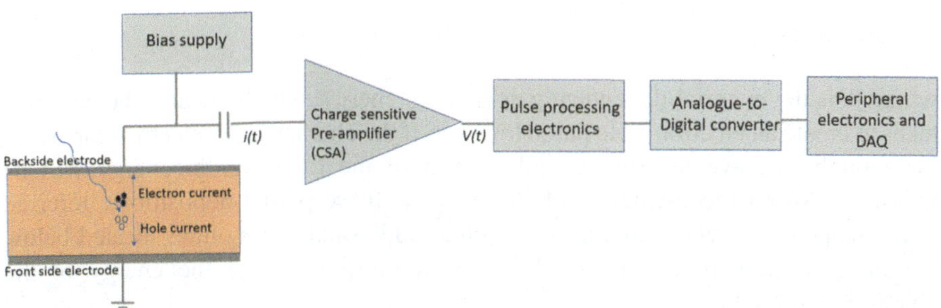

Fig. 3 Basic configuration for the detection of X-rays in a semiconductor detector that might be CdTe or CZT. The sensor and functional blocks making up the electronics chain are illustrated [11]

reference to (w.r.t.) a set time frame (time of flight). We therefore need to be able to identify when the photon is absorbed in the sensor and relate it to the external time frame of interest.

3.2 X-Ray and Gamma-Ray Interactions with CdTe/CZT Sensors

When X-rays enter a semiconductor sensor, it has a finite probability of being absorbed. An X-ray beam entering a sensor will be attenuated according to:

$$\frac{I}{I_0} = e^{-\mu_a \rho x}$$

where I is the intensity of the X-ray beam after it has traveled a distance x into the material, I_0 is the beam's intensity prior to entering the sensor, ρ is the mass density of the material, and μ_a is the attenuation coefficient of the material. The attenuation coefficient of the material is given by adding the attenuation coefficients of each of the X-ray matter interactions that can take place in that sensor. X-rays and γ-rays interact with matter mainly through four basic processes: *photoelectric absorption*, *elastic scattering*, *Compton scattering*, and *pair production*.

In *photoelectric absorption*, all the photon's energy is transferred to an atomic electron, with the highest probability electron being a K-shell electron. The electron that receives the energy from the photon, referred to as a photoelectron, becomes free to move in the sensor and interacts with neighboring atoms to create a charge cloud. The atom that gave up the photoelectron ends up in an excited state. Usually, the excited atom is relaxed by having a higher orbital electron fill the vacancy, emitting a characteristic photon in the process. Under normal conditions, the characteristic photon is reabsorbed in the sensor, creating a second charge cloud that is measured together with the original charge cloud. For a given X-ray photon energy, E, the total number of charge carriers, N, in the charge cloud is given by:

$$N = \frac{E}{W}$$

where W is the material-dependent pair creation energy, which dictates the energy it takes to create an electron-hole pair. However, if the characteristic electron can travel far enough to leave the sensor, only a part of the total photon energy will be measured. When reconstructing photon spectra, these partial depositions, referred to as escape peaks, can sometimes be seen as additional energy lines located below the main peak, with an energy difference corresponding to the energy of the characteristic photon.

Both Rayleigh and Thomson scattering fall under the category of *inelastic scattering*. In this process, the photons interact with the sensor's atoms but do not impart any energy. Instead, the photons are just redirected at an angle with respect to their incoming trajectory. As such, it does not create a charge cloud and is not detected by the detector electronics.

In *Compton scattering*, also known as elastic scattering, the photon will impart some of its energy to a loosely bound electron, known as a recoil electron, and then be scattered at an angle. The amount of energy that is imparted to the recoil electron depends on the photon scattering angle. At 0°, no energy is transferred to the electron, and the maximum energy transfer occurs at 180°. As no energy larger than this maximum energy can be deposited, it creates an upper energy limit referred to as the Compton edge.

Finally, *pair production* can occur if the photon has an energy exceeding 1.022 MeV. This process takes place in an atomic nucleus, and the photon is transformed into an electron-positron pair. Any energy exceeding 1.022 MeV will be imparted as kinetic energy for the newly created pair. After slowing down, the positron annihilates, creating two annihilation photons.

For the described interaction mechanisms, only photoelectric absorption results in the whole photon energy being deposited in the sensor. As such, especially for spectroscopic detectors, it is preferential to try and ensure that this is the most probable interaction mechanism. In Fig. 4, we can see the regions where the different interaction mechanisms dominate as a function of the sensor's atomic number, Z, and the photon energy. From the graph, it is clear that decreasing the photon energy and/or increasing the atomic number of the sensor increases the probability of photoelectric absorption being the dominant interaction mechanism. Due to the high-Z of CdTe/CZT, the Compton range is limited compared to other semiconductor materials, like Ge, gallium arsenide (GaAs), or Si; therefore, the photoelectric effect dominates for a larger energy range. This is the main reason why CdTe/CZT detectors are preferred in most X-ray and gamma-ray applications, as discussed later in this chapter.

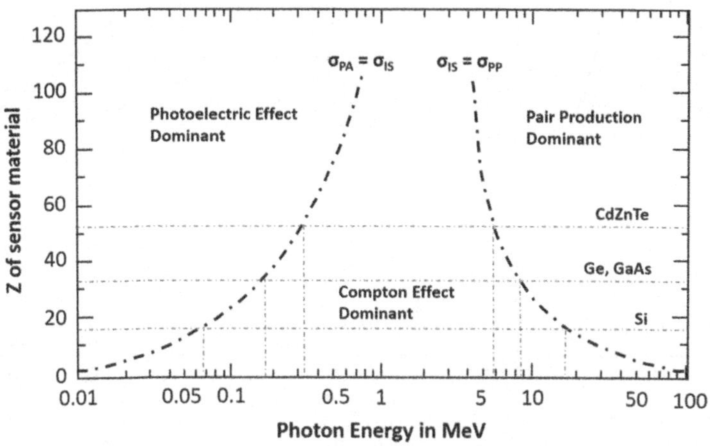

Fig. 4 Illustration of where in phase space the three interaction mechanisms between matter and sensor are dominant, as a function of the atomic number of the sensor and energy of the photon. The atomic numbers for Si, GaAs, Ge, and CdZnTe are highlighted. NIST data [12]

3.3 Sensor Detection Efficiency

As discussed, *the atomic number* of the sensor plays a significant role in the detection efficiency of the sensor. Here, the term detection efficiency refers to the ratio of the number of detected photons vs the total number of incoming photons. Other definitions do exist that take into account the detector response function. Similarly, the *sensor thickness* and the structure of *the entrance window* matter for detector efficiency too. The entrance window is sometimes called the dead layer and refers to any layers or structures on the entrance side of the sensor that do not create a measurable signal for the photons that are absorbed within it.

The effect of an *atomic number* can be seen in Fig. 5, where the detection efficiency at different photon energies is seen for different sensor materials. For ease of comparison, the absorption efficiency for all materials is shown for a sensor thickness of 500 um. It should be noted that these plots assume no entrance window at all. It is seen from the plot that a higher detection efficiency is achieved with a higher atomic number of sensor material. CdZnTe clearly shows superior detector efficiency. However, the absorption edges of the materials also have to be considered for different energy regions. This is seen for CdZnTe in the 27 keV region, where the absorption edges result in the sensor having a worse detection efficiency than what is observed for Ge and GaAs.

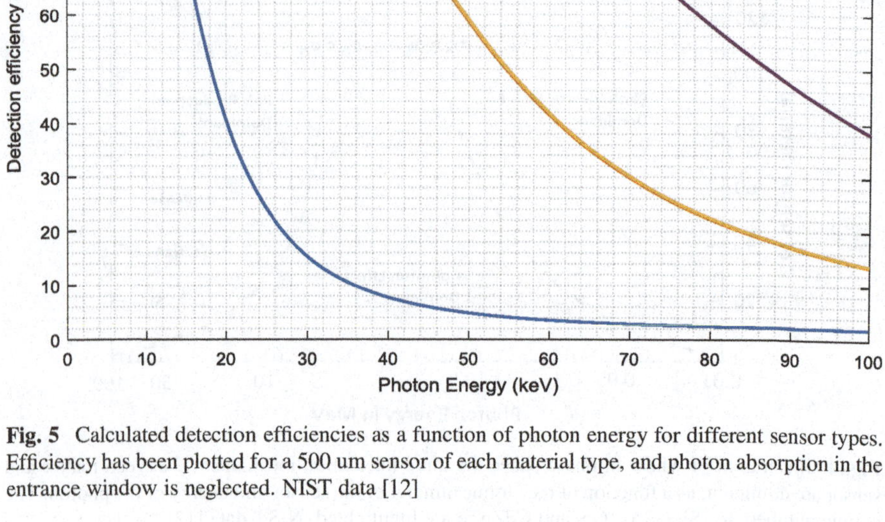

Fig. 5 Calculated detection efficiencies as a function of photon energy for different sensor types. Efficiency has been plotted for a 500 um sensor of each material type, and photon absorption in the entrance window is neglected. NIST data [12]

3.4 Sensor Geometry—Planar and Pixel Detectors

In order to ensure the electrons and holes that constitute the charge cloud separate and induce a signal on the electronics, an electric field has to be established between the sensor electrodes. The most straightforward electrode configuration for a semiconductor sensor can be seen in Fig. 6a. This configuration is referred to a planar configuration and consists of two large electrodes deposited on the opposite sides of the sensor. As a result, the electric field resulting from biasing the electrodes creates a uniform electric field across the sensor‘s active area, as indicated by the dashed blue electric field lines.

While this configuration is the technically easiest to realize, a photon can interact anywhere in the sensor volume, and no finer information regarding the interaction location can be obtained; i.e., the detector does not have any spatial resolution outside of the size of the detector itself. If the detector's targeted application requires spatial resolution, as is the case for imaging, a more complicated electrode configuration is needed. This can be realized by segmenting one of the electrodes into sections, effectively creating a pixelated device, and having each pixelated electrode read out by an electronic channel. This is illustrated in Fig. 6b. As can be seen in the resulting electric field line profiles, any photon absorbed within the voxel volume of a specific electrode will be transported to that electrode. Therefore, when the electronics associated with that electrode measure a signal, the interaction point of the photon can be more accurately determined than in the case of the planar electrode configuration.

Additionally, other considerations are needed to understand the operational differences between these two electrode configurations. As indicated above, it is not the collection of the charge clouds at the electrode that allows a signal to be measured by the pixel electronic channel but the movement by the charge cloud within the sensor. The electrostatic coupling between the charge cloud and the electronics (via the electrode) causes a current to be measured at the charge-sensitive amplifier (CSA). As such, the current starts being induced on the electronics as soon as the created charge cloud starts moving and finishes when the charge cloud reaches

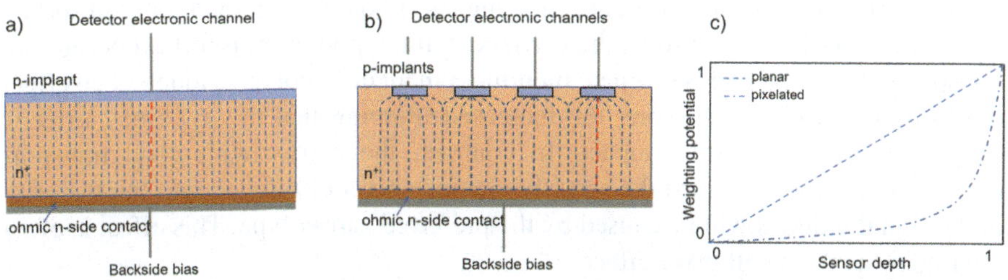

Fig. 6 The electric field line profiles for planar (**a**) and pixelated devices (**b**) [11]. In addition, a 1D cross-section of the weighting field for both detectors is illustrated in (**c**), clearly showing the effect on the weighting field caused by pixelating the electrode (electrode-to-thickness ratio: 0.125). The location of the cross-sections is indicated by the red dashed lines in (**a**) and (**b**)

the electrode. We have to mention that this process is fundamentally different then for example what happens in a CCD detector, where the charge created is transported to the potential pocket created by the pixel electrode, and subsequently transported out via the other pixel electrodes during the readout phase, and finally measured by the readout node when the charge physically arrives at the node.

The process of charge induction on the pixel electronics is dictated by two fields in the detector: the electric field and the weighting field. The electric field has been touched on above, and it dictates how the charge cloud moves in the sensor, with the carriers following the field lines, shown in Fig. 6a, b. As such, an understanding of the electric field allows us to comprehend how the carriers will move in the sensor. On the other hand, the weighting field is a geometrical effect that describes how strongly coupled the charge cloud is to any one electrode when that charge cloud is at any specific point in the detector. As such, the weighting field describes how much current will be induced at the CSA when the charge is moving in a specific part of the sensor. Since the weighting field is directly dependent on the electrode configuration, it is different for the planar and pixel configurations.

This property can be seen in the graph in Fig. 6c, where a one-dimensional (1D) cross-section of the weighting field is illustrated for both configurations. The 1D cross-sections correspond to the dotted lines in Fig. 6a, b. In the graph, sensor depth 1 corresponds to the pixel electrode's position, while position 0 is where the backside contact is located. As can be seen, for the planar configuration, the weighting field has a uniform slope forming a straight line, while for the pixel geometry, the slope increases closer to the pixel electrode. The weighting field slope can be interpreted as a measure of how much charge will be induced in the CSA by the movement of charge in this section of the sensor. Therefore, for the constant slope of the planar electrode, regardless of where the charge is moving along the cross-section, it will induce the same charge on the electronics. For the pixel electrode configuration, on the other hand, the slope is much steeper closer to the pixel electrode. This means that the movement of charge carriers in this region contributes most of the charge of the measured signal on the pixel electronics.

This fact is capitalized in a couple of ways in pixelated detectors. The first one is to optimize energy resolution in compound semiconductors, where one of the two carrier types (usually the holes) has much worse transport properties. By biasing the sensor in the correct direction and ensuring the photon is preferentially absorbed closer to the backside electrode, the carriers with the poorer transport properties are transported toward the region where their movement does not have much of an effect (backside electrode). Therefore, the increased trapping that this type of carrier is subject to will have little impact. In contrast, the carrier with good transport properties will move toward the sensitive region (pixel electrode), and the majority of the signal induced will be caused by the preferred carrier type. This effect is often referred to as the small pixel effect.

The second way the pixel weighting field is capitalized is in the design of detectors that can handle high count rates (quickly measure each photon). For a set electric field strength, the charge cloud will be transported with roughly constant speed across the sensor. As such, for a weighting field with a constant slope, the signal induction time is set by the time it takes for the charge clouds to traverse the entire sensor. For the pixel configuration, however, the signal will be developed during the short period while the charge cloud is moving close to the pixel electrode, making the charge induction time much faster. As such, the electronics can be made faster and still measure a good representation of the full signal when operating a pixelated sensor and consequently be able to handle a higher count rate.

A final point should be made on operating detectors with small pixels. When the sensor absorbs the X-ray, it initially creates a charge cloud, σ_i, with a size of a few um, and as the charge cloud traverses the sensor, the cloud expands through diffusion. The diffusion expansion of the cloud is given by:

$$\sigma_d = \sqrt{2Dt} \sim \sqrt{2\frac{kT}{e}\frac{d^2}{V_b}}$$

where D is the diffusion constant, k is Boltzmann's constant, e is the charge of an electron, d is the thickness of the sensor, and V_b is the applied bias over the sensor. The approximation is valid for an average electric field (i.e., $E = V/d$). As such, for $d = 500$ μm, $T = 300$ K, and $V_b = 100$ V, the charge cloud size (sigma) when reaching the electrode is the initial charge cloud size, σ_i, plus $\sigma_d \sim 11$ μm. Due to this spatial extent of the charge cloud, if the photon is absorbed close to the edge between two pixels, the resulting charge cloud could be measured by two pixels (partial charge induction on both pixels) instead of just one where the photon was absorbed. This effect is referred to as charge sharing and becomes more prevalent in detectors as the pixel size becomes smaller and smaller. When trying to measure the energy of the photon (spectroscopy), which is done by measuring the total amount of charge that the photon has released, charge sharing can have adverse effects even after resuming the measured charge detected by both pixels. However, the sharing of charge between pixels can also be used beneficially by looking at the amount of charge detected by the different pixels in a cluster of affected pixels. By centroiding this cluster, the interaction point of the photon can sometimes be determined to an accuracy higher than the pixel pitch (subpixel spatial resolution).

3.5 *Comparing CdTe/CZT to Other Sensor Materials*

The material properties of interest for the semiconductor crystal to be used as a sensor can be elucidated by walking through this basic detection process. As a set energy X-ray photon impinges on the sensor, the probability for the X-ray to be absorbed, and the absorption profile for multiple photons, is dictated by the *atomic number*, Z, and *density*, ρ, of the crystal. The higher the atomic number and density, the higher the stopping power of the sensor. This is the key advantage of CdTe/CZT.

When a lattice atom in the crystal absorbs the X-ray, many electrons and holes are eventually released. The number of electron-hole pairs that are released is directly related to the size of *the bandgap*, E_g, of the semiconductor. The smaller is the bandgap, the more charge carriers are released and the more accurately the energy of the X-ray can be determined. The reason is as follows. The smaller is the bandgap, the more carrier per unit energy is excited from the valence band to the conduction band. The variation in the number of carriers created for the same photon energy is given by $\sqrt{N}$, where N is the number of created carriers. As such, the higher the number of released carriers is for a set photon energy, the less variation is observed in the number of measured carriers for repeated photons of the same energy.

However, the smaller the bandgap, the larger the sensor's leakage current (thermally generated current). When the leakage current increases, the ability to accurately measure the amount of charge created goes down. The leakage current can be suppressed in several ways, and the most basic method is to cool the sensor to lower temperatures. As such, the bandgap of the sensor material often must be chosen in a trade-off between the number of charge carriers created per unit photon energy, the magnitude of the leakage current, and the restraint imposed if needing to cool the detector. Once the charge clouds have been created, they are separated by the applied electric field and start to drift in opposite directions. The speed with which the carriers can move and the ability of the carriers to make it across the sensor without being recaptured are given by the *carrier mobility*, μ, and *carrier lifetime*, τ, respectively. These two parameters are often lumped together and are referred to as the *mobility-lifetime product*, $\mu\tau$-product, of material when discussing the transport properties of that specific material. Additionally, imperfections of the crystal lattice can introduce electrically active trapping states in the bandgap, resulting in incomplete charge collection. If sufficient trapping occurs, the resulting charge can distort the electric field set up by the bias, resulting in time-dependent variation in response to X-ray photons. The basic material properties that need to be considered for any sensor material are summarized in Table 1.

In addition to these basic material parameters discussed, some additional considerations have to be made when selecting the sensor material: the hardness of the material must be considered when evaluating the ease with which the crystal can be

Table 1 Overview of the basic material parameters of interest when considering a sensor material [11]

Property	Symbol	Unit	Physical consideration
Atomic number	Z	#	Detection efficiency of the sensor
Density	ρ	g/cm^3	Detection efficiency of the sensor
Bandgap	E_g	eV	The number of carriers created per unit photon energy; leakage current
Mobility	μ	cm^2/vs	The ability of charge carriers to move in the sensor under the influence of an electric field
Lifetime	τ	s	The average distance a carrier can survive in the conduction band before being recaptured; not an intrinsic parameter of the material but primarily related to material quality
Mobility-lifetime product	$\mu\tau$	cm^2/V	A measure of the transport quality for a specific material

Table 2 List of some common semiconductor materials used in radiation detection [11]

	Si	Ge	GaAs	$Cd_{1-x}Zn_xTe$	TlBr
Atomic number (Z)	14	32	31,33	48,30,52	81,35
Density (g cm^{-3})	2.33	5.33	5.32	5.4–5.8	7.45
Resistivity (Ω-cm)	1e5	53	10^{10}	$>10^{10}$	10^{12}
Bandgap (eV)	1.12	0.664	1.425	$1.51 + 0.606x + 0.139x^2$	2.68
Pair creation energy (eV)	3.66	2.96	4.18	4.6	6.5
Electron mobility (cm^2V^{-1} s^{-1})	1500	3900	8500	1350	50
Hole mobility (cm^2V^{-1} s^{-1})	480	1900	400	120	4
$\mu_e\tau_e$ (cm^2V^{-1})	>1	>1	1×10^{-4}	1×10^{-2}	3×10^{-4}
$\mu_e\tau_e$ (cm^2V^{-1})	~1	>1	4×10^{-6}	2×10^{-4}	6×10^{-5}

cut into the desired sensor shape; the binding energy between the atoms must be considered to ensure no ionic conduction takes place, changing the structure of the crystal; and the existence of a proper electrode material must be determined that will produce good electric contacts (ohmic or Schottky depending on the desired property) and doping atoms if pn-junctions are required for contacting or sensor operation.

Table 2 lists the key electrical and physical parameters for the most commonly available semiconductor materials that are used as X-ray sensors. The elemental semiconductors Si and Ge are considered the gold standard for sensor materials. This is primarily due to the extremely high-quality crystals available for these materials. As such, they have excellent transport properties and do not suffer from any polarization effects. However, they suffer from some limitations, leading to an interest in alternative material systems. The primary drawback of silicon is its low atomic number, which prevents it from having a high detection efficiency at higher X-ray energies. This drawback in Si has been the primary driver behind the development of high-Z semiconductor materials, namely, GaAs, CdTe, CdZnTe, and thallium(I) bromide (TlBr), for X-ray sensors, the material care clearly won by CdTe and CZT.

While Ge has a reasonably high atomic number, this material also has a very small bandgap. This results in Ge having better intrinsic energy resolution than Si, but it also leads to the need to cool this material, often to cryogenic temperatures, to reduce the thermally generated leakage current and be able to differentiate a photon signal. The need for such significant cooling forces the use of vacuum operation and bulky cooling equipment to operate these detectors. While this can be dealt with for specific applications, it becomes preventative for others, such as space-based applications. As can be seen from Table 2, GaAs offers an alternative to Ge with very similar atomic number and density values but with a much wider bandgap, effectively reducing the need for cooling. The larger bandgap also ensures that GaAs has similar carrier concentrations to depleted Si, meaning it can be operated as "depleted" without pn-junctions. Additionally, GaAs has excellent carrier mobilities, which has led to it being identified as a material of interest for sensor materials that require a swift pulse response (fast timing application). However, the promising features promised by GaAs are currently being prevented by the material's quality, with a large number of impurity states present. This leads to reduced transport properties, as seen from this material's very low mobility-lifetime products. Additionally, deep-level traps (mostly the EL2+ mid-gap defect caused by As substitutionally occupying a Ga lattice site, resulting in the regional distortions of the electric field) have been observed. These problems plaguing GaAs has limited its use to thin sensors, reducing the detection efficiency at higher photon energies. TlBr is available from a few select companies for use as X-ray sensors. However, material quality issues, low hardness, and its nature as a mixed electric-ionic conductor have so far prevented it from being used in commercial products. The most developed high-Z sensors currently available are CdTe and CdZnTe. Both materials have reached a maturity level that has allowed them to be used in several commercial products in medical imaging, nondestructive testing (NDT), and baggage scanning applications.

4 Detector Electronics

The charge pulse that is generated in the CdTe/CZT sensor is induced on the electronics, where it is then processed according to the photon parameter of interest. Two common architectures for electronics design are the pulse processing design, also referred to as the photon counting design, and the charge integrating design. The basic operation of these designs is dictated by the reset methodology used by the first stage in the electronic chain, the charge-sensitive amplifier (CSA). The following sections discuss the differences between these two architectures and what pulse parameter is used to determine the most common photon parameters of interest. Finally, the implementation of these electronic designs into the two application-specific integrating circuit (ASIC) solutions, the hybrid pixel detector design and the monolithic pixel detector design, are discussed.

4.1 Pulse Processing Electronics

Many different amplifier solutions exist to measure the signal generated by detecting an X-ray photon in a semiconductor sensor, including current and voltage amplifiers. However, one of the most used ones is the charge-sensitive amplifier (CSA). The primary reason for the popularity of the CSA is the fact that the gain remains the same regardless of the capacitive loading on its input that the sensor subjects it to, a fact that is not true for many other amplifier solutions.

The basic operation of the CSA can be seen in Fig. 7. Once the X-ray photon creates charge clouds in the sensor and these charge clouds start moving to create a current in the sensor, the charge is induced onto the feedback capacitor of the

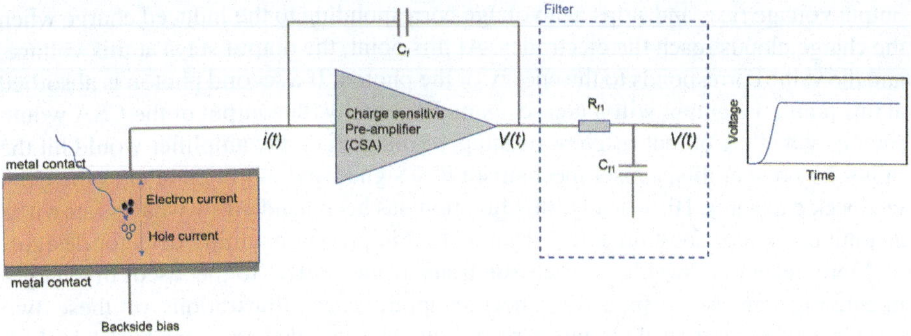

Fig. 7 Illustration of a basic sensor/CSA/filter network used for X-ray detection [11]. As the charge cloud that the X-ray photon has created starts to move, a signal is induced onto the CSA. A filter is often used at the output of the CSA to limit the bandwidth of the network and reduce the system noise. A trade-off between increased noise performance (low bandwidth) and pulse response time (high bandwidth) often has to be done to optimize the application. In the system above, no reset mechanism is present, which is usually added in practice

preamplifier. As this happens, the signal goes from being a current signal to a voltage signal, and as more charge is induced into the CSA feedback capacitor, the voltage at the output of the CSA is driven up. When the full signal has been induced on the CSA, the magnitude of the voltage at the output of the amplifier is directly related to the amount of charge released by the absorbed X-ray photon.

In general, the CSA is designed with a filter attached to its output. The main aim of the filter is to modify the frequency range (bandwidth) of the CSA and filter network to a more suitable one than what is given by the frequency response provided by the CSA alone. As the bandwidth of the frequency response of the CSA/filter network is reduced, the noise is also reduced since signals not belonging to the photon-generated pulse are rejected. This is a simplistic argument highlighting a general rule that reducing the bandwidth reduces the noise. In reality, a detailed study between linear, parallel, and shot noise components in the systems has to be evaluated, and an optimum noise performance occur at a specific shaping time (bandwidth).

However, the bandwidth reduction also results in the rise time at the output of the CSA/filter network becoming slower in the temporal domain (i.e., the signal rise time is increased). As such, the detector needs longer to measure the signal, effectively reducing the maximum count rate the detector can handle accurately. This trade-off between noise level and count rate capability for a detector is one of the design criteria an electronics designer has to take into consideration and can serve as a general example of why a generic electronic channel for all applications cannot be created, but each application has to have its dedicated solution depending on what is the most critical design criteria for that application.

How the voltage at the output of the CSA/filter network behaves as a function of time can be analyzed as follow. Before the photon is absorbed, the output is at the baseline value. As the charge clouds are created and move through the sensor, the output voltage rises and stops at a voltage corresponding to the induced charge when the charge clouds reach the electrodes. At this point, the output stays at this voltage, and the value corresponds to the energy of the photon. If a second photon is absorbed at this point, the output will increase again. Eventually, the output of the CSA would saturate when the output reaches the supply voltage (i.e., the amplifier would hit the rails). To prevent this, a reset mechanism is designed and put in parallel to the CSA feedback capacitor. Historically, this function has been handled by what is known as a continuous reset solution and is what is used in photon-counting detector designs.

More recently, switched reset solutions have started to be used in charge-integrating detector designs for specific applications. Illustrations of these two reset mechanisms and their pulse responses are provided in more detail in Ref. [11]. Continuous reset and switched reset are also sometimes referred to as time-independent reset and time-dependent reset, respectively. This is due to their general nature where the continuous reset will respond to the incoming charge as it comes in, and therefore, it will always have the same response, while for the switched reset, it happens at a set time, and the response to a photon can vary depending on when in time (with respect to the photon arrival time) the pulse characteristic measurement takes place.

4.2 *Monolithic and Hybrid Pixel Detectors*

Many modern pixelated X-ray detectors are implemented using chips from large-scale silicon foundries. These chips are often referred to as application-specific integrated circuits (ASICs) due to their application-tailored nature. Two groupings of these detectors that adhere to the physical principles described above are hybrid pixel detectors and monolithic pixel detectors. The primary difference between the two types of detectors is that the sensor and the chip containing the readout electronics are two different entities for the hybrid pixel detector. In contrast, for a monolithic pixel detector, the sensor and electronics are implemented in the same piece of silicon. While the hybrid pixel detector technology is relatively mature, having been around for several decades, the monolithic pixel detector group is still comparatively early in its development path. As there is no possibility of creating transistors in CdTe/CZT materials, only hybrid pixel detectors are under consideration here.

In hybrid pixel detectors, the sensor and the electronic chip are separate entities that allow for the sensor to be created in a material other than silicon, such as CdTe or CdZnTe. As discussed above, the larger atomic number of these sensors‘atomic constituents allows for a much better detection efficiency at higher energies. Detectors utilizing these high atomic number sensor materials have started to see commercial use in recent years and are poised to become mainstream technology in the coming years.

4.3 *HEXITEC ASICs for CdTe/CZT Readout*

There are numerous application-specific integrated circuits (ASICs) that have been developed for the purpose of reading signals from CdTe/CZT sensors. In this section, we will provide an example of one of the most popular ASICs in the world: HEXITEC ASIC [13–15].

The HEXITEC ASIC was developed for imaging applications where important additional information regarding the objects' properties is contained within the energy spectrum of the absorbed, transmitted, and emitted X-rays. Making use of this additional information requires a detector with excellent spectroscopic performance. The ASIC design was optimized to provide very good hard X-ray spectroscopy but at the expense of readout speed and consists of 80 × 80 pixels on a pitch of 250 um and has dimensions of 20.2 mm × 21.7 mm. Each pixel contains a 52 μm bond pad, which is used for gold stud and silver epoxy bonding to detectors. Figure 8 shows an image of the chip as well as a block diagram of the electronics contained within each pixel of the ASIC. The induced charge formed on the electrode of each pixel is read out using a charge preamplifier feedback circuit, which can compensate for a detector leakage current of up to 250 pA per pixel. The output of each pixel is filtered by a peaking circuit that contains a CR-RC shaper in series with a second-order low-pass filter. A peak-track-and-hold circuit maintains the maximum voltage output of the shaper until the detector is read out using a rolling row readout scheme. The readout is operated with a 20 MHz master clock with separate outputs for four blocks of 20 × 80 pixels; with this configuration, a full frame of 80 × 80 pixels is read out at a rate of 10 kHz. The ASIC is only sensitive to electron readout and has a noise performance of better than 800 eV per channel.

As described earlier in this chapter, the ability to identify and correct charge-sharing events is important for detector performance. In order to be able to correct charge-sharing events without significant degradation of the spectroscopic performance of the detector, only a limited number of pixels can detect an event in a single

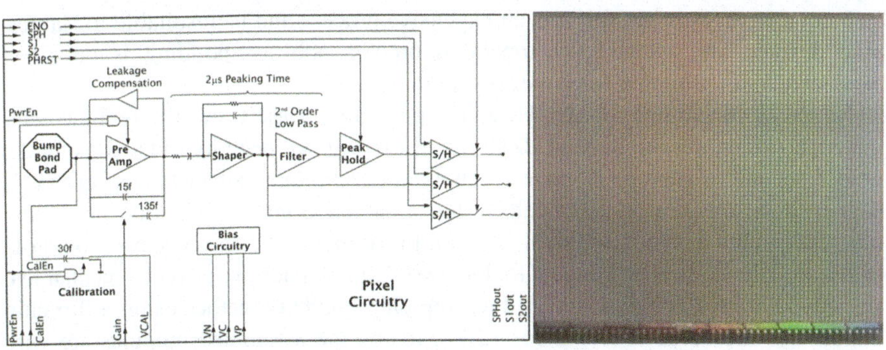

Fig. 8 A block diagram of the HEXITEC ASIC pixel electronics (left). Layout of the 80 × 80 HEXITEC ASIC silicon die; the bottom edge of the ASIC is dedicated to I/O connections (right) [13]

frame. It has been estimated that ten million photons per second per detector (~2.5×10^6 photons s^{-1} cm^{-2}) is the maximum rate that can be detected with the ability to properly correct for charge sharing.

Since the first 20 × 20 pixel device was fabricated in 2008 [13], the ASIC has undergone a number of iterations. The first version of the full 80 × 80 pixel ASIC had high and low gain modes covering energy ranges of 4–200 keV and 40–1200 keV, respectively, while in the second version, the low gain range was adjusted to 12–600 keV to provide better energy resolution for some of the target applications. Until recently, all versions of the ASIC have been limited to tiling in $2 \times n$ arrays due to a 1.5 mm wide area along one edge that contains the I/O pads for wire bonding. A few years ago, a new version of the ASIC was produced where the I/O pads were redistributed using through-via technology (TSV) onto the rear of the ASIC. This allows detectors to be tiled in $n \times n$ arrays with a minimum (~150 um) dead space between individual modules [14]. Very recently, a major upgrade to the HEXITEC ASIC was announced proving a very fast readout capability of 1 MHz with superior low-noise readout properties. The chip is currently in the debugging and testing stage [15].

As mentioned above, materials such as CdTe and CdZnTe are sensitive to the temperature used during bonding processes. Above temperatures of about 150 °C structural changes may occur within the CdTe/CdZnTe material, and impurities and structural defects may become mobile within the crystal lattice; this may lead to the degradation of detector performance. For this reason, conventional solder-reflow flip chip assembly using eutectic tin-lead solder is unsuitable for detector hybridization. At present, two low-temperature flip chip assembly techniques are widely used for the construction of CdTe and CdZnTe detectors. The technique of choice for assembling small pixel detectors with pixel pitches smaller than 100 um, such as the Medipix family of detectors, is cold-weld indium bump bonding. In this process, indium bumps are deposited on both the sensor and the ASIC using standard photolithographic techniques and an indium evaporator. Following the deposition, the photoresist is removed, leaving behind bumps that are then brought together in a flip chip bonder, and a small pressure is applied to form the cold weld. While this technique is very successful, it is an expensive process and is unsuitable for larger-pixel devices. The second low-temperature process, adhesive bonding, is much less expensive and is suitable for larger pixel pitches.

Since the beginning of the HEXITC project, a variety of different detector materials, including Si, GaAs, CdZnTe, and CdTe, have been hybridized using the gold stud and silver epoxy adhesive bonding technique. Silver-loaded epoxy is deposited on the sensor pixel pads using a stencil printing technique producing glue dots with a diameter of about 150 um. In parallel to the epoxy deposition, gold studs are ultrasonically bonded to the ASIC bond pads to provide mechanical stability. The two components are then flip chip bonded together using a flip chip bonder and are cured at 80 °C for 12 h. Finally, the assembled modules are wire bonded to either a stand-alone printed circuit board (PCB) or on an aluminum carrier that allows the detectors to mount in larger detector arrays (see Fig. 9). The testing of detectors hybridized using this technique has consistently produced bonding yields of >99.9%.

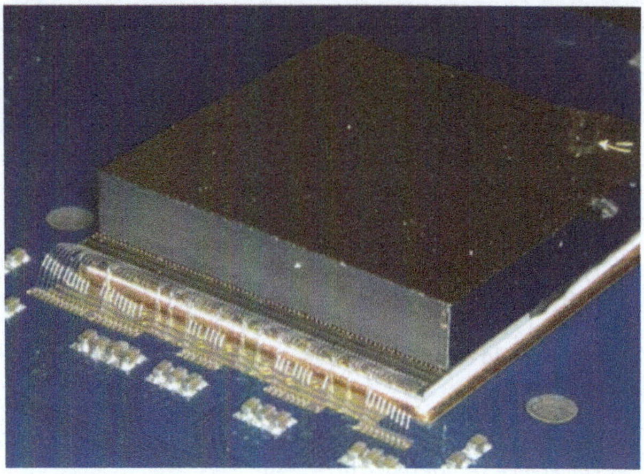

Fig. 9 A single CdZnTe detector flip chip bonded to the HEXITEC ASIC and wire bonded to a PCB

5 Conclusions

Energy-discriminating, photon-counting X-ray detectors (PCDs) enabled by CdTe and CZT materials are a promising new technology for future X-ray detection applications. PCDs have several advantages over traditional, energy-integrating detectors, including higher spatial resolution, retrospective spectral imaging, and higher immunity to electronic noise. Progress in recent years has been rapid. Multiple prototype clinical X-ray scanners have been deployed in nuclear medicine, computed tomography, X-ray diffraction, and baggage scanning. Many of these applications are described in detail in this book.

References

1. McGregor, D. S., & Shultis, J. K. (2021). *Radiation detection concepts, methods, and devices*. CRC Press. https://www.routledge.com/Radiation-Detection-Concepts-Methods-and-Devices/McGregor-Shultis/p/book/9781439819395
2. Spieler, H. (2005). *Semiconductor detector systems*. Oxford University Press. https://global.oup.com/academic/product/semiconductor-detector-systems-9780198527848
3. Owens, A. (2020). *Semiconductor radiation detectors*. CRC Press. https://www.routledge.com/Semiconductor-Radiation-Detectors/Owens/p/book/9780367779689
4. Sze, S. M., & Lee, M. K. (2012). *Semiconductor devices: Physics and technology*. Wiley. https://www.wiley.com/en-us/Semiconductor+Devices:+Physics+and+Technology,+3rd+Edition-p-9780470537947
5. Iniewski, K. (Ed.). (2020). *Advanced Materials for Radiation Detection*. Springer. https://link.springer.com/book/10.1007/978-3-030-76461-6
6. Iniewski, K. (Ed.) (2022). *Advanced X-ray detector technologies: Design and applications*. https://link.springer.com/book/10.1007/978-3-030-64279-2
7. Iniewski, K. (Ed.). (2023). *Advanced X-ray radiation detection: Medical imaging and industrial applications*. Springer. https://link.springer.com/book/10.1007/978-3-030-92989-3

8. Abbene, L., & Iniewski, K. (Eds.). (2023). *High-Z materials for X-ray detection: Material properties and characterization techniques*. Springer. https://link.springer.com/book/10.1007/978-3-031-20955-0
9. Du, J., & Iniewski, K. (Eds.). (2023). *Gamma ray spectroscopy: Technology and applications.* Springer. https://link.springer.com/book/10.1007/978-3-031-30666-2
10. Hansson, C., & Iniewski, K. (Eds.). (2024). *X-ray photon processing detectors: Space, industrial, and medical applications*. Springer. https://link.springer.com/book/10.1007/978-3-031-35241-6
11. Hansson, C., & Iniewski, K. (2024). *"Introduction to detector technologies", in X-ray photon processing detectors: Space, industrial, and medical applications*. Springer.
12. NIST XCOM: https://physics.nist.gov/PhysRefData/Xcom/html/xcom1.html
13. Jones, L. L., Seller, P., Wilson, M. D., & Hardie, A. (2009). HEXITEC ASIC – A pixelated readout chip for CZT detectors. *Nuclear Instruments and Methods in Physics Research A, 604*, 34–37.
14. Seller, P., Bell, S. J., Veale, M. C., & Wilson, M. D. (2012). Through silicon via redistribution of I/O pads for 4-side butt-able imaging detectors. In *IEEE nuclear science symposium conference record* (pp. 4142–2146).
15. Clime, B., et al. (2023, December). Characterisation of HEXITECMHz - A 1 MHz continuous frame rate spectroscopic X-ray imaging detector system. *Nuclear INSTRUMENTS AND METHODS IN PHYSICS RESEARCH SECTION A: Accelerators, Spectrometers, Detectors and Associated Equipment, 1057*, 168,718. https://doi.org/10.1016/j.nima.2023.168718

Dislocation Filtering Layers for Defect Reduction in the Heteroepitaxial Growth of Infrared Optoelectronic Materials

Wenwu Pan, Lorenzo Faraone, and Wen Lei

1 Introduction

1.1 Overview of II–VI HgCdTe and Cd(Zn)Te Semiconductors and Their Applications

II–VI (Hg)(Cd)(Zn)Te thin-film semiconductors (highlighted in red in Fig. 1a) are significant materials with a wide range of applications in high-end infrared sensing/imaging, radiation detectors, and solar cells [1]. For example, II–VI mercury cadmium telluride (HgCdTe), alloyed by closely lattice-matched CdTe and HgTe, is well known as a material system with a widely tunable working wavelength range (1–30 μm), large absorption coefficient, long carrier lifetimes, and high mobilities, which has established it as a leading semiconductor for fabricating high-performance infrared (IR) detectors and imaging focal plane arrays (FPAs) [2]. Apart from IR applications, HgCdTe epilayers grown by molecular beam epitaxy (MBE) have also emerged as attractive materials in developing integrated terahertz photonic systems [3] and topological electronics [4].

Cadmium (zinc) telluride (Cd(Zn)Te) is also a representative II–VI compound with a band gap well suited for solar cell applications [5]. Currently, CdTe is the second most widely used material for thin-film photovoltaic solar cells, after silicon (Si). CdZnTe is also a high-density material with high atomic numbers, making it an excellent candidate for fabricating radiation detectors used for gamma-ray spectroscopy, X-ray imaging, and medical imaging applications [6]. The lattice constant of

W. Pan (✉) · L. Faraone · W. Lei
ARC Centre of Excellence for Transformative Meta-Optical Systems (TMOS), Department of Electrical, Electronic and Computer Engineering, The University of Western Australia, Perth, Australia
e-mail: wenwu.pan@uwa.edu.au; lorenzo.faraone@uwa.edu.au; wen.lei@uwa.edu.au

K. Iniewski (ed.), *CdTe and CdZnTe Materials*,
https://doi.org/10.1007/978-3-031-64521-1_2

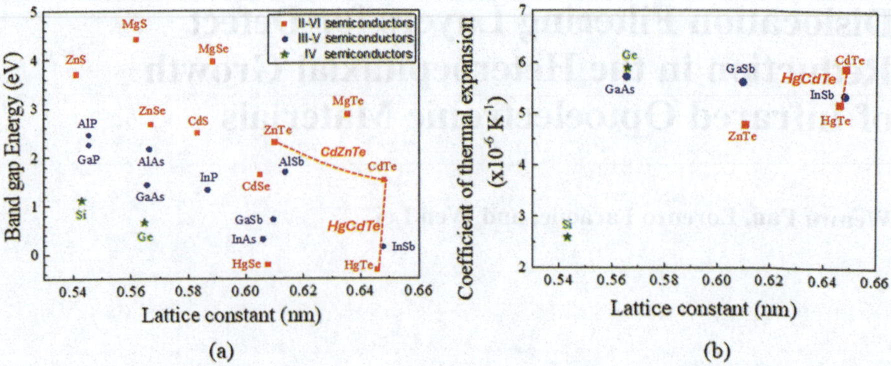

Fig. 1 (**a**) A band gap vs. lattice constant plot of common semiconductor materials [1]; (**b**) lattice constant and CTE (coefficient of thermal expansion) mismatches between HgCdTe and several semiconductor substrates

CdZnTe can be tuned by adjusting the Zn composition of the alloy material, and $Cd_{0.96}Zn_{0.04}Te$ (CZT) has long been the baseline substrate for lattice-matched MBE growth of high-quality HgCdTe for fabricating state-of-the-art IR FPAs [7].

However, II–VI substrates, including CZT, normally are of lower crystal quality, smaller wafer size, and higher unit cost compared to other group IV and III–V substrates. This is primarily because of the soft and brittle nature of II–VI materials, making them more challenging to handle, process, and fabricate compared to more common IV and III–V semiconductor substrates, which are more robust and durable [8]. For instance, the size of CZT substrates is also typically smaller than other standard semiconductor substrates, limiting the number of devices that can be fabricated on them and the ability to produce large-format-size HgCdTe FPAs for high-resolution IR imaging applications. In addition, the quality of CZT substrates can also be a concern as they are often subject to relatively high levels of defects and impurities, which can reduce the yield and reliability of devices fabricated on them. These substrate issues hinder HgCdTe and Cd(Zn)Te from being widely adopted in applications demanding compact, sensitive, and inexpensive optoelectronic devices.

1.2 Advantages and Challenges in the Heteroepitaxy of (Hg)Cd(Zn)Te Films on Alternative Substrates

With the development of IR technologies and their applications, there is a growing demand for high-performance HgCdTe IR detectors with larger array size and lower cost, which has fueled the heteroepitaxial growth of HgCdTe on CdTe buffer layers on large-area, high-quality, cost-effective, and commercially available substrates such as Si [9], germanium (Ge) [10], and gallium arsenide (GaAs) [11]. Compared to the conventional growth on CZT substrates, the growth of II–VI HgCdTe and Cd

(Zn)Te films on these alternative substrates (ASs) offers several advantages, including increased wafer size for the processing of numerous FPAs per wafer and/or larger format size of FPAs. In addition, the AS can offer lower-surface defect density/impurities and better mechanical properties than CZT substrates, thus benefiting detector yield and reducing production costs. This approach aims to increase FPA format size, deliver higher FPA yield and operability, and dramatically lower costs, providing the IR imaging industry with a viable solution to meet the growing demands for advanced IR imaging systems.

Despite these advantages, one of the major challenges associated with the heteroepitaxial growth of single-crystal HgCdTe thin films on ASs is their large lattice mismatch (f_l) with HgCdTe ($f_l = -19.3\%$, -14.3%, and -14.4% for Si, Ge, and GaAs, respectively, as shown in Fig. 1b), which inevitably generates misfit dislocations ($>10^8$ cm^{-2}) in the vicinity of the CdTe/AS interface. The misfit dislocations (MDs) form threading dislocations (TDs) that can propagate into the CdTe buffer layer and the subsequent HgCdTe epitaxial layer, which eventually limits the performance of HgCdTe devices and, therefore, FPA operability. Another concern is their mismatch in CTE (thermal expansion coefficient, f_t), which can also result in strain, defects, or dislocations in HgCdTe thin films grown on these ASs. As evident from Fig. 1b, III–V indium antimonide (InSb) has almost the same lattice constant as CdTe, exhibiting the minimum lattice and CTE mismatches with HgCdTe in comparison to other ASs. Preliminary studies of the MBE growth of HgCdTe on InSb have shown that the crystal quality of (Hg)CdTe grown on InSb substrates can exceed that of (Hg)CdTe grown on CdTe substrates [12–14]. However, a significant complication with the use of InSb substrates is the necessity for complete substrate removal and epitaxial film transfers [13]. This is because InSb absorbs IR light due to its narrow band gap (0.17 eV at 300 K and 0.23 eV at 80 K) and is thus not compatible with the current state-of-the-art backside-illuminated HgCdTe device technology. III–V GaSb is IR transparent and has a relatively smaller lattice mismatch with HgCdTe ($f_l = -6\%$) compared to previously studied ASs (Si, Ge, and GaAs). Experimentally, (Hg)CdTe layers grown on GaSb substrates have generally demonstrated material quality comparable to those on GaAs substrates [15–17]. Further growth studies are hampered by the fact that III Sb substrates are vulnerable to thermal decomposition during the deoxidation thermal process without a background group V Sb flux, which is incompatible with a II–VI MBE system. Higher-quality (Hg)CdTe layers would be expected on III–V ASs by optimizing the thermal cleaning process, such as using a dual-chamber III–V and II–VI MBE system [18, 19].

To reduce the threading dislocation density (TDD) in HgCdTe grown on lattice mismatched ASs, an intermediate CdTe buffer layer is commonly interposed between these ASs and HgCdTe to accommodate their large f_l and f_t. Providing thick buffer layers is unlikely to yield very high-quality material, so different approaches have been studied to reduce the TDD in CdTe and HgCdTe epitaxial layers grown on ASs. These include using low-temperature nucleation layers of ZnTe or CdTe [20] and various forms of cyclic thermal annealing (CTA) [9].

Screening techniques such as X-ray diffraction (XRD) and etch pit density (EPD) analysis are usually used to evaluate the material quality of (Hg)CdTe epilayers [21, 22]. For example, lattice-matched HgCdTe/CZT typically demonstrates XRD full-width-half-maximum (XRD FWHM) of 25–50 arcsec and EPD of ~10^4–10^5 cm^{-2}. For conventional CdTe buffers (5 ~ 10 microns thick) grown on ASs with optimal nucleation layer and cyclic thermal annealing processes, the resultant XRD FWHM and EPD are in the range of 60–80 arcsec and low-10^6 to low-10^7 cm^{-2}, respectively. Moreover, since the direct growth of HgCdTe on CdTe substrates (EPD ~ low-10^4 cm^{-2}) can result in an EPD around low-10^6 cm^{-2} in HgCdTe due to the small lattice mismatch ($f_l = 0.24\%$) between HgCdTe and CdTe [23], it is essential to reduce the HgCdTe/CdTe lattice mismatch in order to fabricate high-performance detectors, especially for long-wave IR (LWIR) detectors since LWIR detectors are much more sensitive to material defects due to their narrower band gap. In the past, one approach to reducing the HgCdTe/CdTe lattice mismatch was to incorporate Zn and/or Se into the CdTe buffer layer. However, high-quality Cd(Zn, Se)Te alloy buffer layers were difficult to obtain by MBE [24].

Overall, due to the limited dislocation confining the efficiency of the CdTe buffer and the residual HgCdTe/CdTe lattice mismatch, none of these approaches yield HgCdTe with material quality, detector performance, and FPA operability comparable to those grown on CZT, especially for LWIR detectors in which EPD levels below 10^5 cm^{-2} is required [25]. This book chapter will discuss a recently developed CdTe buffer layer process based on strained CdZnTe/CdTe superlattices for dislocation filtering aimed at resolving the problem mentioned above [26].

2 Strained CdTe/CdZnTe Superlattice-Based Dislocation Filtering Layers (DFL)

2.1 *Working Principle of DFL*

A strained-layer superlattice (SLS), comprising alternating strained heterolayers, commonly referred to as a dislocation filtering layer (DFL), has been observed to hinder the propagation of TDs and thus mitigate the total TDD in the epitaxial layer above [27–29]. The strain field generated by the SLS induces a lateral force on TDs via Peach-Koehler interaction with the mismatch stress [27], prompting the glide of TD segments. As glide proceeds, the dislocations encounter others and react with them to eliminate the TDs, or the threading segments propagate to the edge of the sample. In either scenario, TDs are transformed into additional MDs confined within the DFL layers.

Although the dislocation filtering efficiency varies with experimental conditions, modeling results indicate that the minimum TDD obtained in any given structure is likely to be limited by kinetic effects to approximately 10^4–10^5 cm^{-2} [29]. The movement of TDs in response to misfit strain combined with thermal energy has been shown to be the best way to enhance dislocation annihilation probability. The

SLS-based DFL technology has been successfully demonstrated in the growth of III–V semiconductors on lattice-mismatched substrates for many years. For example, the TDD has been reduced from ~10^9 cm^{-2} to ~10^5 cm^{-2} for GaAs buffer layers grown on Si by incorporating four sets of 10-nm $In_{0.18}Ga_{0.18}As$/10-nm GaAs SLS-based DFLs, leading to high-performance InAs/GaAs quantum dot lasers [30]. In contrast to the CTA (cyclic thermal annealing) techniques used in the heteroepitaxial growth of (Hg)CdTe on ASs, DFL technology will allow thinner buffer layer thickness, shorter processing time, and lower processing temperature to achieve similar EPD levels due to high dislocation filtering efficiency.

2.2 *Structure Design of CdTe DFL Buffer Layers and MBE Growth*

Figure 3a shows a typical sample structure for a CdTe buffer layer on a 2-inch GaAs (211)B substrate with five sets of $Cd_{1-x}Zn_xTe$/CdTe SLS-based DFL. A detailed description of MBE growth can be found elsewhere [26]. Compared to nonpolar group IV substrates such as Si and Ge, "epi-ready" GaAs is more readily available for the epitaxial growth of II–VI materials and has lower f_l and f_t with (Hg)CdTe. Additionally, while the MBE growth of CdTe on the DFL buffer structure was initially demonstrated on III–V GaSb substrates [31], GaAs substrates are considered more "epi-ready" after thermal cleaning for the subsequent growth of (Hg)CdTe compared to GaSb.

The (211)B orientation has been widely used for growing high-quality HgCdTe since it requires much lower Hg flux than either (001) or (111)A orientations to maintain the same MBE growth rate, and it is less sensitive to microtwin formation in comparison to (111) A/B [32]. As mentioned above, the purpose of the SLS layer structure is to increase the likelihood of dislocation annihilation and inhibit dislocation multiplication and/or propagation. The key to a significant reduction in TDD is TD movement using misfit stress, i.e., in layers with a finite strain thickness product (εh). Moreover, a good rule of thumb for DFL design in heteroepitaxial growth is given by $\varepsilon_c h_c < \varepsilon h < 4\varepsilon_c h_c$, where $\varepsilon_c h_c$ is a constant (experimentally, $\varepsilon_c h_c$ is ~0.24 nm in the GaInAs system) [29]. This can be understood by considering that the stress in the DFL needs to be high enough to divert TDs but not enough to generate additional TDs. In a layer with constant composition, dislocation multiplication is expected to commence when the strain thickness product exceeds ~$4\varepsilon_c h_c$, which places an upper limit on the strain and thickness of a DFL. In addition, $Cd_{1-x}Zn_xTe$ layers that are too thick and/or have more Zn content (x) may increase the risk of Zn segregation. Hence, a constrained Zn content range of $0.08 < x < 0.30$ is estimated for our DFL design. The CdTe DFL buffer layers were grown with different structural designs and annealing conditions to explore their impact on the final EPD of the top CdTe layer. The growth process/parameters consistently resulted in (211)B-orientated CdTe epilayers, while unexpected changes in substrate temperature during growth occasionally resulted in (133)B-orientated layers [33, 34].

2.3 Dislocation Filtering Effect

After MBE growth, there are various methods for characterizing the dislocation filtering effect in CdTe DFL buffer layers. Transmission electron microscopy (TEM) characterizations are straightforward as they can visualize individual defects such as TDs. Apparent TD redirection and termination at the CdZnTe/CdTe heterointerfaces are evident, as indicated by the yellow arrows in Fig. 2b. TEM has a lower detection limit of TDD of 10^7–10^8 cm^{-2} due to the relatively small volume being analyzed. Defect decoration based on chemical etching, known as EPD etching, significantly extends the lower detection limit. For example, 1 min of Everson etching (composition of the etchant: 48% HF: HNO_3: lactic acid = 1:4:25) for the CdTe (211)B surface (etching rate ~1.9 μm min^{-1}) decorates TDs by forming larger surface etch pits, which can be easily detected and counted at lower magnification, thereby facilitating the inspection of larger sample areas. The EPD measurements show that a DFL-based buffer can reduce the EPD of CdTe grown on GaAs from low-10^7 (Fig. 2d) to low-10^6 cm^{-2} (5×10^5 cm^{-2} for the best layer, as shown in Fig. 2c), which approaches the critical EPD level required for fabricating high-performance LWIR HgCdTe detectors.

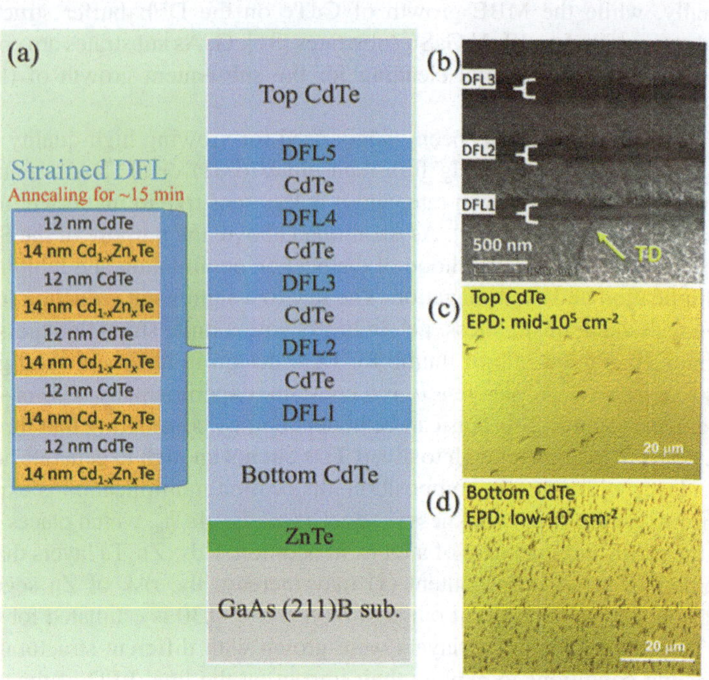

Fig. 2 (**a**) Schematic DFL-based buffer layer structure; (**b**) transmission electron microscopy (TEM) evidence for dislocation filtering by the DFLs; (**c**) and (**d**) CdTe EPD results after and before the incorporation of the DFLs, respectively

Several parameters, such as the x value in $Cd_{1-x}Zn_xTe$ layers, CTA temperature, and the number of DFLs, as summarized in our previous work [26], were found to affect the dislocation filtering efficiency (η) and therefore the final EPD value of the top CdTe layer. In the case of buffer layers with five sets of DFL grown with CTA at 320 °C, an optimal range of $0.15 < x < 0.18$ is observed, which leads to a low EPD level ranging from mid-10^5 to low-10^6 cm^{-2}, and also the highest EPD filtering efficiency of >95% within the Zn composition range of $0.08 < x < 0.27$ is studied. The CdTe DFL buffer layers demonstrate a higher efficiency in TD reduction at much lower CTA temperature in comparison to conventional CdTe/ASs [35–37], which may be beneficial in reducing the risk of structural damage during cooling with substrates having highly mismatched CTE, such as CdTe/Si, as well as reducing the risk of Ga contamination in HgCdTe layers grown on GaAs substrates.

3 Structural and Optical Analysis of CdTe DFL Buffer Layers [38]

A better understanding of the CdTe DFL buffers in terms of their detailed structural properties and associated dislocation filtering efficiency may allow them to be optimized sufficiently to achieve high-quality HgCdTe epilayers. Rapid and nondestructive characterization techniques like XRD, photoluminescence (PL), and cathodoluminescence (CL) spectroscopy have been extensively employed in the development of optoelectronic devices. As will be discussed in the following, these nondestructive techniques have yielded unique insight into both the defect density and lattice strain presented in the CdTe DFL buffer, which is crucial for the subsequent high-quality growth of HgCdTe.

3.1 XRD Characterizations

As shown by Fig. 3, high-resolution XRD characterizations with X-ray beam direction aligned with the $[1\bar{1}\bar{1}]$ direction of the substrate were carried out for reciprocal space mapping (RSM), $\omega-2\theta$, and ω scans for a CdTe DFL buffer structure ($x = 0.178$, EPD $= 5 \times 10^5$ cm^{-2}, 5-μm-thick top CdTe layer). A detailed description of XRD RSM analysis for CdTe DFL buffer layers is provided elsewhere [34]. Based on the RMS analysis, it was found that CdTe layers in/near the DFL region can be compressively strained to partially compensate for the large tensile strain presented in the $Cd_{1-x}Zn_xTe$ layers (Fig. 3b). Similar coherent growth has been previously observed in indium gallium nitride (InGaN)/gallium nitride (GaN) superlattice structures grown on GaN [39]. The in-plane lattice parameter from the coherently grown DFL region shows a lattice mismatch of −0.14% with respect to the top relaxed CdTe, which is close to the target value of −0.24% for the in-plane

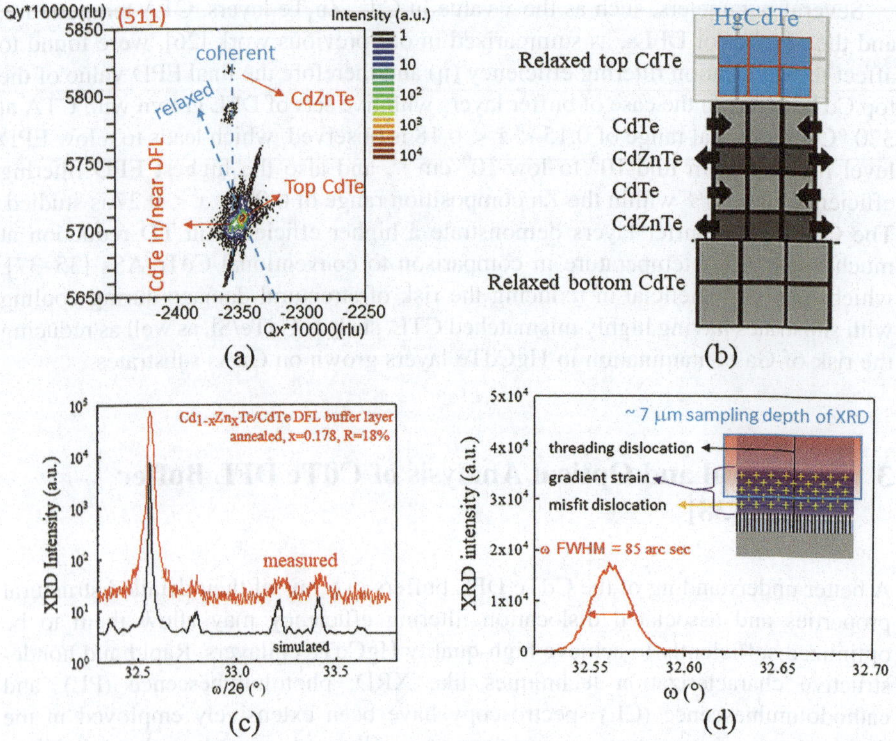

Fig. 3 (**a**) Reciprocal space mapping of the asymmetric (511) reflection of a CdTe DFL buffer. The dominant bottom peak corresponds to the relaxed top CdTe layer, and the nearby subpeak with lower intensity corresponds to the CdTe layers in/near the DFL region. The upper features consist of several interference fringes corresponding to the strained CdZnTe/CdTe DFLs. The tilted line (azure dash) crossing the top CdTe layer peak points toward the origin of reciprocal space and marks the line of fully relaxed (cubic) materials. The vertical line (azure dash-dot) represents reciprocal lattice points where the parallel lattice constant of the CdZnTe superlattice is equal to that of the CdTe in/near the DFL region: that is, a coherently strained superlattice. (**b**) Schematic illustration of stress relaxation in the SLS-DFL region: CdTe layers in/near the DFL region can be compressively strained to partially compensate for the large tensile strain presented in the CdZnTe layers. (**c**) (422) ω−2θ diffraction spectra collected from the CdTe DFL buffer layer and the theoretical simulation. The curves are shifted vertically for better visibility. (**d**) (422) ω diffraction spectrum collected for the CdTe buffer layer, and the inset indicates that three mechanisms are presumably responsible for the XRD FWHM spread: (1) threading dislocations in top CdTe, (2) compressively strained CdTe in/near DFL region, and (3) misfit dislocations in DFLs

lattice-matched growth of mid-wave infrared (MWIR) and long-wave infrared (LWIR) HgCdTe. In principle, the thicknesses of the bottom CdTe as well as the top CdTe can be reduced for the subsequent growth of HgCdTe without affecting dislocation filtering while improving the lattice matching for HgCdTe/CdTe. The coherent growth of the DFL is therefore advantageous in comparison to the conventional pseudomorphic growth in terms of better lattice matching with HgCdTe.

From the RSMs, the relaxation (R) of the CdZnTe layers with respect to the relaxed top CdTe layer is calculated to be $R = 18\%$. For this R value, the structural parameters, such as the Zn content (x) and the thicknesses of the $Cd_{1-x}Zn_xTe$/CdTe in the DFL layers, can be determined by comparing measured and simulated (422) $\omega-2\theta$ diffraction curves, as shown in Fig. 3c. From the fitting for the DFL buffer, the thicknesses of the CdZnTe/CdTe layers and the x value in the $Cd_{1-x}Zn_xTe$ layer are derived to be 14 nm/12 nm and 0.178, respectively.

The XRD FWHM extracted from ω diffraction has been widely used to screen epilayers in terms of TDD based on the empirical observation that a decrease in EPD is generally associated with smaller XRD FWHM, as observed for relaxed CdTe buffers and HgCdTe layers. Figure 3d shows the high-resolution XRD (422) ω diffraction spectrum measured for the CdTe DFL buffer ($x = 0.178$, EPD $= 5 \times 10^5$ cm^{-2}, XRD FWHM of ~85 arcsec) with a top CdTe thickness of 5 μm. It is worth noting that the CdTe DFL buffer layer does not exhibit any noticeable XRD FWHM reduction compared with the reference CdTe buffer layer without DFL (EPD of ~2×10^7 cm^{-2}, XRD FWHM of ~73 arcsec). The main reason is that XRD measurement has a relatively large sampling depth of ~7 μm, so the XRD FWHM essentially samples both the top CdTe layer and the DFL region. As a result, the residual TDs, the gradient strain of the CdTe layers in/near the DFL region, and the formation of MDs in the DFLs contribute to the XRD FWHM spread.

The correlation between the XRD FWHM spread and the strain related to the DFL buffer design has been investigated [26]. Both the CdTe DFL grown with the thickest top CdTe layer (~20 μm) and the CdTe DFL buffer grown with the lowest Zn content of $x = 0.08$ (top CdTe layer thickness ~5 μm) have a narrow XRD FWHM that is very close to the expected range of 40–50 arcsec predicted based on the empirical correlation $\left(\rho = 10^{\left[-14.6+20.8\left(1-e^{-\frac{X_f}{11.3}}\right)+2\left(1-e^{-\frac{X_f}{104.5}}\right)\right]}\right)$ established in our previous work [26]. An XRD FWHM broadening rate of 3 ± 1 arcsec/Zn% is deduced for the DFL buffer layers grown with a 5 μm top CdTe layer and CTA (cyclic thermal annealing) at 320 °C. Other parameters, such as the thickness of the top CdTe layer, the DFL annealing temperature, the number of DFLs, and the thickness of the bottom CdTe layer, are also expected to affect the XRD FWHM values.

3.2 *Optical Properties of CdTe Buffer Layers*

Optical techniques such as scanning electron microscopy-based cathodoluminescence (SEM-CL) and microscopy-based photoluminescence (μPL) also provide a fast and nondestructive approach to evaluating dislocation evolution in the heteroepitaxial CdTe buffer layers with and without DFLs. In contrast to volumetric XRD FWHM measurements, these optical techniques offer much higher spatial resolution due to the much reduced laser and electron beam sampling/

penetrating depth used in μPL and SEM-CL, respectively. Despite nonideal surfaces, interfaces, and background impurities that dominate the overall recombination velocity and screen the impact of bulk TDD, a comparison of luminescence intensities for similar structures grown in the same batch of samples and/or within a short time interval remains relevant to TDD measurements since the MBE growth conditions in terms of background impurities and fluxes can be assumed to be unchanged. In order to evaluate the dislocation filtering effect of the DFLs, both surface PL and cross-sectional SEM-CL spectroscopies were used to examine the direct band gap emission from the CdTe buffer layers at room temperature (RT) and by considering the fact that dislocations usually behave as nonradiative recombination centers that reduce optical emission.

As shown in Fig. 4a, b, a comparison of RT-PL spectroscopy was undertaken for CdTe buffer samples grown within the same MBE growth session. The peak intensities at around 1.5 eV are very well correlated with the EPD values in the range from low-10^7 to mid-10^5 cm^{-2}, suggesting that PL provides a fast and reliable nondestructive approach to evaluating the TDD for CdTe buffer layers with and without DFL. In contrast, a larger XRD FWHM does not necessarily imply higher TDD for the CdTe buffer layers with DFLs due to the gradient strain of the CdTe layers in/near the DFL region.

Figure 4c shows the depth dependence of cross-sectional SEM-CL for a DFL-based buffer layer grown on GaAs. Five CL quenching regions are evident along the growth direction for the CdTe DFL sample, corresponding to the 110 nm thick DFL layers separated by the 500 nm thick CdTe spacers. A blueshift of the CL peak is observed, which indicates the presence of compressive strain for the CdTe layers in/near the DFL region. In addition, the CL intensity of the CdTe peak at around 810 nm (1.530 eV) appears to increase from bottom to top after each DFL, whereas it remains almost unchanged in the CdTe reference sample without DFL [38], suggesting that each DFL acts as a barrier to TD propagation. As depicted in Fig. 4c, it appears that TDs are converted into misfit dislocations (MDs) that are confined within the DFL layers. This, in turn, facilitates nonradiative recombination, decreasing the CL efficiency of the DFL layers while increasing CL emission from the upper CdTe layer. Similar MD-induced CL quenching has been observed in InAs/GaAs quantum dot laser structures on (001) Si [40].

Despite these findings regarding (211)B-oriented CdTe buffers, the effectiveness of the DFL in defect reduction has also been confirmed in (133)B-oriented CdTe buffers [34]. This suggests that the DFL approach can be extended to the growth of CdTe buffer layers in other orientations, such as (001) and (111), which are widely adopted in optoelectronic devices.

4 Conclusions

In conclusion, this book chapter provides a comprehensive overview of the current state of research on defect engineering in epitaxially grown Cd(Zn)Te thin films on lattice-mismatched alternative substrates. We have reviewed our findings concerning

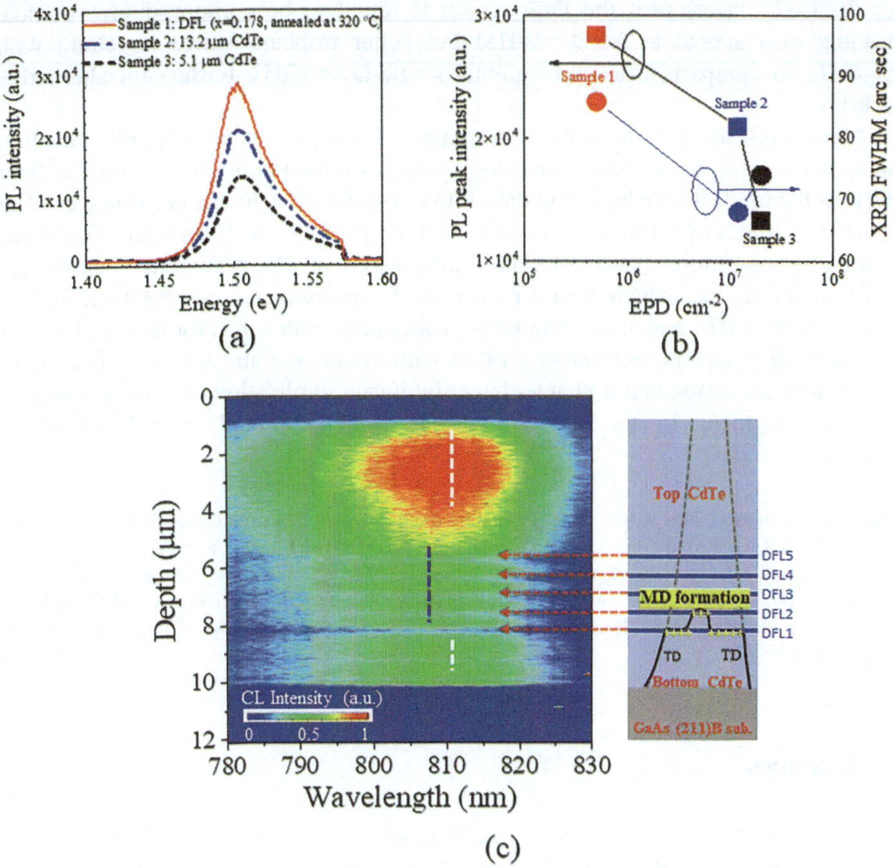

Fig. 4 (**a**) RT-PL spectra of CdTe buffer layers with/without DFL; (**b**) PL peak intensity and XRD FWHM plotted as a function of the measured EPD for different buffer layers; (**c**) left: depth dependence of cross-sectional SEM-CL for a DFL-based buffer layer grown on GaAs; right: schematic representation of approximate dislocation evolution in a CdZnTe/CdTe DFL-based buffer layer

the use of strained CdZnTe/CdTe superlattice layers as dislocation filtering layers (DFLs) for the heteroepitaxial growth of CdTe buffer layers on alternative substrates, such as III–V GaAs. The DFL approach has reduced the routine EPD from low-10^7 cm^{-2} to low-10^6 cm^{-2}, and 5×10^5 cm^{-2} for the best layer, which approaches the critical EPD level required for fabricating high-performance LWIR HgCdTe detectors. The structural and optical properties of the CdTe buffer layers were characterized, and the correlation between the structural/growth parameters and the dislocation filtering efficiency was explored. The dislocation filtering efficiency is found to be highly dependent on the balance between stress and thermal annealing, and even lower EPD can be expected with further optimization of the DFL structural design, growth parameters, and annealing conditions. Moreover, the in-plane lattice of the CdTe layers near the DFL region is found to be compressively strained, leading to a spread in XRD FWHM but better in-plane lattice matching with HgCdTe in comparison to conventional single-layer CdTe buffers on alternative substrates.

More work needs to be done to investigate the impact of growth and annealing temperature and the thickness and composition of superlattice layers on the final EPD of the top HgCdTe layer and their impact on HgCdTe device performance. The beneficial effects of CdZnTe/CdTe DFL are not limited to III–V substrates but could also be applied to the growth of CdTe buffer layers on other alternative substrates, such as Si and Ge, which would be a feasible approach to growing high-quality CdTe and HgCdTe materials on large-area alternative substrates for next-generation HgCdTe IR detectors and imaging FPAs with lower cost and larger array format. The results presented in this chapter have significant implications for future research and development in the field of general heteroepitaxial growth of II–VI semiconductors.

Acknowledgments This work was supported by the Australian Research Council (FT130101708, DP200103188, DP170104562, LP170100088, and CE200100010) and a Research Collaboration Award from the University of Western Australia (UWA). The facilities used in this work are supported by the WA Node of the Australian National Fabrication Facility (ANFF) and the Microscopy Australia Facility at the Centre for Microscopy, Characterization and Analysis (CMCA) at UWA.

References

1. Capper, P. (2017). Narrow bandgap II-VI semiconductors: Growth. In S. Kasap & P. Capper (Eds.), *Springer handbook of electronic and photonic materials* (2nd ed., pp. 343–364).
2. Lei, W., Antoszewski, J., & Faraone, L. (2015). Progress, challenges, and opportunities for HgCdTe infrared materials and detectors. *Applied Physics Reviews, 2*(4), 041303.
3. Ruffenach, S., et al. (2017). HgCdTe-based heterostructures for terahertz photonics. *APL Materials, 5*(3), 035503.
4. Bernevig, B. A., Hughes, T. L., & Zhang, S. C. (2006). Quantum spin Hall effect and topological phase transition in HgTe quantum wells. *Science, 314*(5806), 1757–1761.
5. Metzger, W. K., et al. (2019). Exceeding 20% efficiency with in situ group V doping in polycrystalline CdTe solar cells. *Nature Energy, 4*(10), 837–845.
6. Schlesinger, T. E., et al. (2001). Cadmium zinc telluride and its use as a nuclear radiation detector material. *Materials Science and Engineering: R: Reports, 32*(4–5), 103–189.
7. Berding, M. A., et al. (2000). Critical thickness in the HgCdTe/CdZnTe system. *Journal of Electronic Materials, 29*(6), 676–679.
8. Adachi, S. (2009). *Properties of semiconductor alloys: Group-IV, III-V and II-VI semiconductors* (Vol. 28). Wiley.
9. Wijewarnasuriya, P. S. (2016). Dislocation reduction in HgCdTe grown on CdTe/Si. In *Image sensing technologies: Materials, devices, systems, and applications III*. International Society for Optics and Photonics.
10. Badano, G., et al. (2010). Lattice relaxation and dislocation reduction in MBE CdTe (211) B/Ge (211). *Journal of Electronic Materials, 39*(7), 908–911.

11. Carmody, M., et al. (2012). Recent progress in MBE growth of CdTe and HgCdTe on (211)B GaAs substrates. *Journal of Electronic Materials, 41*(10), 2719–2724.
12. Pan, W., et al. (2023). Structural properties and defect formation mechanisms in MBE-grown HgCdTe on InSb (211) B substrates. *Journal of Applied Physics, 134*(11), 115303.
13. De Lyon, T. J., et al. (2010). MBE growth and transfer of HgCdTe epitaxial films from InSb substrates. *Journal of Electronic Materials, 39*(7), 1058–1062.
14. Liu, S., et al. (2015). Minority carrier lifetime of lattice-matched CdZnTe alloy grown on InSb substrates using molecular beam epitaxy. *Journal of Vacuum Science & Technology B, Nanotechnology and Microelectronics: Materials, Processing, Measurement, and Phenomena, 33*(1), 011207.
15. Lei, W., et al. (2018). Low dislocation density MBE process for CdTe-on-GaSb as an alternative substrate for HgCdTe growth. *Infrared Physics & Technology, 92*, 96–102.
16. Lei, W., et al. (2015). MBE growth of mid-wave infrared HgCdTe layers on GaSb alternative substrates. *Journal of Electronic Materials, 44*(9), 3180–3187.
17. Lei, W., et al. (2014). GaSb: A new alternative substrate for epitaxial growth of HgCdTe. *Journal of Electronic Materials, 43*(8), 2788–2794.
18. Zhao, Y., et al. (2016). Monocrystalline CdTe solar cells with open-circuit voltage over 1V and efficiency of 17%. *Nature Energy, 1*, Article number: 16067.
19. Lassise, M. B., et al. (2018). Growth of II-VI/III-V heterovalent quantum structures. *Journal of Vacuum Science & Technology B, Nanotechnology and Microelectronics: Materials, Processing, Measurement, and Phenomena, 36*(2), 02D110.
20. Jaime-Vasquez, M., et al. (2007). Nucleation of ZnTe on the As-terminated Si (112) surface. *Journal of Electronic Materials, 36*(8), 905–909.
21. Everson, W. J., et al. (1995). Etch pit characterization of CdTe and CdZnte substrates for use in mercury cadmium telluride epitaxy. *Journal of Electronic Materials, 24*(5), 505–510.
22. Yang, J. R., et al. (2008). Traces of HgCdTe defects as revealed by etch pits. *Journal of Electronic Materials, 37*(9), 1241–1246.
23. Benson, J. D., et al. (2012). Growth and analysis of HgCdTe on alternate substrates. *Journal of Electronic Materials, 41*(10), 2971–2974.
24. Chen, Y. P., Brill, G., & Dhar, N. K. (2003). MBE growth of CdSeTe/Si composite substrate for long-wavelength IR HgCdTe applications. *Journal of Crystal Growth, 252*(1–3), 270–274.
25. Johnson, S. M., et al. (1992). Effect of dislocations on the electrical and optical properties of long-wavelength infrared HgCdTe photovoltaic detectors. *Journal of Vacuum Science & Technology B: Microelectronics and Nanometer Structures Processing, Measurement, and Phenomena, 10*(4), 1499–1506.
26. Pan, W. W., et al. (2022). Defect engineering in MBE-grown CdTe buffer layers on GaAs (211) B substrates. *Journal of Electronic Materials, 51*(9), 4869–4883.
27. Peach, M., & Koehler, J. (1950). The forces exerted on dislocations and the stress fields produced by them. *Physical Review, 80*(3), 436.42.
28. Gourley, P. L., Drummond, T. J., & Doyle, B. L. (1986). Dislocation filtering in semiconductor superlattices with lattice-matched and lattice-mismatched layer materials. *Applied Physics Letters, 49*(17), 1101–1103.
29. Ward, T., et al. (2014). Design rules for dislocation filters. *Journal of Applied Physics, 116*(6), 063508.
30. Chen, S., et al. (2016). Electrically pumped continuous-wave III–V quantum dot lasers on silicon. *Nature Photonics, 10*(5), 307.
31. Pan, W., et al. (2020). Strained CdZnTe/CdTe superlattices as threading dislocation filters in lattice mismatched MBE growth of CdTe on GaSb. *Journal of Electronic Materials, 49*(11), 6983–6989.
32. Faurie, J.-P. (1994). Molecular beam epitaxy of Hg1−xCdxTe: Growth and characterization. *Progress in Crystal Growth and Characterization of Materials, 29*(1–4), 85–159.
33. Iso, K., et al. (2019). Vapor phase epitaxy of (133) and (211) CdTe on (211) Si substrates using metallic Cd source. *Journal of Electronic Materials, 48*(1), 454–459.

34. Pan, W., et al. (2023). Structural properties of MBE-grown CdTe (133) B buffer layers on GaAs (211) B substrates with CdZnTe/CdTe superlattice-based dislocation filtering layers. *Journal of Applied Physics, 133*(18), 185301.
35. Chen, Y., et al. (2008). Dislocation reduction in CdTe/Si by molecular beam epitaxy through in-situ annealing. *Journal of Crystal Growth, 310*(24), 5303–5307.
36. Bakali, E., Selamet, Y., & Tarhan, E. (2018). Effect of annealing on the density of defects in epitaxial CdTe (211)/GaAs. *Journal of Electronic Materials, 47*, 4780–4792.
37. Badano, G., et al. (2010). Reduction of the dislocation density in molecular beam epitaxial CdTe (2 1 1) B on Ge (2 1 1). *Journal of Crystal Growth, 312*(10), 1721–1725.
38. Pan, W., et al. (2022). Non-invasive and non-destructive characterization of MBE-grown CdZnTe/CdTe superlattice-based dislocation filtering layers. *Journal of Applied Physics, 131*(20), 205304.
39. Moram, M. A., & Vickers, M. E. (2009). X-ray diffraction of III-nitrides. *Reports on Progress in Physics, 72*(3), 036502.
40. Selvidge, J., et al. (2020). Defect filtering for thermal expansion induced dislocations in III–V lasers on silicon. *Applied Physics Letters, 117*(12), 122101.

Explicit Processing Steps on the Melt Growth of Stoichiometric $Cd_{0.80}Zn_{0.20}Te$ Crystal for High-Energy Radiation Detector

Ching-Hua Su

1 Introduction

The compound semiconductor $(Cd_{1-x}Zn_x)Te$ has been studied extensively in the last three decades for the application of advanced high-energy radiation sensors [1–6]. To be an efficient room-temperature high-energy radiation detector, there are two major requirements for the semiconducting cadmium zinc telluride (CdZnTe) crystal [7–11]. These requirements are (1) high electrical resistivity to reduce the bulk leakage current and (2) the reduction of structural defects that act as trapping and recombination centers. The electrical property of a compound semiconductor is dictated by its impurity contents, deviation from stoichiometry, and doping level. To consistently achieve the high electrical resistivity requirement, besides starting with ultra-high purity materials, the growth processes need to reproduce the stoichiometry of the grown crystals toward the theoretically intrinsic 1 to 1 ratio of the elements on the two sublattices to minimize the electric carriers caused by the native point defects as well as to apply intentionally proper doping for compensation. On the second requirement, the structural defects in the processed CdZnTe crystals, such as grain boundaries, twins, and subgrain boundaries, have been improved recently except that the occurrences of Te-rich precipitates can still deteriorate the X-ray and gamma-ray spectroscopic response of the detectors [12, 13]. When the composition of the initially solidified crystal has a deviation from stoichiometry, the extra element will likely precipitate into the second phase during the cooling of the processed ingot because of its retrograde solubility. Therefore, both requirements can be satisfied by processing $(Cd_{0.80}Zn_{0.20})_{1/2}Te_{1/2}$ crystals with the stoichiometric composition; i.e., $x_{Te} = 0.5000$. However, the precise control of the stoichiometry is difficult because

C.-H. Su (✉)
Materials and Processing Laboratory, EM31 NASA/Marshall Space Flight Center, Huntsville, AL, USA
e-mail: ching.h.su@nasa.gov

K. Iniewski (ed.), *CdTe and CdZnTe Materials*,
https://doi.org/10.1007/978-3-031-64521-1_3

the shift of the melt/solid composition is caused by the loss of different amounts of elements to the vapor phase during growth at elevated temperatures. At the same time, the partial pressures in equilibrium with the solid compound over the narrow homogeneity vary by orders of magnitude from metal-rich to Te-rich boundaries [14] and can be used to accurately control the stoichiometry of the grown crystals by applying controlled partial pressures over the melt during the crystal growth process [15].

Recently, the partial pressures of Cd in equilibrium with the stoichiometric composition of the melt and solid of $Cd_{0.80}Zn_{0.20}Te$, i.e., $x_{Te} = 0.5000$, have been determined as functions of temperature [16], and the directional solidification of $Cd_{0.80}Zn_{0.20}Te$ has been processed with controlled Cd overpressure [17] with several meticulous practices to enhance the yield of bulk CdZnTe single crystals of stoichiometric composition with high crystallinity quality. These practices include (A) initializing the homogenization of CdZnTe from pure elements into the ternary starting material via a traveling zone process, (B) diminishing the contamination of copper (Cu) and other elements by homogenizing the starting material ampoule under external vacuum conditions, (C) reducing the interaction between the sample and the fused silica ampoule during growth through the etching of hydrogen fluoride (HF) with the internal surface of the empty growth ampoule before loading the starting material, (D) promoting the yield of single crystallinity by applying a mechanical pulsed perturbation to the growth ampoule during the nucleation stage of the crystal in the melt, and (E) consistently producing stoichiometric CdZnTe crystals via directional solidification and programming postgrowth furnace cooldown under controlled Cd overpressure.

In general, the ultimate goal is to establish a crystal growth process that is able to reproduce high-purity, stoichiometric CdZnTe crystals of high-quality crystallinity with an insignificant amount of Te precipitates. Without getting into detailed results from the previous basic research, the explicit processing steps on the melt growth of the stoichiometric $Cd_{0.80}Zn_{0.20}Te$ crystal that is consistently qualified for the applications of room temperature high-energy radiation detector will be presented below.

2 Homogenization

Prior to the melt growth processes, there are several important procedures that can improve the quality of the grown crystal during the preparation of the growth ampoules.

2.1 Homogenization Ampoule

During the homogenization process, the small chunks of the starting elements of Cd, Zn, and Te, preferably seven-nines purity or a higher grade, with the single dimension of each chunk less than 15 mm, were weighed to an accuracy level of 10 μg. The homogenization and growth ampoules were first cleaned with HF acid for 30–60 s and then rinsed repeatedly with acetone, methanol, and distilled H_2O. The loaded ampoule was sealed after it had been evacuated to a vacuum lower than 10^{-5} torr. Usually, there were two concerns in the process of homogenizing elements into alloys. Firstly, because of the large mass of the sample, the ampoules were sometimes explored inside the rocking furnace during heating up. It was concluded that the exothermic eutectic reaction between the Group II elements, Cd and Zn, and Group VI element, Te, released a large amount of energy, which rapidly raised the sample temperature and caused the instantaneous jump of vapor pressure inside the ampoule. Secondly, the adhesion of the sample to the silica wall, or “wetting,” was often observed after the homogenization and crystal growth process. Not only did wetting cause the sample to fracture during cooling, but it also contaminated the sample by the incorporation of silicon (Si) impurities. In an experiment on cadmium sulfide (CdS) crystal growth by vapor transport [18] and other investigations [19], it was found that the reaction of Cd with silica might be the cause of the wetting [20]. Even in graphitized, a.k.a. carbon-coated, ampoules, which are commonly used to prevent wetting, Si peaks, although much weaker than in the ungraphitized case, were still found in the EDX (energy dispersion of X-rays) spectra [18]. To reduce the occurrence of these two incidents, the loaded homogenization ampoule was designed to first go through a localized eutectic reaction. The ideal procedure is similar to a zone melting process in that a single heated zone, with a maximum temperature of 520 °C to induce localized Te eutectic reaction (about 450 °C), was translated over the vacuum-sealed ampoules at a translation rate of 15–20 cm/h to initiate the eutectic exothermic reaction locally. The formation of Cd/Te and Zn/Te spongy mixture from this reaction lowers the chemical potential of Cd and reduces the reactivity between Cd and fused silica during high-temperature processing.

2.2 *Homogenized Under External Vacuum Environment*

In the early stage of the project, the sealed ampoule, filled with elements, was loaded directly inside a tubular rocking furnace to melt and mix the elements without zone melting. The ampoule was heated up to 1070 °C (about 20 °C lower than the melting temperature); soaked for 36 h to improve its uniformity by solid-state diffusion; raised to 1160 °C, i.e., about 40 °C above its melting point; and then rocked horizontally for 3–5 h before casting by lowering the ampoule to about 2 cm outside the bottom of the furnace, with the furnace holding at an angle of 45° lower than the horizon. However, the electrical resistivity of the grown crystals was low, and the impurity level was high. As shown in the first column of Table 1, the Hall measurements on a slice of sample cut at a distance of 2.5 cm from the tip of the grown crystal of $(Cd_{0.8}Zn_{0.2})Te$, CZT-2, resulted in a low electric resistivity of 100 ohm-cm. The levels of impurities in the slice were measured by GDMS (glow discharge mass spectroscopy), and the elements of high impurity and their measured levels are shown in the first two columns of Table 1, which indicate high levels of carbon (C),

Table 1 The measured electrical resistivity, ρ (given in the first row), and impurity levels (in ppb, atomic, i.e., part per billion atoms, given in three columns on the right with the elements listed in the left column) from three disks cut from two as-grown ($(Cd_{0.8}Zn_{0.2})Te$) ingots

Sample	CZT-2; 2.5 cm ($\rho \sim 100$ ohm-cm)	CZT-9; 2.5 cm ($\rho \sim 10^{10}$ ohm-cm)	CZT-9; 5.3 cm
Element	ppb, atomic	ppb, atomic	ppb, atomic
C	980	20	24
N	45	15	20
O	980	35	40
Na	12	54	60
Mg	170	27	59
Al	26	7	<5
Si	11	8	<1
S	230	280	130
Cl	56	49	20
Cr	28	67	580
Mn	<10	60	140
Fe	480	93	300
Ni	23	46	80
Cu	3300	110	130
Se	160	210	220

The disks were (given in the first row) one slice cut at 2.5 cm from first-freeze (FFF) tip of the CZT-2 ingot and two slices cut at 2.5 cm and 5.3 cm FFT tip of the CZT-9 ingot

oxygen (O), and iron (Fe) contents and an extremely high level of Cu, 3300 ppb, atomic. Hence, the process was modified to insert the sealed ampoule first inside another closed fused silica tube, which was evacuated to 10^{-3} torr continuously and was then placed inside the tubular rocking furnace for homogenization. With the extra layer of vacuum provided by the outer fused silica tubing, the contamination from Cu and other elementals was minimized, as shown in Table 1. The impurity levels of the slices cut at 2.5 cm and 5.3 cm from the CZT-9 crystal, grown from the starting material produced by the modified homogenization, are given in columns 2 and 3, respectively, and have improved with lower contaminations in C and O and a much lower Cu content from 3300 to 110 ppb, atomic. This improvement has caused the electrical resistivity measured on the 2.5 cm slice of CZT-9 to increase to 10^{10} ohm-cm. Without the "wetting" of the ampoule's inner wall, the cast ingots usually slid inside the ampoule, and the homogenized samples were retrieved as starting materials for crystal growth.

The other simpler procedure, without going through the zone melting procedure, has also been practiced by controlling the localized eutectic reaction slowly during the homogenization process. The whole idea of eliminating "wetting" is to decrease Cd activity in the sample so as to reduce the rate of reaction between Cd and the silica wall. This may be done by first heating the sample to an intermediate temperature. This temperature should be low enough so that the activity of pure Cd is too low to cause a serious reaction between Cd and the silica wall but high enough to start a homogenization reaction between the elements. Once the elements react, the Cd activity in the sample is greatly reduced from its value of pure Cd. Thus, further heating to a temperature above the liquidus temperature will not cause significant reactions between Cd and the silica wall. Hence, after the ampoule was loaded with pure elements and then evacuated and sealed, it was loaded inside the rocking furnace with the extra fused silica tube and heated to around 500 °C slowly (in about 2 h) and was held there for another 20 h or longer. Then the same homogenization process described above was used for the rest of the procedure. No wetting was observed either after the sample was cooled down.

3 Crystal Growth Ampoule Preparation

During the processing of crystal growth at elevated temperature, the CdZnTe sample has a long time to interact with the fused silica container and form chemical bonding, the so-called "wetting" described above, which makes the sample attached to the ampoule wall. During the cooldown of the sample after growth, as the bulk of the sample goes through a larger thermal contraction than that of the fused silica, the wetting area remains adhered to the fused silica wall, which, consequently, causes the sample to internally separate, leading to cracks and other structural defects. These defects are detrimental to the electronic/optic performance of materials. A thin graphite coating for the inner wall of the ampoule is not recommended because it can cause chemical contamination of the grown crystals as patches of the thin carbon film sometimes disappear during growth due to its reduction by oxide impurities.

The severity of the chemical interaction between the sample and the fused silica inner wall depends on several factors: (1) the processing temperature, (2) the duration at the temperature, (3) the chemical activity of the sample, and (4) the contact area of interaction between the elements and the fused silica. However, once the material system and the crystal growth technique are decided, factors 1 and 2 are rather inflexible. In factor 3, the chemical activity of the sample, especially Cd, has been reduced through zone melting or by soaking the sample at low temperatures, as discussed above. To further reduce the interaction between the samples and the fused silica ampoules, procedures to promote factor 4, i.e., to reduce the contact area of interaction, have been devised. The total surface area that the grown crystal contacting the container depends on the surface features of the inner wall. A very flat and smooth surface provides a large contact area, whereas a rough surface, with peaks and valleys, limits the contact area to the peaks due to the surface tension of the melt/crystal. With this concept, a simple method to modify the surface morphology of the fused silica inner wall was developed by etching it with HF acid before crystal growth activity. Typically, HF rinsing for 30 s to 2 min has been the widely adopted method for cleaning the surface of fused silica inner tubing before any processing activities. However, it was found that long-time HF etching for 30 min can modify the surface features of fused silica into peak and valley and can affect the interaction between the sample and its container [21]. The effects of HF etching have also been studied recently on the surface micromorphology, roughness, and wettability of dental glass ceramics, such as lithium disilicate [22–24]. HF etching is only needed for the lower section of the ampoule where the sample, either melt or solid, is in contact with the ampoule's inner wall. The method is applicable to all processes involving the interaction between materials and fused silica containers at elevated temperatures. After HF etching, the empty growth ampoule went through the same procedures for cleaning and was baked out overnight under vacuum procedures, as reported in Sect. 2.1.

4 Directional Solidification

4.1 *Tapping to Induce Nucleation*

During the measurement of the differential thermal analysis (DTA) of a CdTe ampoule, the sample was heated up rapidly to above its melting point of 1090 °C and was soaked for 9 h before starting a furnace cooling rate of about 10 °C/min. During the cooling of the furnace, the sample temperature remained at 1090 °C and then suddenly started to decrease when the furnace reached 1040 °C, i.e., a supercool of 50 °C [25]. In the process of directional solidification, a high supercool not only results in a large section of ingot solidifying uncontrollably with spontaneous nucleation but also prohibits the ideal growth condition of small single-crystal nuclei forming at the very tip of the ampoule and growing into large single grains and, consequently, also results in the low yield of single crystals. To prevent the undesired formation of a large multigrained spontaneous nucleation, i.e., to promote nucleation under the condition of small supercooling, a short-time mechanical perturbation was applied to the growth ampoule when the melt at the ampoule tip is just below the liquidus temperature, as described below.

4.2 *Growth Process*

The starting CdZnTe compound was retrieved from the homogenized ampoule and was broken into chunks with a linear dimension smaller than 5 mm. These chunks from the same homogenization ampoule need to be mixed well to maintain uniform compositions. After the desired amount of starting material was weighed and loaded inside the growth ampoule, a fused silica cup holding about 2–3 g of pure Cd was fixed by two small, restricted bumps made inside the tube at the top of the ampoule just below the seal-off cup. It served as the Cd reservoir to control Cd overpressure during growth. The exact location of the Cd reservoir relative to the sample will be determined from the thermal profile of the crystal growth furnace, as discussed below. The ampoules were then sealed under a vacuum lower than 10^{-5} torr.

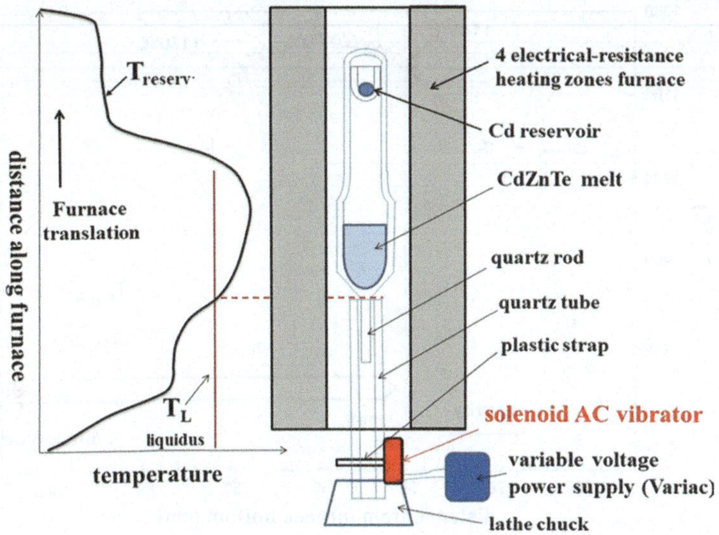

Fig. 1 Schematics of the tapping arrangements at bottom to provide mechanical perturbation, four zones growth furnace with growth ampoule inside, thermal profile, the temperatures of liquidus, T_L, and the Cd reservoir at temperature T_{reserv}

The mechanical tapping arrangement is shown schematically in Fig. 1. The tapered tip of the growth ampoule has an extended fused silica rod supported by a fused silica tube, which was fixed vertically by a lathe chuck on the floor. The mechanical perturbation provided for the growth experiments was a commercially available 60 Hz solenoid AC vibrator, which was coupled to the base of the supporting fused silica tube by a plastic strap. The magnitude of the perturbation was adjusted by varying the power input of a variable voltage power supply, Variac, from 0 to 120 V. The vertical Bridgman growth furnace, as shown in Fig. 1, had four independently controlled electrical-resistance heating zones. They are, starting from top to bottom, the Cd reservoir, the hot, the booster, and the cold zones. Figure 1 also illustrates schematically the thermal profile inside the furnace bore and the initial position of the growth ampoule, with the sample staying above the liquidus temperature, T_L.

The initial positions of the growth ampoule, including the Cd reservoir, relative to the thermal profile of a typically vertically directional solidification crystal growth furnace is given in Fig. 2. With an isothermal furnace liner, or heat pipe, at the top zone, the temperature of the Cd reservoir, T_{reserv}, is situated at the upper end of a 20 cm long "Cd reservoir zone" of constant temperature, usually kept between 805 and 820 °C, in this case about 810 °C. The hot zone and booster zone provided a maximum of 1170 °C for the melt and, together with the cold zone, a thermal gradient of about 12 °C/cm between the roughly linear thermal profile of 1140 and 1040 °C.

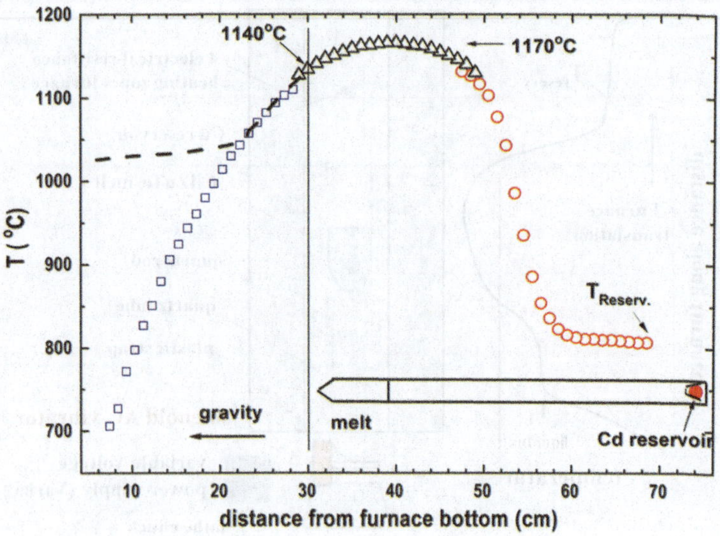

Fig. 2 The thermal profile data and the initial ampoule position for typical crystal growth by vertically directional solidification

To design a crystal growth ampoule for the thermal profile, two things need to be paid attention to. Firstly, the melt needs to be in a vertically thermally stabilized condition, i.e., the melt temperature monotonically increases from the bottom to the top of the melt, to prevent bubbling in the melt, which generates extremely violent fluid flows. In the case shown in Fig. 2, with the ampoule tip at a 31 cm location and the thermal maximum around 40 cm, the total length of the sample melt should be less than 9 cm. Secondly, the distance between the top surface of the sample and the seal-off location for the ampoule needs to be adjusted for each growth furnace. In the case shown in Fig. 2, this distance was designed to be about 40 cm so that after the 8 cm long sample is completely solidified, the Cd reservoir temperature can still be under control in the isothermal "Cd reservoir zone."

It was reported by Rudolph et al. [26, 27] that in compounds of high ionicity, such as II–VI compounds of CdTe, ZnTe, and zinc selenide (ZnSe), the melt retains a high degree of molecular association at the melting point. Therefore, overheating the melt to about 20–30 K above the melting temperature was needed to reduce the melt association and thus improve crystal quality by lowering the number of large-angle grain boundaries and twins [27]. It has also been reported that during the DTA measurements on the melting of CdTe [28, 29], the initial fragmentation of solid into clusters dominated in the small superheating range, about 5 °C. Therefore, further heating of the sample up to 20–40 °C above the melting temperature resulted in the dissociation of the cluster into separate atoms and molecules.

A detailed procedure for the growth of CdZnTe crystals with mechanically induced nucleation is as follows:

(i) Position the growth ampoule along the thermal profile inside the furnace and heat the furnace such that the complete sample is 20–40 °C above the liquidus temperature (1119 °C) when the furnace temperature settings have been reached, as shown in Fig. 2.
(ii) Soak the melt for at least 24 h.
(iii) Start the furnace translation moving upward at a rate of 1.0–1.25 mm/h.
(iv) When the ampoule tip is at a position in the furnace where the temperature is 5–8 °C below the liquidus temperature, apply mechanical tapping by running the solenoid vibrator for 20–30 s at a voltage of about 50 V.
(v) Continue with the growth by translating the furnace at a rate of 1.0–1.25 mm/h until the sample section is completely below the solidus temperature (1093 °C).
(vi) Program the cooldown of the grown ingot and Cd reservoir.

4.3 Cd Reservoir Temperature

With the measured liquidus temperature of 1119 °C from the partial pressure measurements of Cd [16], a best fit to the measured P_{Cd} of the stoichiometric $(Cd_{0.80}Zn_{0.20})_{1-x}Te_x$ melt (x = ½) shows a straight line in the plot of log (P_{Cd}) vs. 1/T as:

$$\log_{10} P_{\mathrm{Cd}}(\mathrm{atm}) = -10,560/T(\mathrm{K}) + 7.702 \quad \text{for } T > 1119\ {}^{\circ}\mathrm{C} \tag{1}$$

Similarly, the best fit for P_{Cd} in equilibrium with the stoichiometric (Cd0.80Zn0.20)1-xTex solid (x = ½) is given by [16]:

$$\log_{10} P_{\mathrm{Cd}}(\mathrm{atm}) = -13,010/T(\mathrm{K}) + 8.928 \quad \text{for } T < 1093\ {}^{\circ}\mathrm{C} \tag{2}$$

These two equations for P_{Cd} over the stoichiometric melt and solid are plotted in Fig. 3 as red and black lines, respectively, and connected by a dotted black line. For comparison, the measured values of P_{Cd} in equilibrium with the stoichiometric $(Cd_{0.96}Zn_{0.04})Te$ crystals, to minimize the Te inclusions/precipitates for the application as the substrate for the growth of HgCdTe infrared (IR) detector, reported by Franc et al. [30] is plotted as a blue dashed line in Fig. 3.

The temperature of the Cd reservoir during crystal growth is dictated by the temperature of the $Cd_{0.8}Zn_{0.2}Te$ melt/solid. The Cd pressure over pure Cd, P^{o}_{Cd}, can be expressed as a function of temperature [31] by:

$$\log_{10} P^{o}_{\mathrm{Cd}}(\mathrm{atm}) = -5,317/T(\mathrm{K}) + 5.119 \tag{3}$$

To correlate the temperatures of the Cd reservoir to those of the melt, we set the value of P_{Cd} in Eq. 1, calculated from the melt temperature, T_M, equal to the pressure over pure Cd, given by Eq. 3, calculated by using the temperature of the Cd reservoir, T_{R}. Then the value of T_{R} can be related to T_{M} as:

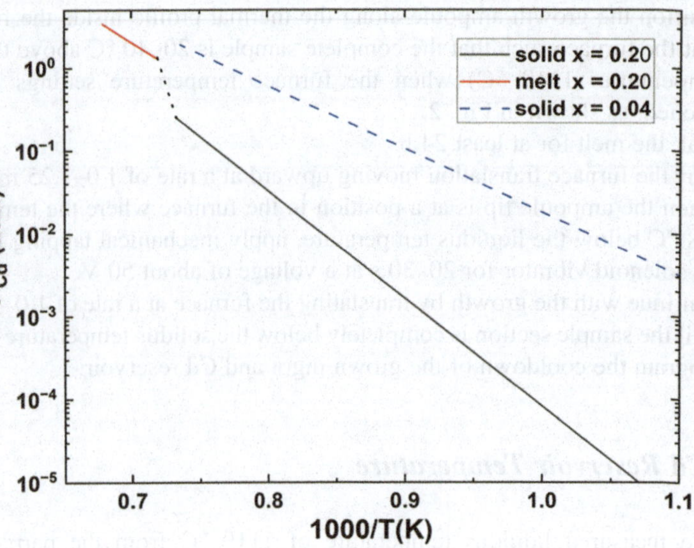

Fig. 3 The partial pressure of Cd, P_{Cd}, over the stoichiometric $(Cd_{0.80}Zn_{0.20})Te$ melt and solid and $(Cd_{0.96}Zn_{0.04})Te$ solid as functions of temperature in 1000/*T*(K)

$$T_R(^\circ C) = \frac{5317 \times T_M(K)}{10560 - 2.583 \times T_M(K)} - 273.1 \quad \text{for } T_M > 1119 \ ^\circ C \tag{4}$$

Similarly, the values of T_R as a function of the temperature of a growing solid crystal, T_S, can be derived from Eqs. 2 and 3 as:

$$T_R(^\circ C) = \frac{5317 \times T_S(K)}{13010 - 3.809 \times T_S(K)} - 273.1 \quad \text{for } T_S < 1093 \ ^\circ C \tag{5}$$

From Eqs. 4 and 5, the temperature of the Cd reservoir is plotted as functions of the sample temperature of the melt, in red line, as well as the temperature of the solid, in black line.

5 Cooling of the Grown Sample and the Cd Reservoir

The furnace translation needs to stop at the end of growth, i.e., when the temperature at the top of the ingot decreases to below 1093 °C, and then the Cd reservoir temperature, T_R, needs to be programmed as a function of the solid sample's temperature, T_S, given in Eq. 5 and shown in Fig. 4. The temperature of the whole grown crystal, T_S, usually covers a range from the first-freeze tip to the top of the crystal. It is recommended to provide a flat furnace thermal profile along the crystal,

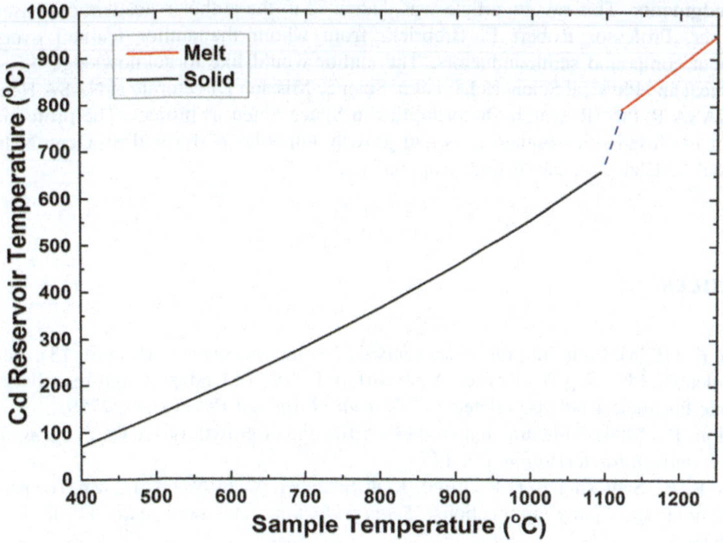

Fig. 4 The temperature of the Cd reservoir plotted as functions of the sample temperature of the melt, in red line, and the temperature of the solid, in black line

shown as a black dashed line in Fig. 2, to narrow down this range with the help of an isothermal furnace liner so that the whole grown ingot can be cooled down roughly isothermally under the same Cd overpressure. The solidified sample was usually cooled down to room temperature with constant rates over a period of 96–144 h. Even without the wetting of the crystal via the practice of HF etching, described in Sect. 3, the thermal gradient inside the crystal can generate a serious strain from the uneven thermal expansion caused by the large nonlinear internal thermal gradient. The slow cooling rate will reduce the internal thermal gradient and minimize the strain.

6 Discussions

The processing details presented above will improve $Cd_{0.80}Zn_{0.20}Te$ crystals grown by directional solidification with controlled Cd overpressure, contamination control under vacuum, yield of high-quality crystallinity, and consistent stoichiometry of $x_{Te} = ½$. The production of high electrical resistance will depend on the impurity contents of the starting elements employed for the growth. From the experience of our practice with the seven-nines purity elements, the doping of In, 4–6 ppm, atomic, will compensate the acceptor levels from the residual impurity or the native acceptors and result in high electrical resistivity.

Acknowledgments This review article was dedicated to the author's academic adviser and life-long teacher, Professor Robert F. Brebrick, from whom the author learned everything he knows about compound semiconductors. The author would like to acknowledge the support of the Biological and Physical Sciences Division, Science Mission Directorate at NASA Headquarters, and the NASA ROSS (Research Opportunities in Space Science) project. The professional glass blowing of the fused silica optical cells and growth ampoules performed by Gene Nelson of the A. G. Scientific Glass Co. Inc. is highly appreciated.

References

1. Zanio, K. (1978). Cadmium telluride. *Semiconductors and semimetals* (Vol. 13). Academic.
2. Triboulet, R., Marfaing, Y., Cornet, A., & Siffert, P. (1974). Undoped high-resistivity cadmium telluride for nuclear radiation detectors. *Journal of Applied Physics, 45*, 2759.
3. Rudolph, P. (1994). Fundamental studies on Bridgman growth of CdTe. *Progress in Crystal Growth and Characterization, 29*, 275.
4. James, R. B., Schlesinger, T. E., Lund, J., & Schieber, M. (1997). $Cd_{1-x}Zn_xTe$ spectrometers for gamma and X-ray applications. *Semiconductors and semimetals* (Vol. 43, p. 335). Academic.
5. Schlesinger, T. E., Toney, J. E., Yoon, H., Lee, E. Y., Brunett, B. A., Franks, L., & James, R. B. (2002). Cadmium zinc telluride and its use as a nuclear radiation detector material. *Materials Science and Engineering, 32*, 103.
6. Del Sordo, S., Abbene, L., Caroli, E., Mancini, A. M., Zappettini, A., & Ubertini, P. (2009). Progress in the development of CdTe and CdZnTe semiconductor radiation detectors for astrophysical and medical applications. *Sensors, 9*, 3491.
7. Bolotnikov, A. E., Black, M., Camarda, G. S., Carini, G. A., Cui, Y., Kohman, K. T., Li, L., Salomon, M. B., & James, R. B. (2006). The effects of Te precipitates on measurable characteristics of CdZnTe detectors. *SPIE Proceeding, 6319*, 631903.
8. Rudolph, P., Engel, A., Schentke, I., & Grochocki, A. (1995). Distribution and genesis of inclusions in CdTe and (Cd,Zn)Te single crystals grown by the Bridgman method and by the travelling heater method. *Journal of Crystal Growth, 147*, 297.
9. Szeles, C., & Driver, M. C. (1998). Growth properties of semi-insulating CdZnTe for radiation detector applications. *SPIE Proceeding, 3446*, 1.
10. Amman, M., Lee, J. S., & Luke, P. N. (2002). Electron trapping nonuniformity in high-pressure-Bridgman-grown CdZnTe. *Journal of Applied Physics, 92*, 3198.
11. Elshazly, E. S., & Tepper, G. (2008). Correlation of tellurium inclusions and carrier lifetime in detector grade cadmium zinc telluride. *Applied Physics Letters, 93*, 042112.
12. Carini, G. A., Bolotnikov, A. E., Camarda, G. S., Wright, G. W., James, R. B., & Li, L. (2006). Effect of Te precipitates on the performance of CdZnTe detectors. *Applied Physics Letters, 88*, 143515.
13. Szeles, C. (2004). Advances in the crystal growth and device fabrication technology of CdZnTe room temperature radiation detectors. *IEEE Transactions on Nuclear Science, 51*, 1242.
14. Fang, R., & Brebrick, R. F. (1996). CdTe I: Solidus curve and composition-temperature-tellurium partial pressure data for Te-rich CdTe(s) from optical density measurements. *Journal of Physics and Chemistry of Solids, 57*, 443.
15. Su, C.-H., & Lehoczky, S. L. (2011). Melt growth of high-resistivity CdZnTe crystals by controlling Cd over-pressures. *Journal of Crystal Growth, 319*, 4.
16. Su, C.-H. (2022). Partial pressure of Cd in equilibrium with stochiometric melt and solid of $(Cd_{0.8}Zn_{0.2})Te$ between 300 °C and 1250 °C for melt growth. *Journal of Crystal Growth, 596*, 126837.

17. Su, C.-H. (2020). Growth process of $Cd_{0.8}Zn_{0.2}Te$ crystal to enhance its performance as detector for high-energy radiation. *Current Smart Materials, 5*, 1.
18. Su, C.-H. (1987). Growth rate of CdS by vapor transport in a closed ampoule. *Journal of Crystal Growth, 80*, 333.
19. Taguchi, T., Shirafuji, J., & Inuishi, Y. (1978). Growth by travelling heater method and characteristic of undoped high-resistivity CdTe. *Japanese Journal of Applied Physics, 17*, 1331.
20. Ching-Hua, S., Lehoczky, S. L., & Szofran, F. R. (1986). A method to eliminate wetting during the homogenization of HgCdTe. *Journal of Applied Physics, 60*, 3777.
21. Su, C.-H. (2020). Improving quality of materials by reducing its interaction with fused silica container during processing. *Crystal Research and Technology, 55*, 1900208.
22. Puppin-Rontani, J., Sundfeld, D., Costa, A. R., Correr, A. B., Puppin-Rontani, R. M., Borges, G. A., Sinhoreti, M. A. C., & Correr-Sobrinho, L. (2017). Effect of Hydrofluoric acid concentration and etching time on bond strength to lithium disilicate glass ceramic. *Operative Dentistry, 42*, 606.
23. Ramakrishnaiah, R., Alkheraif, A. A., Divakar, D. D., Matinlinna, J. P., & Vallittu, P. K. (2016). The effect of hydrofluoric acid etching duration on the surface micromorphology, roughness, and wettability of dental ceramics. *International Journal of Molecular Sciences, 17*, 822.
24. Sundfeld Neto, D., Naves, L. Z., Costa, A. R., Correr, A. B., Consani, S., Borges, G. A., & Correr-Sobrinho, L. (2016). The effect of hydrofluoric acid concentration on the bond strength and morphology of the surface and interface of glass ceramics to a resin cement. *Operative Dentistry, 40*, 470.
25. Su, C.-H. (2015). A method of promoting single crystal yield during melt growth of semiconductors by directional solidification. *Journals of Crystal Growth, 419*, 35.
26. Saucedoa, E., Rudolphb, P., & Dieguez, E. (2008). Modified Bridgman growth of CdTe crystals. *Journal of Crystal Growth, 310*, 2067.
27. Rudolph, P. (2024). Contributions to the development of crystal growth technologies. *Journal of Crystal Growth, 625*, 127456.
28. Shcherbak, L., Feichouk, P., & Panchouk, O. (1996). Effect of CdTe "postmelting". *Journal of Crystal Growth, 161*, 16.
29. Shcherbak, L. (1999). Pre-transition phenomena in CdTe near the melting point. *Journal of Crystal Growth, 197*, 397.
30. Franc, J., Grill, R., Hlidek, P., Belas, E., Turjanska, L., Hoschl, P., Turkevych, I., Toth, A. L., Moravec, P., & Sitter, H. (2001). The influence of growth conditions on the quality of CdZnTe single crystals. *Semiconductor Science and Technology, 16*, 514.
31. Yu, T.-C., & Brebrick, R. F. (1992). The Hg-Cd-Zn-Te phase diagram. *Journal of Phase Equilibria, 13*, 476.

Advancements in CdZnTe Detectors with the Drift Strip Method

I. Kuvvetli, S. R. H. Owe, and C. Budtz-Jørgensen

1 Introduction

Despite the promising properties of compound CdZnTe (CZT) material as a room temperature semiconductor detector, the material suffers from inefficient hole collection, which compromises its performance. Addressing this challenge, modifications in electrode design offer a solution by isolating the electron drift contribution to signal formation. At the Danish Space Research Institute (DSRI), today DTU Space, significant efforts have been made in the 3D CZT drift strip detector technology, yielding high-intrinsic spatial and spectral resolution, while keeping the number of electronic readout channels at a minimum. This chapter reviews the evolution of this technology from early development up until today, highlighting its unique capabilities, applications, and future research directions.

2 Early Development Phase: Proof of Concept

The detector group at DTU Space) initiated a CZT room temperature semiconductor detector development program in the beginning of 1996. In this period, the detector group was heavily involved in the development of the Joint European X-Ray Monitor (JEM-X) for the ESA INTEGRAL space mission [1]. The imaging detector was based on the gas microstrip technology covering the energy range from 3 to 35 keV [2]. The gas detector program was terminated with the successful launch of JEM-X on-board the INTEGRAL satellite in 2002. JEM-X is in fact still

I. Kuvvetli (✉) · S. R. H. Owe · C. Budtz-Jørgensen
DTU Space, Technical University of Denmark, Kongens Lyngby, Denmark
e-mail: irfan@space.dtu.dk

K. Iniewski (ed.), *CdTe and CdZnTe Materials*,
https://doi.org/10.1007/978-3-031-64521-1_4

operative.[1] The change to solid state detector development was primarily caused by the advent of so-called room temperature detectors. The novel semiconductor materials, especially CZT offered many advantages compared to gas-based systems, such as:

- Extended energy range into the MeV region due to better stopping power at higher energies.
- Improved energy resolution (<1% @ 662 keV) due to minimum ionization energy.
- Compact detector bulk, with potential sub-millimeter 3D interaction position resolution using orthogonal novel strip electrode geometry.
- Stable performance without sparking and breakdowns common for proportional counters.

Moreover, CZT detectors do not require cryogenic cooling as is the case for conventional semiconductor materials, which is especially favorable for space applications.

An important goal for the DSRI detector development was to develop focal plane detectors for the novel type of hard X-ray telescope [3] which was being studied at DSRI. With the advent of the Danish Small Satellite program, it was furthermore recognized that CZT detectors were very well suited for two proposed missions: eXCALIBur and AXO [4].

2.1 *Improving Spectroscopic Performance of CZT Detectors*

The semiconductor detector research program at DSRI was in the beginning, focused on improving the spectroscopic properties of CZT detectors. Although the quality of CZT semiconductor material had improved substantially at the time, it was still a drawback that most single crystal CZT semiconductors contain defects, which can trap the charge carriers generated by the ionizing radiation. This will perturb the signal generation and, as a consequence, reduce the energy resolution significantly of the detector. The charge transport problem is most severe for holes, which have trapping probabilities that are more than an order of magnitude higher than the one of electrons. Various methods already existed to diminish the effects caused by hole trapping. These methods are divided into two groups:

1. *Signal rejection*, where the detector signals with large charge contribution from the holes are identified and rejected, for example, by rise-time discrimination [5] and dual shaping time [6]. With these techniques, many events are rejected to obtain improved spectral resolution.

[1] As of April 2024.

2. *Reducing hole contribution* to the detector signal, for example, by the coplanar technique [7]. This is achieved by subtracting the signals on two sets of coplanar strips yielding a signal that is nearly independent of hole contribution.

The detector group at DSRI looked into methods on how to reduce the hole contribution to the signal, with inspiration from solutions found in gas detectors, resulting in the Frisch strip solution.

2.2 Frisch Strip Solution

The situation of slow-moving holes and fast electrons is analogous to the situation found in gas detectors [7], wherein the positive ions move much slower than the electrons and contribute minimally to the detector signal. For gas detectors, the problem is solved by using the so-called Frisch grid. The Frisch grid screens the readout electrodes from the influence of the slow-moving positive ions, making the detector device sensitive to the fast moving electrons alone. It has been shown that Fano-factor-limited energy resolution can be obtained with Frisch-gridded ionization chambers [8]. The method presented reduces sensitivity to hole trapping by utilizing Frisch strips as an analog to the Frisch grid. These strips consist of broad electrode segments placed between narrow ones, with the narrow segments serving as readout electrodes, also known as anodes. While the broad Frisch strips shield the readout electrode from slow-moving holes, the bias voltage between the Frisch strips and the readout electrode ensures the drift and collection of charge carriers by the readout electrode. Figure 1 illustrates a cross section of a detector using the principle

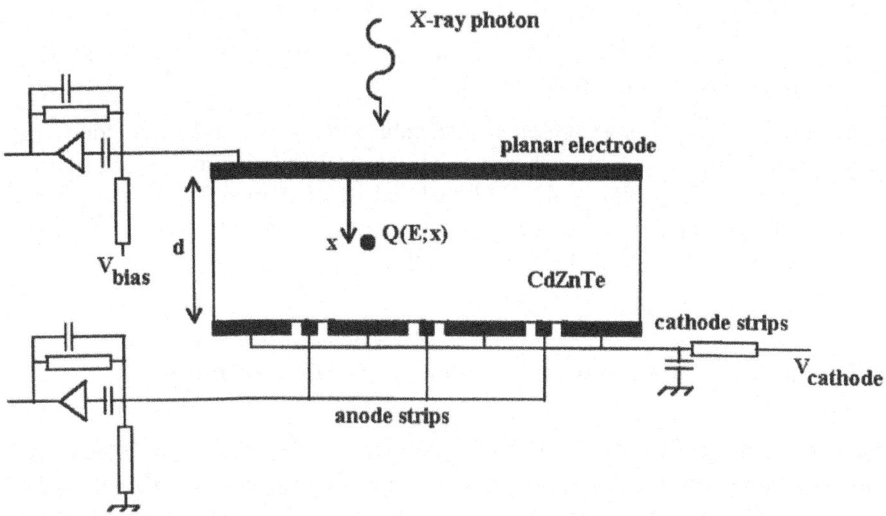

Fig. 1 Schematic view of the Frisch strip semiconductor detector principle. (Figure reprinted from Ref. [10] with permission from Elsevier.)

of Frisch strips. In fact, this layout was inspired by the design of the gas micro strip detectors as referred to above [2]. Here, the purpose of high voltage difference between anode and Frisch strips (in the figure referred to as cathode strips) is to generate gas amplification of the primary electrons. However, we knew that a gas microstrip plate operating at low voltage difference (no amplification) between anode and cathode strips behaves as a Frisch-gridded ionization chamber [2]. For more details on the theory behind the signal generation in the Frisch strip detector, the reader is referred to [9].

A pattern of monolithic thin anode strip electrodes and large size cathode electrodes in between was deposited on a CZT crystal to demonstrate the effect of Frisch strips, as illustrated in Fig. 1. The anode strips were 15 μm wide and had a pitch of 300 μm. The cathode strips were 150 μm wide. The dimension of the crystal was $7 \times 15 \times 1.5$ mm^3. The Frisch strips were interconnected on the crystal.

Two adjustable voltage sources supplied the detector bias, $V_{\mathrm{bias}} = -200$ V, to the planar electrode and the bias of the Frisch strips, $V_{\mathrm{Frisch}} = -30$ V. The readout strips (anodes) were held at ground potential. The detector was illuminated through the planar electrode with different sources: ^{241}Am, ^{137}Cs, and ^{133}Ba, covering a photon energy range from 20 to 662 keV. The main conclusions of these investigations were [9]:

- The proof of concept demonstrated that utilizing Frisch strips significantly improved the spectral resolution at energies above 40 keV. The energy resolution (Full Width at Half Maximum (FWHM)) of the ^{241}Am 59.6 keV line improved from 25% at the cathode to 5% at the anode strip.
- The spectral resolution could be improved further by correcting for hole trapping with the ratio of the signals on the planar electrode and the readout strips.
- Compared to the coplanar technique, the readout electronics is less complex, since no signals were subtracted.
- No events were rejected to improve the spectral resolution, as is normally done with rise-time discrimination techniques.

Although the resolution increased considerably, there was still place for further improvement. The obtainable energy resolution was mainly limited by the electronic noise and the poor quality of the CZT material which also suffered from electron trapping. Therefore, it was decided to investigate the concept with segmented Frisch strips employed for high-quality CZT crystals.

2.3 *Segmented Frisch Strips: The Drift Strip Solution*

Because of the results and difficulties experienced with our first prototype Frisch strip detector, it was decided to proceed with using high-quality CZT material of so-called spectral grade. We decided to employ monolithic strip electrodes of identical dimensions replacing the Frisch strip, with a segmented electrode solution, now referred to as drift strips. This had the advantage that we could test different

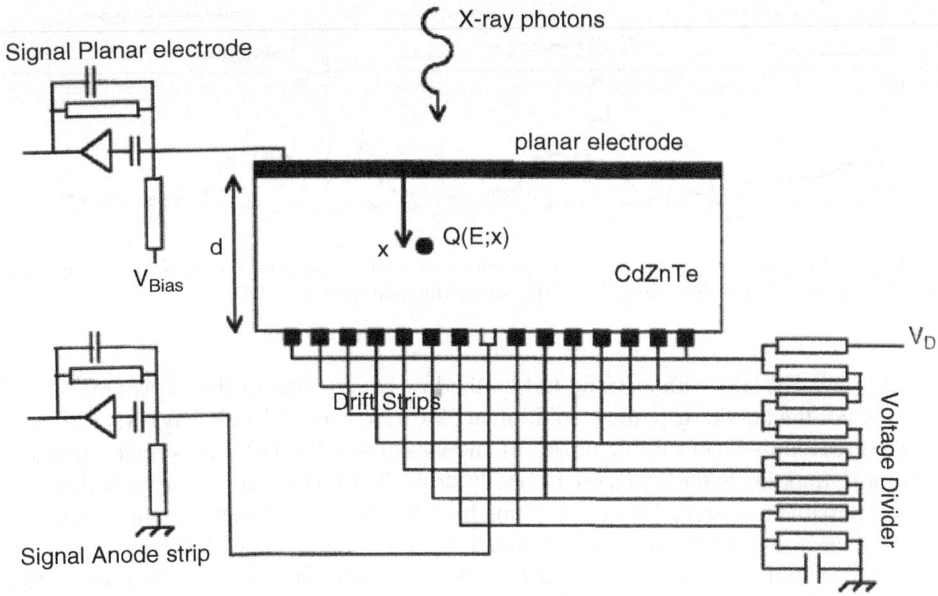

Fig. 2 Schematic view of the drift strip detector in a single anode strip configuration. (Figure reprinted from Ref. [10], with permission from Elsevier)

strip configurations using only one detector by configuring it by different electrode readouts. It soon became clear that the so-called drift strip configuration, where one strip is used for charge (electron) collection and the neighboring strips are used to steer the electrons to the anode, offered many advantages [11]. The principle of the drift strip electrode geometry applied on a CZT detector bulk is shown in Fig. 2.

In the first prototype CZT drift strip detector, the drift strips were applied to only one side of the crystal, while using a single planar electrode on the opposite side. When the electrons approach the side with the drift strips, they are directed toward the anode strip. The small area of the anode strip reduces the electronic noise while the drift strips act as the previous Frisch strips making the detector insensitive to the hole influence. A signal is also generated on the planar electrode. This signal can be used to correct for hole trapping. Such a drift strip detector was tested by depositing 3 mm long strips with a width of 40 μm and a pitch of 100 μm on a 1.5-mm thick CZT crystal. A gold layer was deposited on the back side. The bias of the planar electrode was −200 V, while the voltage difference between two adjacent strips was −20 V. The drift strip detector was tested with ^{241}Am, ^{137}Cs, and ^{133}Ba, covering a photon energy range from 20 to 662 keV. Due to the small area of the anode strip, the electronic noise (FWHM) reduced from 4.68 keV for the planar electrode to 3.10 keV for the anode strip. Thus, low-energy lines were much better resolved with the anode strip. At energies above 40 keV, the energy resolution for the planar electrode became strongly influenced by hole trapping, while it is still electronic noise limited for the anode strip. For the planar electrode, the resolution was

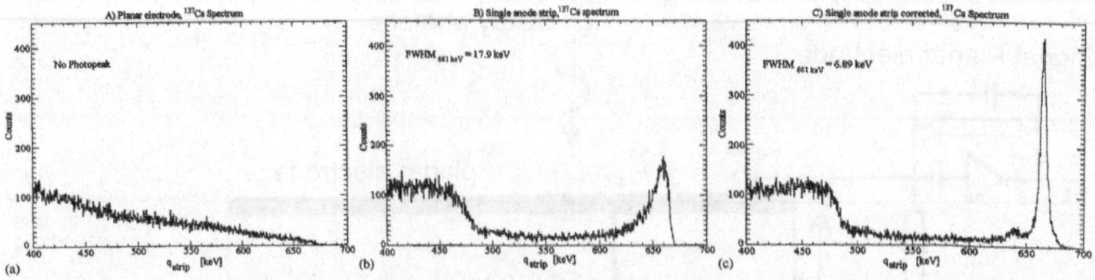

Fig. 3 ^{137}Cs spectra showing the different improvements that were achieved for the CZT drift strip detector. (Figure reprinted from Ref. [10], with permission from Elsevier)

11.0 keV FWHM with a strong tail on the low-energy side of the ^{241}Am 59.6 keV line. For the anode strip, the resolution at 59.6 keV was 3.21 keV FWHM, which is comparable to the electronic noise. At energies above 100 keV, a residual influence of hole trapping to the signal on the anode strip existed. However, as described in [9], it is possible to correct this hole trapping by using the ratio, *R*, between the signals on the planar electrode and on the anode strip. In Fig. 3, the different stages of improvements that were achieved for the ^{137}Cs peak at 661 keV are shown. No peak could be observed with the planar electrode (Fig. 3a), while the peak has a width of 17.9 keV with the anode strip (Fig. 3b). In Fig. 3c, the resolution is improved further to 6.9 keV after correction for hole trapping, with a peak-to-valley ratio of 40:1 and a peak-to Compton ratio of 4:1. For further details, the reader is referred to [11].

With the gained experience from the proof-of-concept studies, the detector group at DTU Space continued the development program. In 2013, the group successfully designed and manufactured prototype 3D CZT drift strip detectors based on this experience in a European Space Agency funded ESA R&D development project. In the next section, the 3D CZT drift strip detector state of the art will be described in terms of the drift strip method, followed by a discussion of noteworthy publications and applications of the technology.

3 The Drift Strip Method

The drift strip method (DSM) encapsulates a specific electrode design together with associated pulse shape analysis procedures. The specific electrode geometry is based on the principle of the segmented Frisch strips, referred to as drift strips, in a configuration now known as the 3D CZT drift strip detector. This design is based on the expertise gained from the studies described in the previous section. Additionally, a pulse shape analysis procedure enables event characterization, sub-millimeter spatial resolution, and excellent spectral resolution. This section provides an overview of the method.

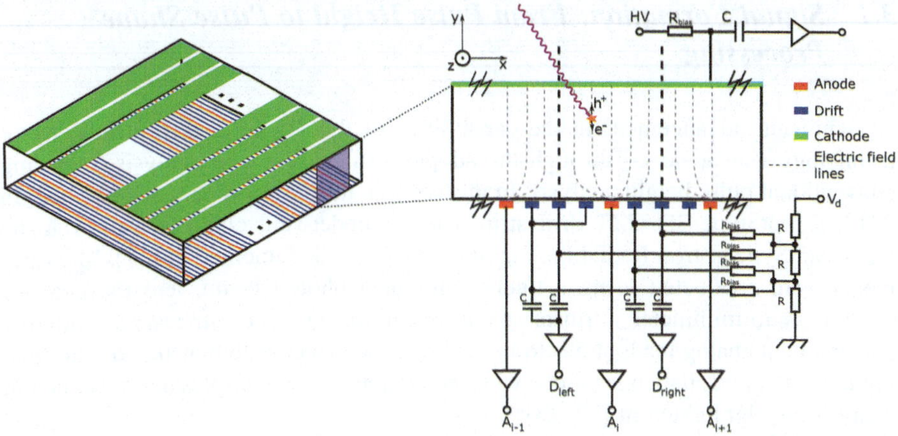

Fig. 4 Schematic of the 3D CZT drift strip detector (left) and a single drift cell (right). (Figure obtained from Ref. [16])

As described in the preceding section, we uncovered that the small area of the anode strip serves to minimize electronic noise due to low capacitance and leakage current. Subsequently, the small pixel effect ensures that the weighting potential of the anode is compact, as described by the Shockley-Ramo theorem [12, 13]. The drift strips screen the anodes from hole movement, while the negative bias directs the electrons toward collection at the anode. Following the initial proof-of-concept studies, the detector group at DTU Space refined the method, focusing on electrode geometry alongside the development of advanced position algorithms and energy correction methods. Notably, two patents were secured related to the technology [14, 15]. The current electrode geometry concept of the 3D CZT drift strip detector is illustrated in Fig. 4.

The distinct electrode geometry used in the DSM is designed to shield the collecting anodes from the inefficient movement of holes in CZT, and at the same time allow for 3D spatial sensitivity. The method relies on a geometrical weighted electrode configuration carefully selected to ensure that the weighting potential of the anode remains compact and insensitive to hole movement. The drift strip detector design was improved by adding segmented cathodes on one side of the CZT bulk and perpendicular anodes and drift strips on the other side, as shown in Fig. 4. This configuration is now known as the 3D CZT drift strip detector principle. The figure provides a schematic depiction of

the 3D CZT drift strip detector geometry (left) alongside an illustration of a drift cell (right) together with the coordinate system related to the detector. When a photon interacts in the detector bulk, a photoelectron is ejected, leading to the creation of an electron charge cloud and associated holes. Here, it is shown how the bias configuration of anodes, cathodes, and drift strips directs electrons toward collection at the anode, while holes migrate toward the cathode.

3.1 Signal Formation: From Pulse Height to Pulse Shape Processing

A significant advancement in the capabilities of the 3D CZT drift strip detector technology was achieved through the adoption of pulse shape analysis, replacing conventional pulse height analysis. In this section, this development is described. In 2013, a prototype 3D CZT drift strip detector underwent characterization at the European Synchrotron Radiation Facility (ESRF), as documented in [17]. Measurements were conducted using a finely collimated photon beam, and the detector exhibited sub-millimeter intrinsic spatial resolution in three dimensions, utilizing conventional analog readout electronics. While the detector demonstrated a promising performance, challenges arose with the readout system employing pulse height analysis, as highlighted in [17, 18]:

- There existed a strong correlation between the x- and y-position determination in the detector.
- The determination of z-position became uncertain for events originating close to the cathode plane, where the spatial resolution approached the cathode pitch.
- Additionally, there was no event-type characterization within a single drift cell.

In response to the limitations and challenges experienced with the conventional analog readout system, the detector was equipped with a high-speed digitizer to allow for pulse shape analysis [18]. An example of the raw data acquired from the high-speed digitizers for a single photon interaction in a 3D CZT drift strip detector is given in Fig. 5. The reader should note that the number of anodes, cathodes, and drift strips depends on the specific detector size and configuration. In the given

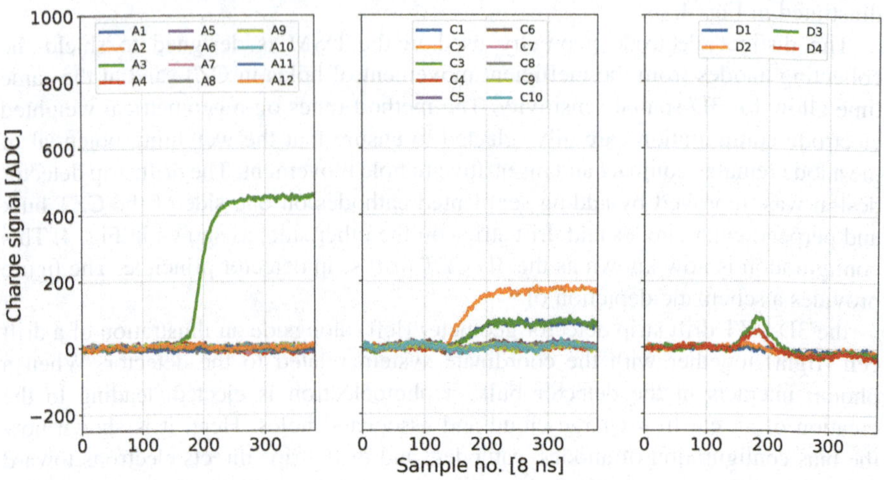

Fig. 5 26 Raw data pulse shapes for an arbitrary event in a 3D CZT drift strip detector. (Figure obtained from Ref. [16])

example, the detector has 26 readout channels, and thus for a measurement of N interactions, the raw data would consist of $N \times 26$ pulse shapes for a given measurement. During charge collection in the detector, the movement of electrons and holes will induce a signal on all electrodes in the detector; anodes, cathodes, and drift strips. The pulse shapes generated for each electrode provides a unique fingerprint of the interaction that took place in the detector. Introducing pulse shape analysis enables event-type characterization, uncorrelated sub-millimeter position determination, and a sophisticated energy correction procedure.

In the following sections, the pulse shape procedures for the 3D CZT drift strip detector are summarized, followed by a presentation of some of the noteworthy applications enabled by the 3D CZT drift strip detector.

Event-Type Characterization

The output signal from anodes, cathodes, and drift strips will look different depending on what kind of interaction took place and where it occurred in the detector. Specifically, the pulse shape of the anode signal can reveal the event type. This is illustrated in Fig. 6. If the anode signal displays a single sharp rise and only one triggered anode, a single cell event takes place (Fig. 6a). In this case, a single electron charge cloud is generated and collected. This event can, for example, be a photoelectric absorption event or a Compton scattering followed by the scattered photon escaping the material. If a Compton scatter interaction occurs in the same drift cell, all generated electron charge clouds are collected by the same anode; it is referred to as a single-cell multiple event (Fig. 6b). In this case, the anode pulse shape will display a staircase-like signal, with a steep rise for each electron

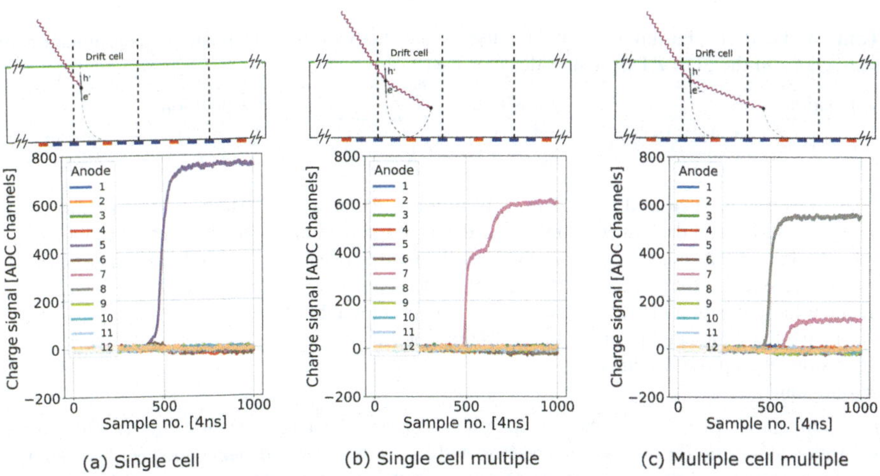

Fig. 6 The pulse shapes of the anode signal hold valuable information on which type of interaction occurs in the detector volume. (Figures modified from Ref. [16])

charge cloud collected. If a similar interaction occurs, but the Compton scattered photon interacts in a different drift cell in the detector, and thus the electron charge clouds are collected by different anodes, a signal similar to Fig. 6c will occur. In this case, it is referred to as a multiple-cell multiple event, and we will see a steep rise in different anode signals. A combination of the mentioned events can occur, and all can either be a full absorption or terminated by an escaping photon.

Intrinsic Position Determination

To attain sub-millimeter intrinsic spatial resolution in three dimensions, specialized algorithms are employed to determine the interaction position of a photon. These algorithms leverage the unique pulse shapes generated by the 3D CZT drift strip detector. While these algorithms have been previously outlined [18–20, 16], a concise summary will be provided here specifically for single interaction events. If the reader is interested in understanding how these algorithms are applied to Compton interactions, they are referred to [19].

Table 1 provides an overview of position algorithms related to the 3D CZT drift strip detector, and Fig. 7 illustrates the dimensions of the detector related to the coordinate system, where the x-axis is perpendicular to the anodes, the y-axis is between the anode and cathode plane, and the z-axis is perpendicular to the cathodes. The method applied to determine the z-position has its shortcomings for events close to the cathode plane, where the pulse heights of neighboring cathodes will be vanishing, and full-induced signal on the central cathode of the event in question is also verified by the Shockley-Ramo theorem [12, 13]. However, pulse shape analysis of the digitized signal allows to utilize the information of transient events, to correct this issue. For a description of this method, the reader is referred to [18].

Table 1 Position algorithms used in pulse shape analysis for 3D position determination of interactions in the 3D CZT drift strip detector [18]

x-position	y-position	z-position
$x_{pos} = [x_{trig} + F \cdot \frac{A_{DR} - A_{DL}}{A_{DR} + A_{DL}}] \cdot P_{an}$	$y_{pos} = \frac{\sum_{j=1}^{J} C_j}{A_m} \cdot d_{CZT}$	$z_{pos} = \frac{\sum_{j=1}^{J} j \cdot C_j}{\sum_{j=1}^{J} C_j} \cdot P_{ca} - \frac{P_{ca}}{2}$
The x-position is calculated using the anode and drift strip signals. The activated anode number provides a rough estimate of the x-position of an interaction, and the neighboring drift strips then provide a finer positioning	The y-position is calculated using the depth-of-interaction method, by the ratio between the sum of cathode pulse heights and the pulse height of the collecting anode	The z-position is calculated by the center-of-gravity method of cathode pulse heights

Pan is the anode pitch, *Pca* is the cathode pitch, F is an adjustable scaling factor to calibrate the x-position algorithm for the specific detector, and x_{trig} is the activated anode number. A_{DL} and A_{DR} are the amplitudes of the inverted current signals of the left and right drift strips of the collecting anode. *Am* ($m = 1, 2, ..., M$) is the pulse height of the collecting anode, where M is the total number of anodes. C_j ($j = 1, 2, ..., J$) is the cathode pulse height, where J is the total number of cathodes. d_{CZT} is the detector thickness

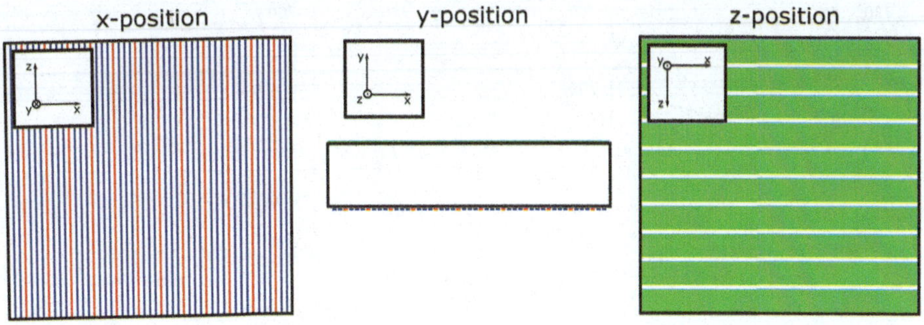

Fig. 7 Illustration of the orientation of x-, y-, and z-positions in the 3D CZT drift strip detector. (Figure obtained from Ref. [16])

Spectral Sensitivity and Correction

The 3D-position-sensitive nature of the 3D CZT drift strip detector allows to determine and correct for position-dependent pulse height variations throughout the detector bulk, and thereby correct the pulse heights and improve the overall energy resolution.

The pulse heights in the 3D CZT drift strip detector is dominantly affected by the depth-of-interaction (DOI) given by the y-position, where two opposing effects contribute to the reduction in resolution [11]:

1. *The anode weighting potential distribution* is compact close to the anode side; however, for events occurring close to the anode side, the electron charge cloud will not pass through the entire weighting potential, resulting in a reduced induced signal.
2. *Electron trapping* will result in reduced pulse heights, increasing with longer drift paths of the electron charge cloud.

In Fig. 8, the measured pulse heights as a function of x-, y-, and z-position are given. The sinusoidal dependence of the x-position shows the maxima at the anode position, due to longer drift path along the edges. The z-position shows variations related to CZT material inhomogeneity effects. Figure 9 shows a ^{137}Cs spectra before and after correction, where the FWHM was improved from 3.7% to 1.6% after spatial correction. This specific detector was dominated by electronic noise; for other detectors, this has been improved even further, for example, 1.14% reported in [18].

4 Noteworthy Results and Applications

While the 3D CZT drift strip detectors were initially designed to be intended for use in X-ray and gamma-ray astronomy, it became apparent that these detectors could also provide significant benefits in various applications. In this section, a short review of some noteworthy results and applications from the drift strip method will be presented, followed by an outlook on the future.

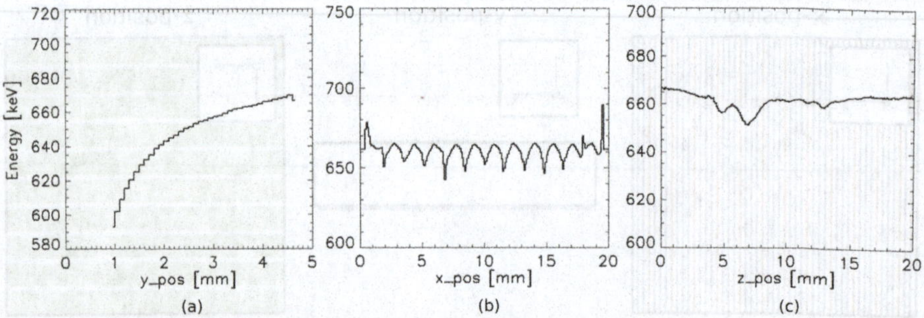

Fig. 8 Pulse height dependence on the x-, y-, and z-position for ^{137}Cs 662 keV photons detected by photoelectric absorption. (Figure obtained from Ref. [21])

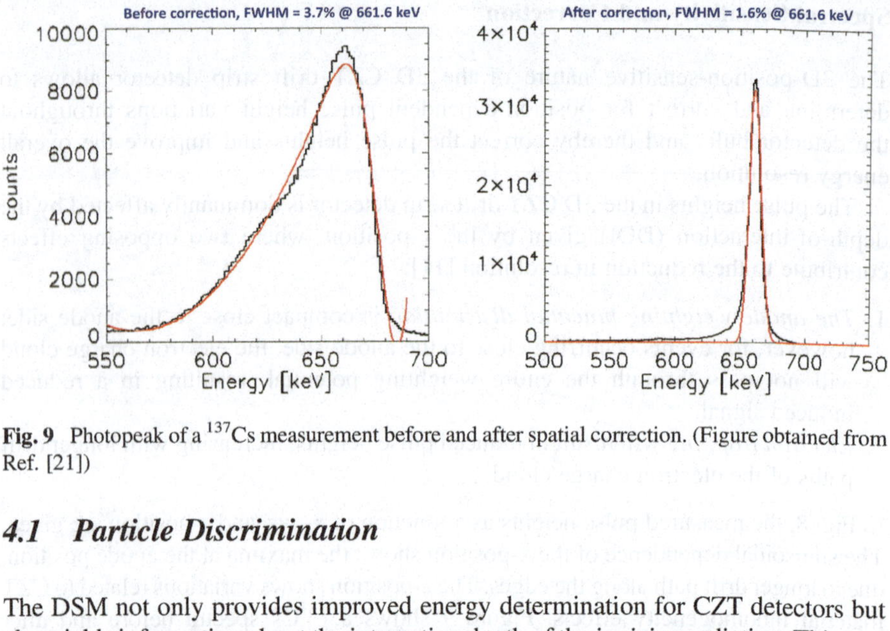

Fig. 9 Photopeak of a ^{137}Cs measurement before and after spatial correction. (Figure obtained from Ref. [21])

4.1 *Particle Discrimination*

The DSM not only provides improved energy determination for CZT detectors but also yields information about the interaction depth of the ionizing radiation. This was demonstrated in a study done by the DTU Space detector group [22], using an early version of the 3D CZT drift strip detector, namely the geometry illustrated in Fig. 2. The study could also be applied to more recent versions of the 3D CZT drift strip detector.

DOI information serves a dual purpose in the field of particle physics. It can be used to correct and compensate for energy measurements affected by charge trapping and can additionally be used to differentiate between gamma rays and charged particles. The DOI information provides the ratio between the signal induced on the planar cathode, Q_{planar}, and the collecting anode strip, Q_{strip}, given by $R = \sim Q_{\text{planar}}/Q_{\text{strip}}$. It will be close to 1 for interactions near the planar electrode and a value close to 0 for interactions near the strip electrodes.

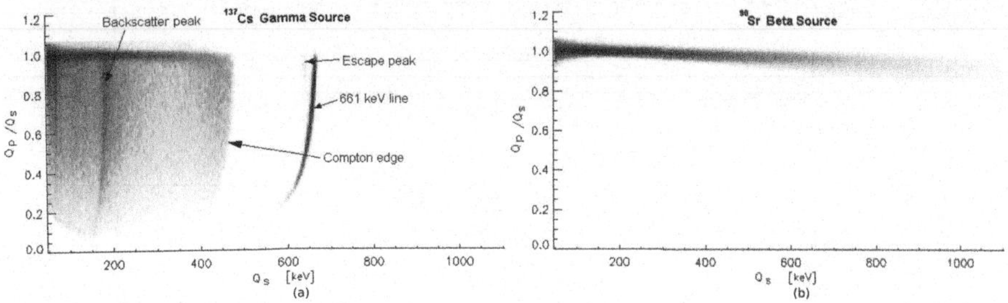

Fig. 10 Depth of interaction derived from the ratio between a planar electrode and the collecting strip signal plotted as a function of measured energy for **(a)** ^{137}Cs gamma rays and **(b)** ^{90}Sr beta rays. The distinct interaction patterns observed for each type of radiation demonstrate the utility of this method in differentiating between these two types of ionizing radiation. (Figure obtained from Ref. [22])

In Fig. 10, the DOI is plotted as a function of measured energy for (a) ^{137}Cs gamma rays and (b) ^{90}Sr beta rays. The figures demonstrate that the electrons, up to energies of 1 MeV, are stopped close to the planar electrode, contrary to the gamma rays which have randomly distributed interactions along the depth in the detector bulk [22]. This allows for a particle discrimination capability, which is especially important for space missions. It will allow gamma-ray bursts to be distinguished from electron bursts, as their DOI distributions will be significantly different. This capability could enhance the potential of any space mission sensitive to particle background and possibly suppress the background, thus enhancing telescope sensitivity.

4.2 *Radiation Damage Measurements*

The spectroscopic properties of CZT drift strip detectors are largely independent of the material hole collection characteristics. However, as electron-only devices, they are significantly dependent on the efficiency of electron collection. A major concern for the use of CZT detectors in space is the impact of particle-induced radiation damage, which is known to affect the material charge trapping. This aspect is critical for the detector performance and presents a significant challenge for space missions.

The effects of particle radiation on CZT drift strip detectors were experimentally tested on a CZT drift strip detector with a planar cathode [23]. The study was conducted in collaboration with the cyclotron facility at Copenhagen University Hospital [23]. The experiment involved exposing a well-characterized 2.7-mm thick CZT drift strip detector to 30 MeV protons, irradiated from the cathode side in several steps, to achieve the planned proton fluences up to $6.0 \times 10^9 p^+/cm^2$. The range of protons with energy 29.7 ± 0.7 MeV is in CZT 2.89 ± 0.12 mm [24] which covers the full detector thickness in this experiment. The experiment focused on the

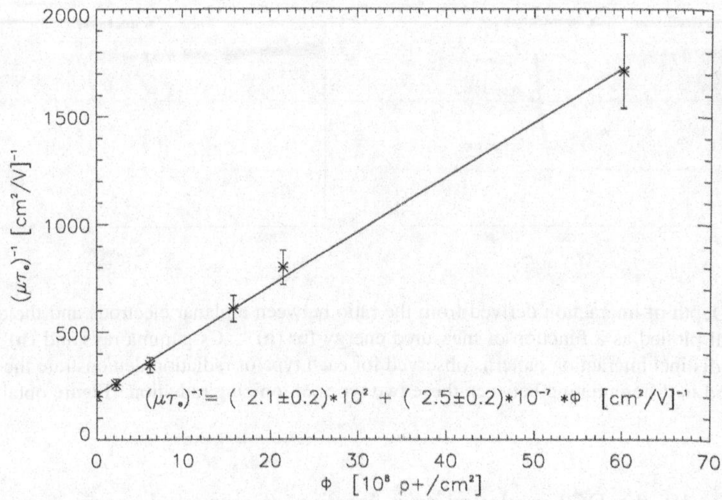

Fig. 11 The derived average charge trapping parameter $(\mu\tau_e)^{-1}$ values as a function of the proton fluencies as a function of fluence. (Figure reprinted from Ref. [23], with permission from Elsevier. Copyright reference to be delivered)

particle fluence dependence of the detector charge trapping properties, the detector energy resolution, and the main features of the activation measurements. The analysis of the detector charge collection properties was based on the detector response measured with a ^{137}Cs source by evaluating the mobility-lifetime product $\mu\tau$.

As demonstrated in Fig. 11, the experimental results provide a comprehensive summary of the study on radiation effects. The analysis revealed a proportional increase in charge trapping with proton fluence, with the contribution to electron trapping following the relation

$$(\mu\tau)_{\text{rad}}^{-1} = (2.57 \pm 0.2) \times 10^{-7} \times \Phi \quad (V/\text{cm}^2) \tag{1a}$$

with Φ in p^+/cm^2. The equation can be expressed as a function of the received proton dose, D, assuming an initial proton energy of 30 MeV

$$(\mu\tau)_{\text{rad}}^{-1} = (0.8 \pm 0.1) \times 10^3 \times D \quad (V/\text{cm}^2) \tag{1b}$$

with D in krad.

Equations (1a) and (1b) offer a framework for assessing proton radiation effects on CZT detectors. An annealing procedure, involving 22 h of heating at 100 °C in 2-h increments, enabled recovery of radiation-induced electron trapping and determination of $\mu\tau$. CZT drift strip detectors exhibited resilience under high-energy proton exposure, making them ideal for long-term space missions. Their charge-trapping properties and energy resolution remained stable, with electron trapping increasing predictably with proton fluence. The annealing process further enhances their suitability for space radiation detection by allowing performance recovery.

4.3 Fano-Factor-Limited Energy Resolution

The DSM enables a significant reduction in hole trapping sensitivity for the anode, resulting in a significant improvement in achievable energy resolution. The best achievable spectral response was presented in [25], where the detectors were characterized using custom-designed ultralow-noise readout electronics as the first-stage preamplifiers. This readout electronic could achieve as low as 0.8 keV FWHM total electronic noise at −10 °C. The theoretical resolution limit for the CZT drift strip detectors was investigated, and it was shown that they exhibit some of the best performance figures yet reported for full illumination of a wide energy range using ^{241}Am, ^{109}Cd, ^{133}Ba, and ^{137}Cs calibration sources [25].

The aim of the spectral measurements was to determine the detector's spectral response across a wide energy range and to establish the intrinsic energy resolution as a function of photon energy. For instance, at 356 keV, the FWHM energy resolution was 2.51 keV (see Fig. 12a). At 80.8 keV, the corresponding resolution was 1.64 keV FWHM. This should be compared to the calculated Fano resolution of 500 eV FWHM at 80.8 keV. The measurements show that our energy resolution was limited by the total electronic noise below energies of 100 keV. However, at medium and high energies, the Fano noise became an appreciable fraction of the resolution function. The measured intrinsic energy resolution as a function of photon energy is shown in Fig. 12b for the best of our CZT drift strip detectors. The energy resolutions shown here have been corrected for the electronic noise contribution, which was 0.8 keV. The solid line indicates the statistical limit from the Fano noise using a Fano factor of $F = 0.1$ for CZT. This result shows that DSM can achieve energy resolutions which are within a factor of 2–3 of the CZT Fano-limited resolution.

4.4 Compton Camera

When observing radiation in the MeV-band (0.1–100 MeV) emanating from celestial sources, several challenges emerge, including low flux, limited interaction probability, three energy loss processes, and elevated background radiation rates. In the poorly observed MeV-domain, Compton interactions are dominant, and future observatories will likely be based on the Compton imaging technique. Therefore, distinguishing event types and extracting energy and position information from Compton scatter interactions within the same detector bulk will be crucial qualities for future detectors intended for space applications in this energy domain.

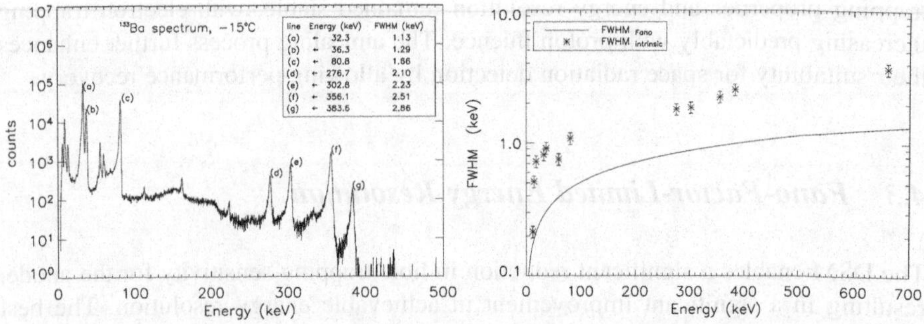

(a) (b)

Fig. 12 (**a**) ^{133}Ba spectrum measured with a CZT drift strip detector. The spectrum is energy calibrated and corrected. The total electronic noise is 0.8 keV FWHM for the anode strip. The operation temperature was −15 °C. (**b**) The intrinsic energy resolution as a function of photon energy, for the CZT drift strip detector, corrected for electronic noise. The solid line indicates the statistical limit from Fano noise. (Figures reprinted with permission from Ref. [25])

With the introduction of high-speed digitizers and pulse shape analysis, the inclusion of timing information facilitates the identification of Compton events in the 3D CZT drift strip detector, including Compton interactions in the same drift cell, rendering these features highly advantageous for a spectral imager utilized as instrumentation in astrophysical studies in the MeV-domain.

In a study, we illustrated that it is possible to operate a single 3D CZT drift strip detector of size $2 \times 2 \times 0.5$ cm^3 as a Compton camera with limited angular resolution [19]. The imaging capability of a prototype detector was tested as a proof of concept, and utilizing the toolbox MEGAlib,[2] we successfully reconstructed an image of a 662 keV source from two-site Compton scatter interactions in a single detector bulk, as illustrated in Fig. 13. This study highlighted the imaging capabilities of the 3D CZT drift strip detector enabled by pulse shape analysis and underlined the suitability of the 3D CZT drift strip detectors in future Compton camera space missions, either on larger scale telescopes or on small satellite Compton cameras [19, 26].

4.5 3D Mapping of Material Nonuniformities

Despite its larger band gap, which enables room temperature operation and makes CZT valuable for a wide range of applications, the charge trapping mentioned earlier within the material results in ineffective charge collection and poses challenges in producing large, defect-free single crystals. Additionally, the crystalline structure of the material can exhibit inhomogeneous quality.

---

[2]https://megalibtoolkit.com/home.html

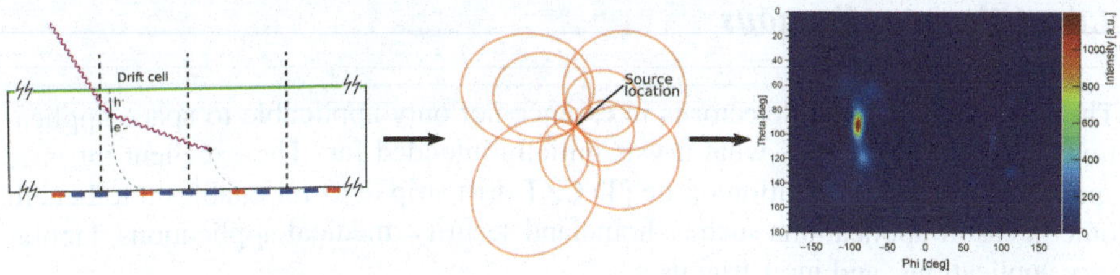

Fig. 13 Position and energy information of Compton interactions in a single 3D CZT drift strip detector are extracted, and the Compton cones reconstructed to determine the source location with an ARM of 16° [19]. (Figures from Refs. [16, 21])

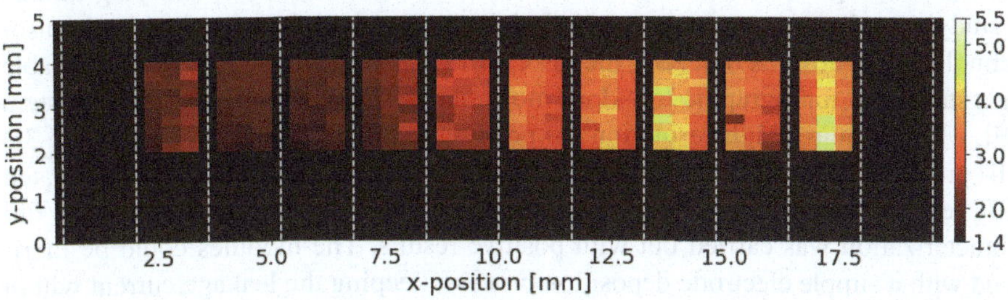

Fig. 14 Map of the electron mobility and lifetime in a 3D CZT drift strip detector, for a 1 mm slice along the z-direction. A clear nonuniformity in the detector material is evident. (Figure reprinted with permission from Ref. [20])

The consequence of the poor charge mobility together with nonuniformity of the material will greatly affect the generated pulse shapes from the charge collection. Conventional methods for evaluating the $\mu\tau$-product assess one single value for the entire detector bulk, not taking into consideration the nonuniformity of the detector material. In a study, we illustrated that it is possible to utilize the 3D sensitive nature of the 3D CZT drift strip detector, to map the $\mu\tau_e$-product of the electrons in 3D [20]. In Fig. 14, a 3D map of the $\mu\tau$-product of electrons in a 3D CZT drift strip detector is given. The plot represents a 1-mm slice along the z-direction. From the plot, it is evident that the material is not uniform and that the $\mu\tau$-product of the electrons changes throughout the detector material. Furthermore, it underlines how the 3D spatial sensitivity of the 3D CZT drift strip detector enables this kind of material analysis, where the material properties can be analyzed and mapped in 3D.

The study clearly illustrated the nonuniformity of the CZT material. Furthermore, it also supplies an outlook for the detector technology, where such a material map can be included in advanced theoretical detector models or used for signal corrections in 3D.

4.6 Other Applications

The 3D CZT drift strip detector is in essence not only applicable to space applications, even though that is what it was initially intended for. The excellent intrinsic spatial and spectral resolution of the 3D CZT drift strip detector enables it to benefit other areas of applications such as homeland security, medical applications, laboratory applications, and industrial use.

In this section, we would like to highlight one specific application which was recently investigated in a project collaboration between DTU Space and Kromek group plc [27]. In the molecular breast imaging (MBI) project, we investigated what medical imaging and astronomy have in common. Whether it is identifying tumors or studying colliding neutron stars, both domains can benefit from the same sensor technology. As described in [28], the 3D CZT drift strip detector could be applicable for a stationary tomographic low-dose molecular breast imaging (LD-MBI) system [29], which requires high-resolution 3D position sensitive CZT detectors. In the 3D MBI project carried out in a collaboration between DTU Space and Kromek [27], ten MBI test modules were fabricated based on the drift strip method, and a full characterization was carried out with positive results. The modules could be fabricated with a simple electrode deposition process, keeping the leakage current within the desired range. Furthermore, the spatial resolution met the requirements of LD-MBI systems.

This study emphasizes the technology transfer that often occurs between technology developed for space applications and relevant technologies on Earth, as the requirements for space technology push the boundaries and call for state-of-the-art technology development.

5 Future Outlook

As presented in this chapter, many important steps have been taken over time to improve the 3D CZT drift strip detector technology to its current state. The technology has consistently exceeded expectations in various studies. The detector can achieve sub-millimeter spatial resolution, excellent spectral resolution, and event-type discrimination with significantly fewer readout channels compared to a conventional pixel detector. However, its current implementation is limited to a laboratory setup on an optic board, and significant efforts must be made to elevate the technology to a space-ready or commercialized state.

The advancement of high-resolution 3D CZT drift strip detectors for photon-by-photon detection and radiation measurements is hindered by limitations in computational power and the size of digital data generated during position algorithms and offline analysis. This restricts the achievable flux and obstructs the real-time processing of detector signals. To address this challenge, a new radiation detector readout technology development program was initiated at DTU Space in 2017. This program aims to pioneer a novel readout method capable of handling large data volumes while achieving digital electronic-limited fluxes. By leveraging innovative readout techniques and advanced signal processing algorithms, we seek to enhance the efficiency and accuracy of photon detection and radiation measurements in 3D CZT drift strip detectors. Through collaborative research and interdisciplinary approaches, this initiative holds the potential to revolutionize detector readout technology and unlock new frontiers in radiation detection and imaging applications.

References

1. Lund, N., Budtz-Jørgensen, C., Westergaard, N. J., Hornstrup, A., Lundgaard Rasmussen, I., Laursen, S., Kristensen, R., Mogensen, P. B., Harbo Andersen, K., Rasmussen, I., et al. (1999). Jem-x the x-ray monitor on integral. *Astrophysical Letters and Communications, 39*, 339.
2. Budtz-Jørgensen, C. (1992). Features of the microstrip proportional counter technology. *Review of Scientific Instruments, 63*(1), 648–654.
3. Christensen, F. E. (1996). Multilayer mirrors for future x-ray missions. In *Proceedings of the NGXO Meeting*, 1996. NGXO Meeting; Conference date: 01-01-1996.
4. Budtz-Jørgensen, C., Kuvvetli, I., Westergaard, N. J., Jonasson, P., Reglero, V., Eyles, C., & Neubert, T. (2001). The czt x-ray imager on axo. *Science with Minisat, 01*, 281–289.
5. Ivanov, V. I., Garbusin, V. A., Dorogov, P. G., Loutchanski, A. E., & Kondrashov, V. V. (1995). Spectro- metric characteristic improvement of cdte detectors. *IEEE Transactions on Nuclear Science, 42*(4), 258–262.
6. Lund, J. C., Olsen, R., Van Scyoc, J. M., & James, R. B. (1996). The use of pulse processing techniques to improve the performance of cd/sub 1-x/zn/sub x/te gamma-ray spectrometers. *IEEE Transactions on Nuclear Science, 43*(3), 1411–1416.
7. Luke, P. N. (1995). Unipolar charge sensing with coplanar electrodes-application to semiconductor detectors. *IEEE Transactions on Nuclear Science, 42*(4), 207–213.
8. Budtz-Jørgensen, C., Knitter, H.-H., Straede, C., Hambsch, F.-J., & Vogt, R. (1987). A twin ionization cham- ber for fission fragment detection. *Nuclear Instruments and Methods in Physics Research Section A: Accelerators, Spectrometers, Detectors and Associated Equipment, 258*(2), 209–220.
9. Van Pamelen, M. A. J., & Budtz-Jørgensen, C. (1998). Novel electrode geometry to improve performance of cdznte detectors. *Nuclear Instruments and Methods in Physics Research Section A: Accelerators, Spectrometers, Detectors and Associated Equipment, 403*(2–3), 390–398.
10. Van Pamelen, M. A. J., Budtz-Jørgensen, C., & Kuvvetli, I. (2000). Development of cdznte x-ray detectors at dsri. *Nuclear Instruments and Methods in Physics Research Section A: Accelerators, Spectrometers, Detectors and Associated Equipment, 439*(1), 625–633.
11. Van Pamelen, M. A. J., & Budtz-Jørgensen, C. (1998). Cdznte drift detector with correction for hole trapping. *Nuclear Instruments and Methods in Physics Research Section A: Accelerators, Spectrometers, Detectors and Associated Equipment, 411*(1), 197–200.
12. Shockley, W. (1938). Currents to conductors induced by a moving point charge. *Journal of Applied Physics, 9*(10), 635–636.
13. Ramo, S. (1939). Currents induced by electron motion. *Proceedings of the IRE, 27*(9), 584–585.
14. Kuvvetli, I., & Budtz-Jørgensen, C. (2015). X-ray and gamma-ray radiation detector, June 2015. Also registered as: WO2014EP75643, EP20130194445; WO2015078902; G01T1/24.

15. Kuvvetli, I., & Budtz-Jørgensen, C. (2018). Semiconductor detector with segmented cathode, April 2018. WO2018065024; G01T 1/ 24 A I.
16. Owe, S. R. H. (2023). *Development of 3D imaging detectors for high energy astronomy instrumentation*. PhD thesis, Technical University of Denmark.
17. Kuvvetli, I., Budtz-Jørgensen, C., Zappettini, A., Zambelli, N., Benassi, G., Kalemci, E., Caroli, E., Stephen, J. B., & Auricchio, N. (2014). A 3d czt high resolution detector for x- and gamma-ray astronomy. *SPIE – International Society for Optical Engineering. High Energy, Optical, and Infrared Detectors for Astronomy VI; Conference date: 22-06-2014 Through 25-06-2014.*
18. Budtz-Jørgensen, C., & Kuvvetli, I. (2017). New position algorithms for the 3-d czt drift detector. *IEEE Transactions on Nuclear Science, 64*(6), 1611–1618.
19. Owe, S. H., Kuvvetli, I., Budtz-Jørgensen, C., & Zoglauer, A. (2019). *Evaluation of a compton camera concept using the 3d cdznte drift strip detectors* (Vol. 14). IOP Publishing. 20th International Workshop on Radiation Imaging Detectors; Conference date: 24-06-2018 Through 28-06-2018.
20. Owe, S. H., Kuvvetli, I., & Budtz-Jorgensen, C. (2021). Carrier lifetime and mobility characterization using the dtu 3d czt drift strip detector. *IEEE Transactions on Nuclear Science, 68*(9), 2440–2446.
21. Owe, S. H. (2018). *Characterization of new generation 3d czt high resolution x- and gamma-ray detector for high energy space mission concept.*
22. Kuvvetli, I., Budtz-Jørgensen, C., Jørgen, N. Westergaard, S., Jonasson, P., van Pamelen, M. A., Reglero, V., Eyles, C. J., & Neubert, T (1999). Development of large-area czt-detector systems. In *Proceedings of SPIE – The International Society for Optical Engineering*, volume 3765, 1999. 1999 EUV, X-Ray, and Gamma-Ray Instrumentation for Astronomy X; Conference date: 21-07-1999 Through 23-07-1999.
23. Kuvvetli, I., Budtz-Jørgensen, C., Korsbech, U. C. C., & Jensen, H. J. (2003). Radiation damage measurements on czt drift strip detectors. *Nuclear Instruments and Methods in Physics Research, 512*(1–2), 98–105.
24. Ziegler, J. F., Biersack, J. B., & Littmark, U. (1999). *The stopping and range of ions in solids*. Pergamon Press.
25. Kuvvetli, I., & Budtz-Jørgensen, C. (2005). Pixelated cdznte drift detectors. *IEEE Transactions on Nuclear Science, 52*(5), 1975–1981.
26. Owe, S. R. H., Kuvvetli, I., Zoglauer, A., & Budtz-Jørgensen, C. (2024). *Feasibility of using 3d czt drift strip detectors for small compton camera space missions* (Vol. 19). IOP Publishing. 24th International Workshop on Radiation Imaging Detectors; Conference date: 25-06-2023 Through 29-06-2023.
27. Project: Czt drift strip detectors for 3d molecular breast imaging, 2018.
28. Owe, S. R. H., Kuvvetli, I., Cherlin, A., Harris, B., Tomita, H., Baistow, I., Tcherniak, D., & Budtz-Jorgensen, C. (2023). Evaluation of czt drift strip detectors for use in 3d molecular breast imaging. *IEEE Transactions on Radiation and Plasma Medical Sciences, 7*(2), 113–123.
29. Cherlin, A., Wirth, A., Baistow, I., Erlandsson, K., Thielemans, K., & Hutton, B. F. (2021). A new concept for a low-dose stationary tomographic molecular breast imaging camera using 3d position sensitive czt detectors. IEEE NSS MIC (RTSD).

Study of Transient Current Distortion in CZT Detectors Based on Sentaurus TCAD

Chen Xiang

1 Introduction

The cadmium zinc telluride (CdZnTe or CZT) crystal is well known for its high resistivity, large γ/X-ray cross-section, high mobility-lifetime product, wide bandgap, etc. Due to all these advantages, CZT detectors have been widely applied in the fields of nuclear medicine, space exploration, and industrial inspection [1, 2]. However, the further application of the CZT crystal in pulse γ-ray detection has been largely restricted due to high bulk defect concentration, causing performance inadequacy (such as the "space charge effect") in CZT γ-ray detectors [3–5]. With the accumulation of space charges generated by the carrier trapping effect in deep-level defects, the internal electric field distribution will be gradually changed by the intense space charge, and a polarization effect occurs. The carrier collection efficiency will be further reduced, and the output of the CZT detector will deteriorate [6, 7].

Many studies have focused on deep-level defects in CZT crystals. Franc et al. found that under high radiation fluxes, three deep levels participated in the polarization and depolarization process in CZT crystals, including $E_V + 0.75$ eV, $E_C - 0.88$ eV, and $E_C - 1.13$ eV [8]. Rejhon et al. presented a method for evaluating deep levels in CZT crystals through the electro-optic effect. Four main defect levels were identified, namely, $E_V + 0.41$ eV, $E_V + 0.77$ eV, $E_V + 0.94$ eV, and $E_C - 0.90$ eV [9]. All these defects were responsible for the trapping and detrapping of the charge carriers, causing the distortion of the internal electric field distribution and the performance deterioration of the CZT detector under high γ-ray flux rate irradiation (higher than 10^8 γ·mm^{-2}·s^{-1}). Chen et al. investigated the current distortion of CZT detectors existing under high γ-ray flux, including sharp peak and flat current

C. Xiang (✉)
China Institute of Atomic Energy, Beijing, China

Northwest Institute of Nuclear Technology, Xi'an, China

K. Iniewski (ed.), *CdTe and CdZnTe Materials*,
https://doi.org/10.1007/978-3-031-64521-1_5

nonlinearity [10]. Theoretical and experimental analyses showed that the decay constants after the sharp peaks are consistent with the detrapping lifetimes of deep-level defects on the time scale [10]. This result primarily verified the relationship between the distorted current signal and the deep-level defect in CZT detectors. But no systematic proof has been given. However, there is little systematic study focusing on the relationship between the deep-level defects and the current distortion of CZT detectors under high flux rate γ rays.

In this chapter, our concentration is mainly focused on providing convincing evidence for the relationship between the current distortion and the specific defect level in CZT detectors. Firstly, the temporal response model for CZT detectors, constructed by Sentaurus TCAD software, is used for this study. Secondly, with the response model and γ-ray irradiation experiment, the relationship between the defect effect (trapping, detrapping, and recombination) and the current distortion is analyzed. Finally, the relationship between the current distortion and the defect level in CZT detectors is specified. This work can explain the current distortion phenomenon for CZT detectors under high flux-rate γ-ray radiation and will further provide important support for the techniques used for suppressing distortion and optimizing performance.

2 Theoretical Analysis by Sentaurus TCAD

2.1 Research Idea

In this paper, the research idea involves three parts. Firstly, the response model for CZT detectors, which is constructed by Sentaurus TCAD, is used in this study, with the necessary crystal parameters such as bandgap, resistivity, etc. Secondly, the effects of the different defects on the current CZT detector are explored by this response model. Lastly, the relationship between the crystal defects and the current distortion from CZT detectors under γ-ray radiation is analyzed.

2.2 Response Model

The CZT detector model, shown in Fig. 1, is constructed by Sentaurus TCAD. This simulation model is based on the basic drift-diffusion equation. In addition, carrier mobility, recombination (including Shockley-Read-Hall (SRH) and Auger recombination), and defect models are all considered in the CZT response model. The dimension of the CZT crystal is 5 mm × 5 mm × 2 mm. The parameters for the CZT crystal are listed in Table 1, such as the spatial distribution of charge carriers, resistivity, mobility, etc. The electrode material is gold, which is set on parallel surfaces of 5 mm × 5 mm. The thickness of the gold electrode is 50 nm, and the work function of gold is 5.1 eV. Contact types between the CZT crystal and electrodes are

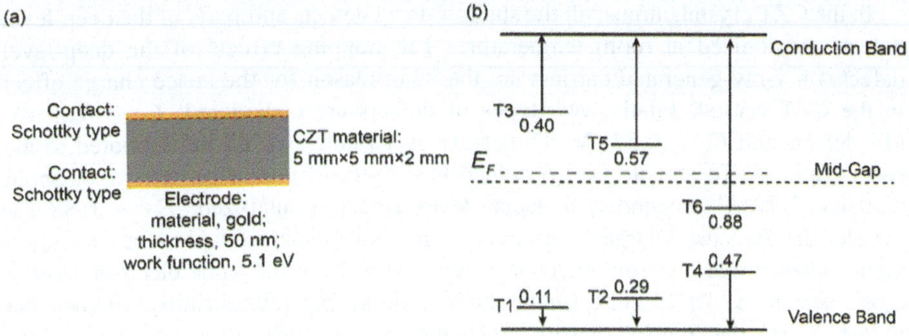

Fig. 1 Theoretical model illustration of a CZT detector. (**a**) Detector model. (**b**) Defect parameters

Table 1 Parameters for the CZT crystal used in the TCAD model

Parameter	Unit	Value
Average atomic number	–	49.1
Density, ρ_m	g/cm^3	5.78
Bandgap, E_g	eV	1.57
Ionization energy, w	eV	4.64
Resistivity, ρ_R	Ω·cm	3.00×10^{10}
Electron mobility, μ_e	cm^2/V/s	1000
Electron mobility-lifetime product, $\mu_e\tau_e$	cm^2/V	$(3–5) \times 10^{-3}$
Hole mobility, μ_h	cm^2/V/s	50 – 80
Hole mobility-lifetime product, $\mu_h\tau_h$	cm^2/V	5.00×10^{-5}

Table 2 Defect parameters in the CZT crystal used in the TCAD model

No.	Level (eV)	Density (cm^{-3})	Type	Method
T1	$E_V + 0.11$	10^{13}	Acceptor	TSC and TEES
T2	$E_V + 0.29$	10^{12}	Acceptor	TSC and TEES
T3	$E_C - 0.40$	10^{11}	Donor	TSC and TEES
T4	$E_V + 0.47$	10^{11}	Acceptor	TSC and TEES
T5	$E_C - 0.57$	10^{10}	Donor	TSC and TEES
T6	$E_C - 0.88$	–	Donor	Ref. [8]

set as Schottky contacts. Considering that the energy of γ-ray is high enough to penetrate through the CZT crystal, the distribution of γ-ray-generated carriers in the CZT crystal follows a bulk uniform order. A theoretical model illustration is shown in Fig. 1a.

Defect distribution is another important parameter for the CZT detector model. The typical defect parameters (T1 ~ T5) of a CZT crystal (from Northwestern Polytechnical University, China) are listed in Table 2 and Fig. 1b, which is experimentally tested through the thermally stimulated current (TSC) method and thermoelectric effect spectroscopy (TEES) [11]. T6 defect ($E_C - 0.88$ eV) is given by Ref. [8].

In the CZT crystal, almost all the shallow-level defects and parts of the deep-level defects are ionized at room temperature. The trapping effects of the deep-level defects on γ-ray-generated carriers are the main reason for the space charge effect in the CZT crystal. Firstly, two types of defects are considered: $E_V + 0.29$ eV (T2 defect) and $E_C - 0.57$ eV (T5 defect). A T5 defect could be attributed to the second ionized Te antisite (Te_{Cd}), which has a second ionization energy of about 0.59 eV [11–13]. Secondly, a deeper-level defect is analyzed: $E_C - 0.88$ eV (T6 defect). Because a large uncertainty exists in the electron capture cross-section values, these three electron defects are assumed to have the same electron capture cross-section σ_{et} 10^{-14} cm^2. Under this condition, the relationships between the defect level, the defect density, and the current distortion can be clearly explained [8, 9].

2.3 *Temporal Response Simulation*

Effects of T2 Defects and T5 Defects

Firstly, the effects of T2 and T5 defects on the current distortion of CZT detectors are analyzed by this model. The defect densities are 10^{10} cm^{-3}, 10^{11} cm^{-3}, 10^{12} cm^{-3}, and 10^{13} cm^{-3}, respectively. The outer voltage is set at 600. The flux rates of γ-ray are set from 5 R/min to 85 R/min, and the intervals are 10 R/min. The process of radiation flux rate can be divided into five states: (a) no radiation flux rate for 5 seconds, (b) a change from no radiation to radiation state for 1 millisecond, (c) radiation flux rate for 20 seconds, (d) a change from radiation to no radiation state for 1 millisecond, and (e) no radiation flux rate for 5 seconds. The temporal currents of a CZT detector under different flux rates of γ-ray radiation and different defect densities are shown in Figs. 2 and 3. For the existence of T2 and T5 defects, no sharp peak appears even if the defect density approaches 10^{13} cm^{-3}, as shown in Figs. 2d and 3d.

To analyze the mechanism of flat current nonlinearity, the relationship between T2 and T5 defect density and the current distortion is studied based on the CZT response model. Taking the flat current as the y-axis value, the relationship between the flat current in Figs. 1 and 2 and the flux rate is shown in Fig. 4; Fig. 4a is for the T2 defect, and Fig. 4b is for the T5 defect.

When the concentration of T2 defects (or T5 defects) in the CZT crystal remains constant, the peak current of the output signal of the CZT detector shows a linear increase trend with the increase in the radiation dose rate. Conversely, when the radiation dose rate remains constant, the peak current of the output signal of the CZT detector gradually decreases with the increase in the concentration of T2 defects (or T5 defects) in the CZT crystal. Thus, under conditions of different defect types (T2 and T5) and levels, the variation process of the peak current of the detector output signal is essentially consistent, indicating that the difference in defect levels is not the main cause of the decrease in the peak current. As the defect concentration

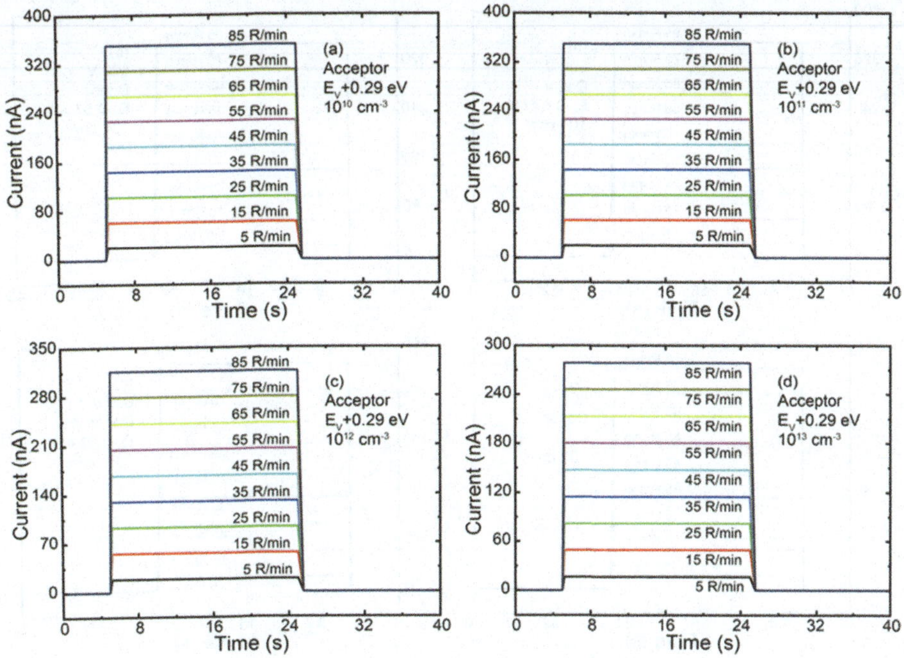

Fig. 2 Effect of T2 defects on the stable current of current-mode CZT detectors (the defect densities are 10^{10} cm^{-3}, 10^{11} cm^{-3}, 10^{12} cm^{-3}, and 10^{13} cm^{-3}, and the working voltage is 600 V)

increases, more defects become activated. Since defects do not affect the collection efficiency of charge carriers due to their capture and release processes, based on the physical principles considered by the current-type CZT detector response model, it is believed that the loss of charge carriers is primarily due to the activation of defects enhancing the polarization of the electric field within the crystal, resulting in more charge carriers undergoing recombination and loss. The possible reasons for the distortion of the CZT detector output signal will be explored next. It is clear that the result shows that T2 and T5 defects cannot cause the nonlinearity of flat currents, even if the defect density reaches 10^{13} cm^{-3}.

Effect of T6 Defects

In this section, the effect of T6 defects on the current distortion will be analyzed with the detector response model. The output currents of a CZT detector under different defect densities are shown in Fig. 5. As shown in Fig. 5a, b, T6 defects with low density (such as 10^{10} cm^{-3}) cannot cause the sharp peak either when the defects with low density (such as 10^{11} cm^{-3}). When the density of a T6 defect reaches 10^{12} cm^{-3} or even higher, a sharp peak appears at the beginning of γ-ray radiation, as shown in Fig. 5c, d. This theoretical result confirms that the trapping and detrapping effects of

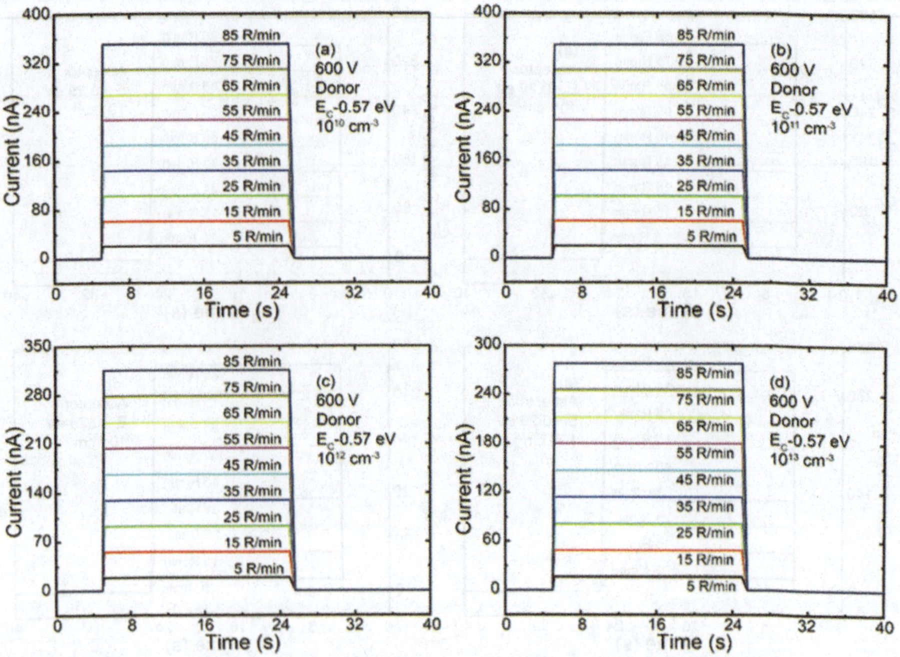

Fig. 3 Effect of T5 defects on the stable current of current-mode CZT detectors (the defect densities are 10^{10} cm^{-3}, 10^{11} cm^{-3}, 10^{12} cm^{-3}, and 10^{13} cm^{-3}, and the working voltage is 600 V)

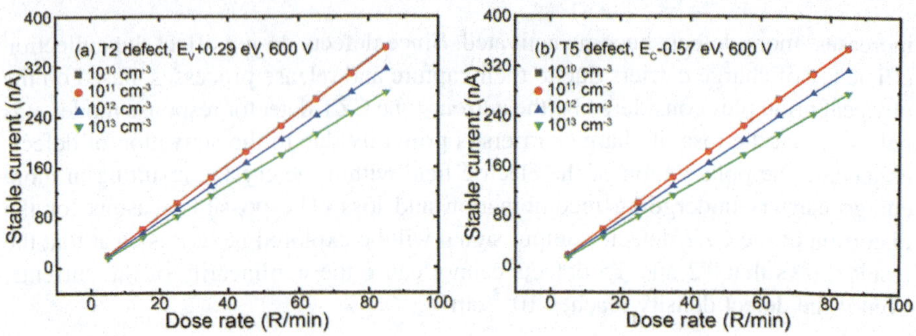

Fig. 4 Effect of T2 and T5 defects on the relationship between the dose rate and the stable current of current-mode CZT detectors (the defect densities are 10^{10} cm^{-3}, 10^{11} cm^{-3}, 10^{12} cm^{-3}, and 10^{13} cm^{-3}, and the working voltage is 600 V)

deep-level defects on γ-ray-generated carriers have significant contributions to the sharp peak of the temporal response at the beginning of γ-ray radiation.

The relationship between the flat current and the flux rate for a T6 defect is analyzed, which is illustrated in Fig. 6. As the result shows, under low defect density, a T6 defect cannot cause flat current nonlinearity either (Fig. 6a, b). However, when

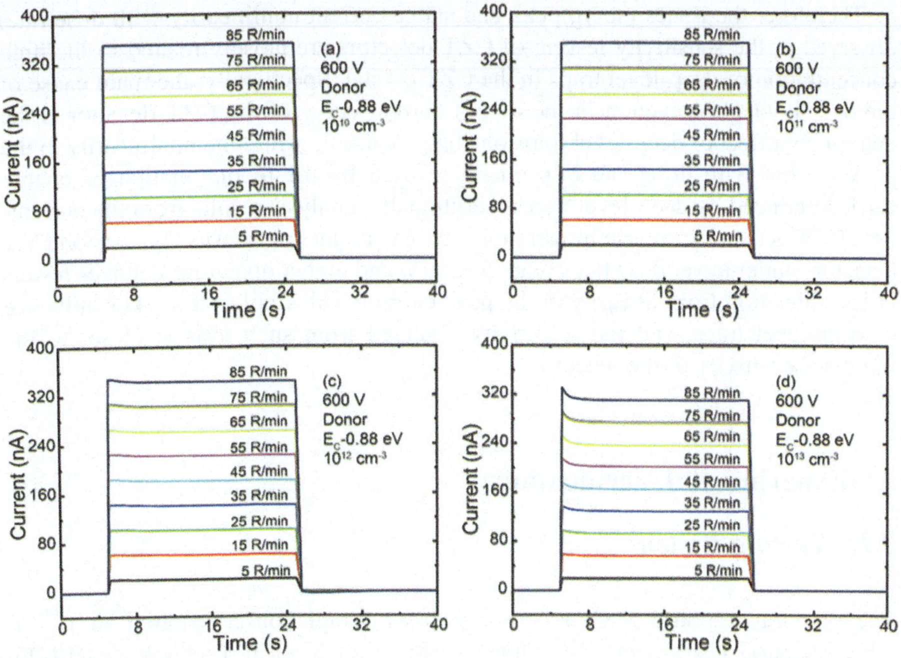

Fig. 5 Effect of E_C − 0.88 eV defects on the relationship between the dose rate and the stable current of current-mode CZT detectors (the defect densities are 10^{10} cm^{-3}, 10^{11} cm^{-3}, 10^{12} cm^{-3}, and 10^{13} cm^{-3}, and the working voltage is 600 V)

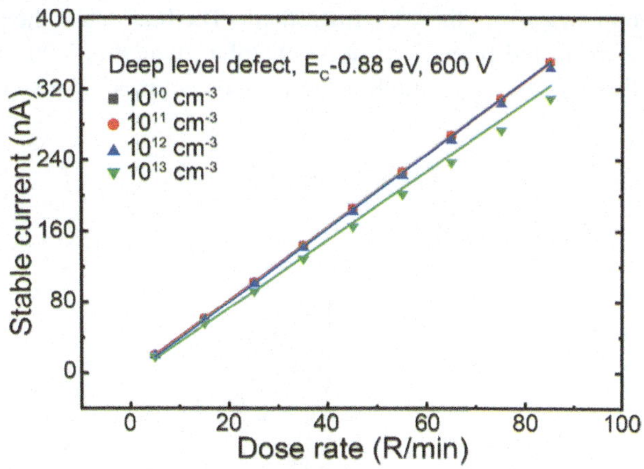

Fig. 6 Effect of E_C − 0.88 eV defects on the relationship between the dose rate and the stable current of current-mode CZT detectors (the defect densities are 10^{10} cm^{-3}, 10^{11} cm^{-3}, 10^{12} cm^{-3}, and 10^{13} cm^{-3}, and the working voltage is 600 V)

the defect density reaches 10^{12} cm^{-3} or higher, nonlinearity of the flat current occurs, as shown in Fig. 6c, d. This theoretical result indicates that flat current nonlinearity occurs with an accompanying sharp peak, meaning that flat current nonlinearity is also related to deep-level defects in the CZT crystal.

Therefore, the issues (sharp peak and stable current nonlinearity with dose rate) observed in the sensitivity testing of CZT detectors are closely related to the high concentration of deep-level traps in the CZT crystal. Specifically, the main cause of the overshoot phenomenon in the output current curve of the CZT detector is the detrapping effect of deep-level traps on charge carriers, while the nonlinearity in the peak current with dose rate is primarily caused by the recombination of charge carriers induced by deep-level traps. Additionally, analysis results from the current-type CZT detector response model indicate that conducting sensitivity tests on CZT detectors under lower dose rates (e.g., 5 R/min) and higher operating voltages results in less deviation from linearity in the peak current. This implies a weaker influence of deep-level traps, and the sensitivity obtained from such tests is closer to the intrinsic sensitivity of the detector.

3 Experimental Verification

3.1 Testing System

The temporal response of CZT detectors to γ-ray radiation is acquired on a ^{60}Co γ-ray radiation instrument at Northwest Institute of Nuclear Technology (NINT), Xi'an, China. The activity of ^{60}Co source is about 2000 Ci. The output current of CZT detectors is recorded by a microelectrometer, which can acquire current signals that are smaller than pA. The outer voltage is supplied by a high-voltage source. The layout of the temporal current test system is shown in Fig. 7. The CZT detector is set 6 m away from the ^{60}Co instrument. The flux rate at the position of the CZT detector is measured by an ionization chamber, which is 3.903 R/min. Based on Eq. 1, the flux rate at the position of the CZT detector is 1.33×10^8 MeV·cm^{-2}·s^{-1} [10].

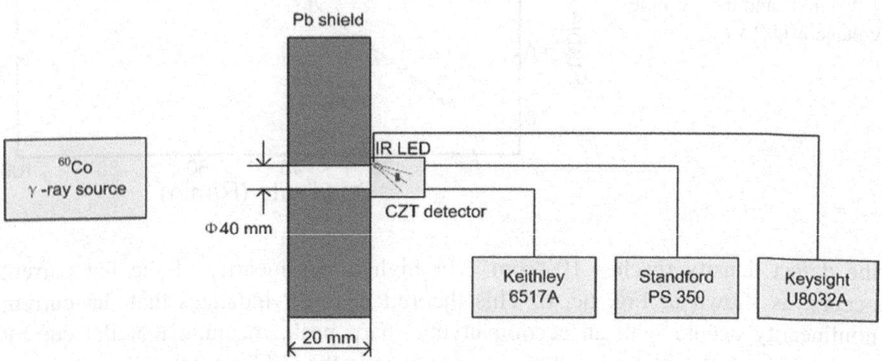

Fig. 7 Layout of a CZT detector sensitivity calibration system

$$\frac{d\phi}{dt}\left(\text{MeV}\cdot\text{cm}^{-2}\cdot\text{s}^{-1}\right) = I\left(\text{R}\cdot\text{min}^{-1}\right) \times 3.405 \times 10^7 \quad (1)$$

3.2 Experimental Result

The temporal current curves of a CZT detector under different outer voltages are shown in Fig. 8. When the outer voltage is not higher than 300, the flat current of the CZT detector remains stable, even though a sharp peak exists at the beginning of

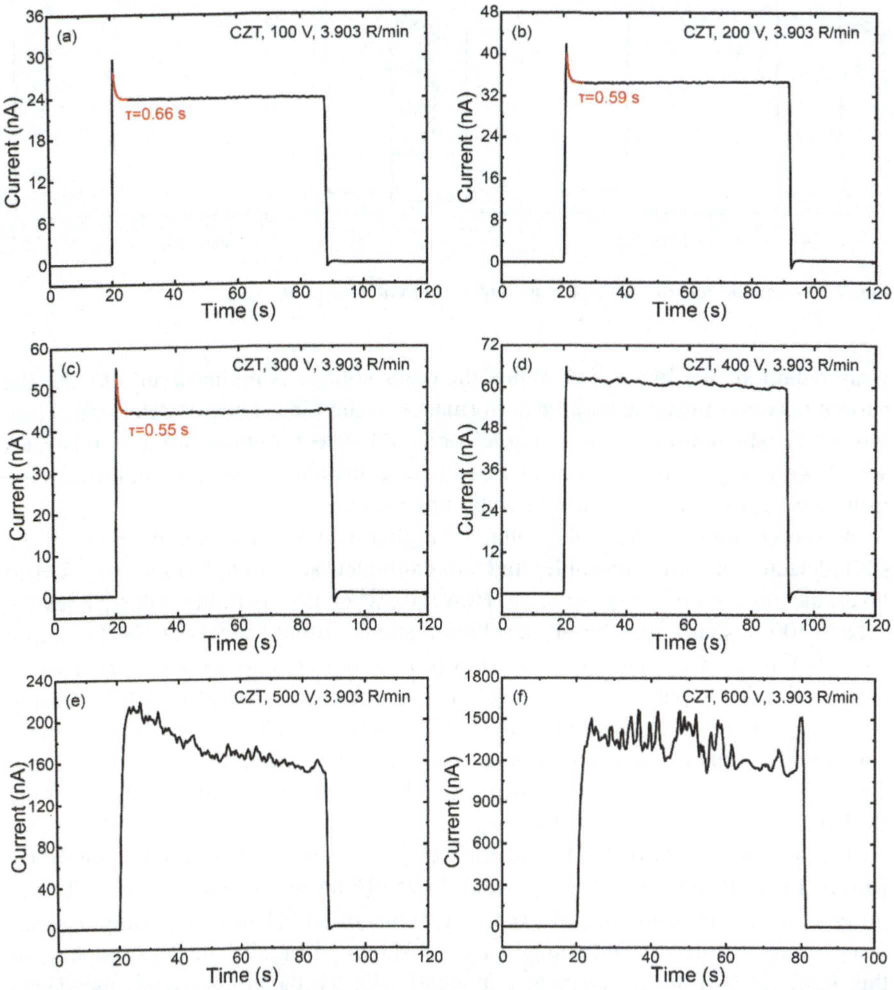

Fig. 8 Output currents of CZT detector change with the working voltages under the dose rate of 3.903 R/min

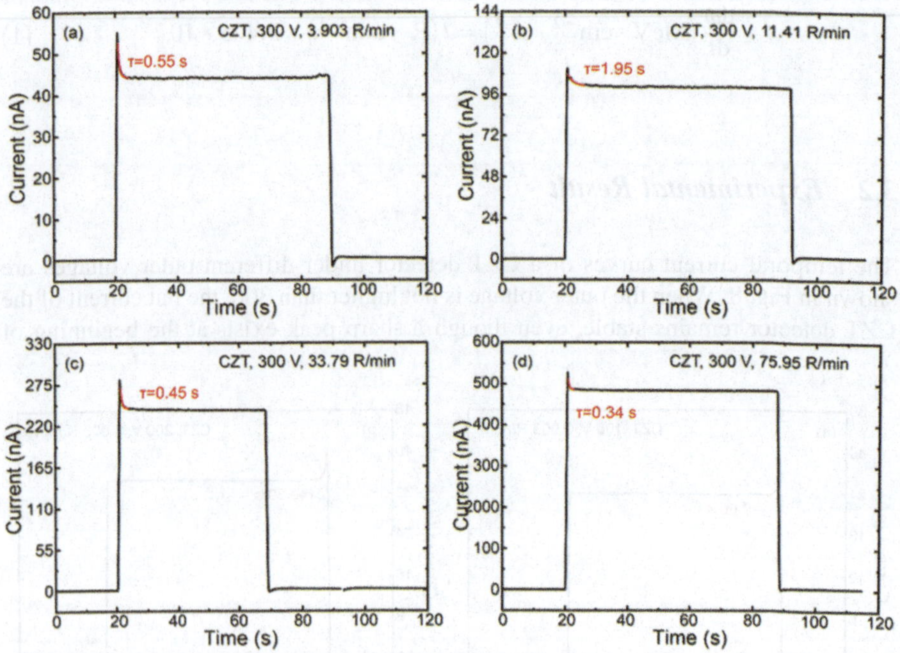

Fig. 9 Output currents of CZT detectors under different dose rates

γ-ray radiation (see Fig. 8a–c). When the outer voltage is higher than 300, the flat current becomes unstable and the performance of the CZT detector deteriorates (see Fig. 8d–f). The temporal current curves of a CZT detector under different flux rates are shown in Fig. 9. It should be clarified that all the above experimental results are from one CZT detector and from similar phenomena.

As shown in Fig. 8, when the voltage is higher than 300, the output current of the CZT detector becomes unstable, and the calibrated sensitivity cannot be used to invert the intensity of γ-ray radiation. However, when the operating voltage does not exceed 300, a sharp peak before reaching a stable output state exists in the output current of the CZT detector, and the decay process of this sharp peak follows a single exponential decay pattern. The decay constant decreases gradually with increasing operating voltage, as shown in Fig. 8a–c. Considering that higher operating voltages correspond to smaller decay constants along the sharp peak, the detector's output signal can reach a stable value more quickly. Therefore, the operating voltage for the CZT detector should be set to 300.

The output signals of CZT detectors when the front-end dose rates are 3.903 R/min, 11.41 R/min, 33.79 R/min, and 75.95 R/min are shown in Fig. 9. Under different dose rate conditions, the output currents of a CZT detector exhibit a sharp peak before reaching a stable output state, as shown in Fig. 9. The decay process of this sharp peak follows a single exponential decay pattern. Overall, the decay constant obtained from fitting the output signal of a CZT detector decreases gradually with an increasing dose rate.

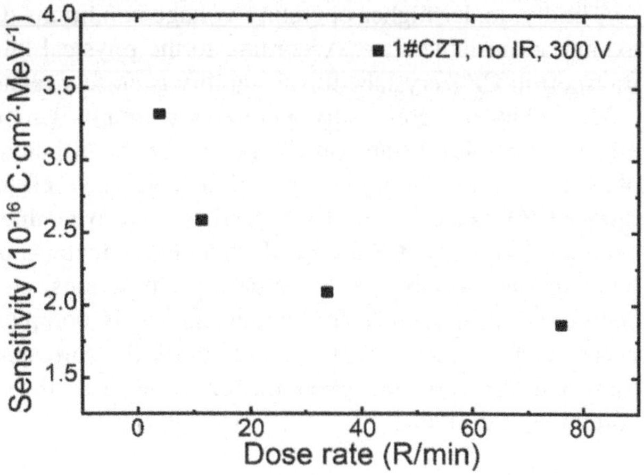

Fig. 10 Sensitivity values of CZT detectors under dose rate conditions

Based on the definition of steady-state sensitivity, the peak current is used as the output current when the CZT detector operates stably. The relationship between the sensitivity of CZT detectors and the dose rate is shown in Fig. 10. It can be observed that there is significant nonlinearity in the sensitivity of CZT detectors with respect to the dose rate. For instance, when the dose rate increases from 3.903 R/min to 75.95 R/min, the sensitivity of the CZT detector decreases from 3.32×10^{-16} C·cm^2/MeV to 1.87×10^{-16} C·cm^2/MeV, representing a decrease of 43.8%.

In a sensitivity testing experiment, the influence of operating voltages and dose rates on the sensitivity of CZT detectors is systematically analyzed, as well as the variation patterns of the detector output signal distortion and its relationship with operating voltage and dose rate. The specific findings can be summarized as follows.

Firstly, in both nonradiation and radiation conversion states, CZT detector output signals exhibit a sharp peak at the beginning of radiation and a decay time after the sharp peak on the order of seconds. With an increase in the dose rate, the decay time gradually shortens, and the operating voltages have little influence on the decay time. These characteristics indirectly reflect that the transport process of charge carriers is not the main cause of the sharp peak. The detrapping of the charge carriers by deep-level traps is an inherent factor in generating a sharp peak.

Secondly, the peak value of CZT detector output signals should linearly vary with the dose rate, resulting in a constant detector sensitivity if no defect exists. However, in actual tests, as the dose rate increases, the CZT detector sensitivity gradually decreases, primarily due to increased charge carrier recombination. Charge carrier recombination mainly originates from two sources: near-Fermi level recombination centers and the accumulation of space charges caused by the polarization of the electric field distribution. Combining the sensitivity analysis results obtained from the current-type detector response model, it is observed that the effect of recombination centers on sensitivity variation with dose rate is minimal. Therefore, charge carrier recombination in the low electric field region formed by space charge effects is an inherent factor in the variation of CZT detector sensitivity with dose rate.

Thirdly, under higher operating voltage conditions, CZT detector output signals exhibit an unstable state. According to the physical laws governing charge carrier transport in CZT crystals, this instability is closely related to deep-level traps.

This chapter theoretically and experimentally demonstrates that the detrapping effect of deep-level traps on charge carriers is an inherent factor in the sharp peak phenomenon and the nonlinear variation of stable current with dose rate observed during CZT detector sensitivity testing. However, due to the complex origins of deep-level traps in CZT crystals, including metal-semiconductor interface band bending and, possibly, intrinsic defects, impurities, etc., the contribution of these deep-level traps to detector output signals is complex. In future research, it is necessary to further investigate and clarify the correspondence between deep-level traps and the overshoot phenomenon, as well as the nonlinear variation of peak current with dose rate.

4 IR Improvement Analysis

Both the sharp peak and the flat current nonlinearity are mainly contributed by deep-level defects. To suppress the current distortion of CZT detectors, it is significant to lower the density of deep-level defects and their effects on γ-ray-generated carriers.

Subbandgap illumination (infrared (IR) illumination) refers to those light, who has a larger wavelength value than ~817 nm, which corresponds to the bandgap of CZT crystals (~1.52 eV). Since the energy of IR illumination is lower than the bandgap of CZT crystals, it will penetrate through the CZT crystal and directly react with the deep-level defects. It has been verified that subbandgap illumination can suppress the current distortion (including the sharp peak and flat current nonlinearity) of CZT detectors under a temporal response [14]. The parameters for IR light-emitting diode (LED) are a peak wavelength of 1200 nm and full width at half maximum (FWHM) of ~45 nm. The reason for choosing 1200 nm wavelength (~1.03 eV) is that most defect levels are lower than ~1 eV, and 1200 nm wavelength can almost excite all the defects in CZT crystals. Since the increased dark current irradiated by 1200 nm IR illumination is mainly contributed by 1200 nm wavelength exciting carriers from defect level, the dark current level is used to describe the effect of IR illumination on CZT detector current. While no IR illumination is operated, the temporal current curve of the CZT detector shows an obvious sharp peak. With an increase in IR LED driving voltage, the sharp peak gradually decreases and finally disappears. The flat current value decreases by ~12% with an increase in the dark current from 10 nA to 340 nA 1200 nm.

Here, the reasons for the IR optimization will be analyzed through the CZT detector model. Theoretically, the main reason for the effect of IR illumination on deep-level defects is the trapping and detrapping cross-sections for the defect. The lifetimes for the trapping (τ_{et}) and detrapping (τ_{ed}) process are listed as Eqs. 2 and 3 below, where N_t is the total electron trapping state density with an energy level of E_{et}, n_t is the trapped electron density at this state, p_t is the trapped hole density at the

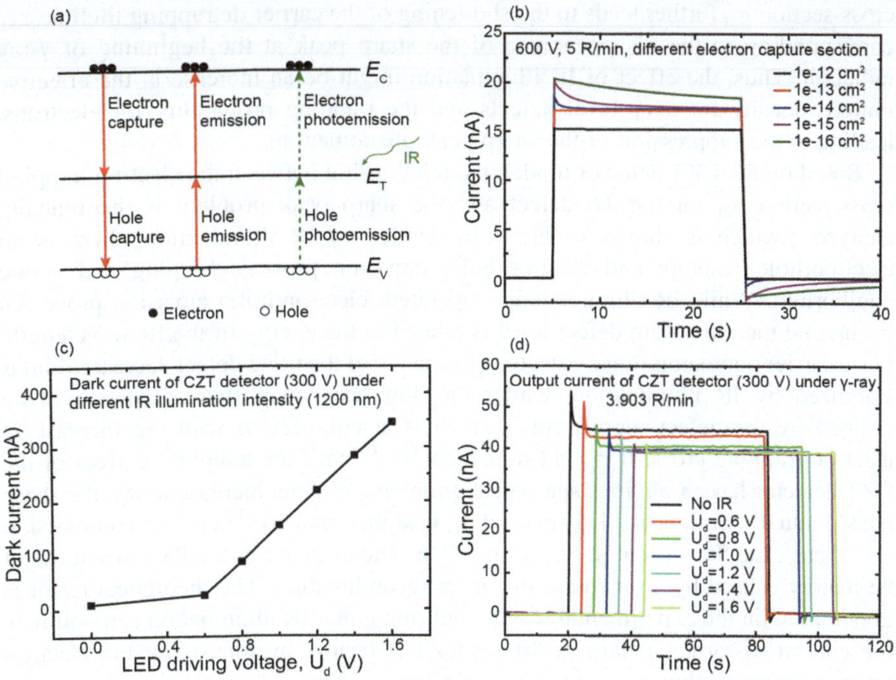

Fig. 11 IR optimization analysis. (**a**) Six carrier transit processes in the defect. (**b**) Different electron cross-sections in the CZT detector model. (**c**) Dark current and (**d**) output current of CZT detectors under different LED driving voltages

hole trapping state, v_{th} is the thermal velocity, N_c is the available state density for the conduction band, σ_{et} is the electron trapping cross-section, σ_{er} is the recombination cross-section between the free electron and the trapped holes, k is the Boltzmann constant, and T is the Kelvin temperature [15].

$$\tau_{et} = \frac{1}{(N_t - n_t)\sigma_{et} v_{th}} \tag{2}$$

$$\tau_{ed} = \frac{1}{N_C \sigma_{et} v_{th} \exp(-(E_C - E_{et})/kT)} \tag{3}$$

Under a certain intensity of IR illumination, large amounts of trapped carriers are detrapped. For example, holes are excited from the valence band to the defect level (hole photoemission), and electrons are excited from the defect level to the conduction band (electron photoemission), as shown in Fig. 11a. Carrier photoemission increases the dark current, as shown in Fig. 11b. Since more defects are ionized, more carriers will be trapped by these newly ionized defects, equivalent to the increase of the electron trapping cross-section σ_{et}. An increase in electron trapping cross-section σ_{et} further leads to the shortening of the carrier detrapping lifetime τ_{ed}, corresponding to the disappearance of the sharp peak at the beginning of γ-ray radiation. Thus, the effect of IR illumination might be an increase in the effective ionized density for deep-level defects and the trapping probability for electrons, leading to the suppression of the sharp peak phenomenon.

Based on the CZT detector model, the relationship between the electron trapping cross-section σ_{et} for the T6 defect and the sharp peak problem is theoretically analyzed, which is shown in Fig. 11a. Under no IR illumination, there is an electron(hole) capture and electron(hole) emission process, keeping a dynamic equilibrium. While IR illumination is operated, electron(hole) emission processes occur, and the maximum defect level is related to the energy of the IR wavelength. Once carrier photoemission exists, the occupancy of the filled defect decreases and is stabilized by IR illumination. Under this low-occupancy state, carriers will be trapped by the defect more easily, which is at equilibrium with the increase in electron trapping cross. If σ_{et} is lower than 10^{-16} cm^2, the temporal current of the CZT detector has an obvious sharp peak problem. With an increase in σ_{et}, the sharp peak gradually disappears. However, if σ_{et} is higher than 10^{-12} cm^2, as compared to 10^{-13} cm^2, the flat current decreases by 11%. The decrease in the flat current might be mainly caused by an increase in carrier recombination. The theoretical result is good based on the experimental result, confirming that IR illumination can optimize the current distortion (sharp peak) for CZT detectors by improving the electron trapping cross-section σ_{et} of deep-level defects.

5 Conclusion

In this paper, the temporal current distortion of CZT detectors under γ-ray radiation is analyzed, including the sharp peak and the flat current nonlinearity. Sentaurus TCAD software and γ-ray radiation instrument are used in this study. The theoretical and experimental results show that the sharp peak and the flat current nonlinearity of the temporal response for CZT detectors under γ-ray radiation are mainly contributed by the trapping and detrapping effects of deep-level defects on γ-ray-generated carriers. The improvement effect of IR illumination on the current distortion is the increase in the effective ionized density for deep-level defects and the trapping probability for electrons, leading to the suppression of the sharp peak phenomenon. These results can help us better understand the current distortion of CZT detectors under high flux-rate radiation, which can provide significant theoretical support for further research on distortion suppression and performance optimization.

References

1. Henric, S. K., Daniel, S., Fiona, A. H., et al. (2016). X-ray polarimetry with the Polarization Spectroscopic Telescope Array (PolSTAR). *Astroparticle Physics, 75*, 8–28.
2. Zha, G. Q., Yang, J., Xu, L. Y., et al. (2014). The effects of deep level traps on the electrical properties of semi-insulating CdZnTe. *Journal of Applied Physics, 115*(4), 1–4.
3. Camarda, G. S., Bolotnikov, A. E., Cui, Y. G., et al. (2008). Polarization studies of CdZnTe detectors using synchrotron X-ray radiation. *IEEE Transactions on Nuclear Science, 55*(6), 3725–3730.
4. Dedic, V., Rejhon, M., Franc, J., et al. (2017). Space charge oscillations in semi insulating CdZnTe. *Applied Physics Letters, 111*(10), 1–5.
5. Washington, A. L., II, Teague, L. C., Duff, M. C., et al. (2011). Effect of sub-bandgap illumination on the internal electric field of CdZnTe. *Journal of Applied Physics, 110*(7), 1–5.
6. Dedic, V., Franc, J., Sellin, P. J., et al. (2012). Study on electric field in Au/CdZnTe/In detectors under high fluxes of X-ray and laser irradiation. *Journal of Instrumentation, 7*(2), 1–10.
7. Guo, Y., Zha, G. Q., Li, Y. R., et al. (2019). Effect of transient space-charge perturbation on carrier transport in high-resistance CdZnTe semiconductor. *Chinese Physics B, 28*(11), 1–10.
8. Franc, J., Dedic, V., Rejhon, M., et al. (2015). Control of electric field in CdZnTe radiation detectors by above-bandgap light. *Journal of Applied Physics, 117*, 165702.
9. Rejhon, M., Franc, J., Dedic, V., et al. (2016). Analysis of trapping and de-trapping in CdZnTe detectors by Pockels effect. *Journal of Physics D: Applied Physics, 49*, 375101.
10. Chen, X., Song, Z. H., Lu, Y., et al. (2021). Improvement on the temporal response of CZT γ-ray detector by infrared illumination. *IEEE Transactions on Nuclear Science, 68*(10), 2533–2538.
11. Xu, L. Y., Jie, W. Q., Zhou, B. R., et al. (2015). Effects of crystal growth methods on deep-level defects and electrical properties of CdZnTe: In crystals. *Journal of Electronic Materials, 44*(1), 518–523.
12. Yang, F., Jie, W. Q., Zha, G. Q., et al. (2020). The effect of indium doping on deep level defects and electrical properties of CdZnTe. *Journal of Electronic Materials, 49*(2), 1243–1248.
13. Babentsov, V., Franc, J., Hoschl, P., et al. (2009). Characterization of compensation and trapping in CdTe and CdZnTe: Recent advances. *Crystal Research and Technology, 44*(10), 1054–1058.
14. Plyatsko, S. V., Petrenko, T. L., & Sizov, F. F. (2009). Defects instability under IR laser treatment in p-CdZnTe crystals. *Infrared Physics & Technology, 52*, 57–61.
15. Du, Y. F., Lebanc, J., Possin, G. E., et al. (2012). Temporal response of CZT detectors under intense irradiation. *IEEE Transactions on Nuclear Science, 50*(4), 480–484.

Evaluation on the Performance of CZT Detectors in Alpha Particle Detection with Experiment and Simulation

Yu Xiang, Xiaopan Jiang, and Long Wei

1 Introduction

Alpha particle detection is widely applied in many fields, such as in the determination of radioactive materials, radiation protection, environment protection, nuclear fuel facility monitoring, and radionuclide therapy [1, 2]. Since an α particle is heavy and charged, it loses energy mainly through inelastic collisions with atoms that belong to the target material. The collisions result in the excitation and ionization of the atoms. Many kinds of detectors are utilized for α particle detection. They can be simply categorized as gaseous detectors, scintillator detectors, and semiconductor detectors. In a gaseous detector, after the collision between an α particle and an atom, there is a high probability that one of the orbital electrons of the atom escapes from the gas atom and becomes a free electron. An ion is also created while the electron is liberated. The transport of the electron-ion pair in the electric field leads to a change

Y. Xiang · L. Wei (✉)
Beijing Engineering Research Center of Radiographic Techniques and Equipment, Institute of High Energy Physics, Chinese Academy of Sciences, Beijing, China

School of Nuclear Science and Technology, University of Chinese Academy of Sciences, Beijing, China

Jinan Laboratory of Applied Nuclear Science, Jinan, China

CAEA Center of Excellence on Nuclear Technology Applications for Nuclear Detection and Imaging, Beijing, China
e-mail: weil@ihep.ac.cn

X. Jiang
Beijing Engineering Research Center of Radiographic Techniques and Equipment, Institute of High Energy Physics, Chinese Academy of Sciences, Beijing, China

Jinan Laboratory of Applied Nuclear Science, Jinan, China

CAEA Center of Excellence on Nuclear Technology Applications for Nuclear Detection and Imaging, Beijing, China

K. Iniewski (ed.), *CdTe and CdZnTe Materials*,
https://doi.org/10.1007/978-3-031-64521-1_6

in the induced charge on the electrode, which is then processed by the connected electronics and finally becomes an electrical signal representing the energy deposition of the α particle.

In a scintillator detector, the generation process of the electrical signal representing the energy deposition of the alpha particle is quite different and more complicated than that in gas detectors. The incident alpha particle interacts with a scintillator atom, and the atom is therefore excited by the interaction. Then some visible lights are generated during the deexcitation of the atom. The lights are named fluorescence, and they are transformed into photoelectrons on the photocathode of a photomultiplier. Finally, the photoelectrons propagate rapidly in the photomultiplier and are subsequently collected by the anode.

In a semiconductor detector, the generation of a signal is almost the same as that in a gaseous detector. The only difference between them is that electron-hole pairs rather than electron-ion pairs are created in semiconductor detectors. The holes are the vacancies left by the electrons that get rid of the atoms and become free. If an electron comes to fill a hole, a relative electrical current will be generated in this process, which is the same as the current from the opposite movement of a positron, and that is how holes contribute to the electrical signal of semiconductor detectors.

Thanks to the large atomic number and high density of semiconductors, they can serve as efficient absorbers for photons and charged particles. The range of an α particle is usually only a few tens of microns in a semiconductor, while in gas, the value is three orders of magnitude higher, and this is the reason that the volume of semiconductor α particle detectors is permitted to be far smaller than that of gas detectors. Besides, in general, the ionization energy of a semiconductor is rather low, only a few electron volts, meaning that a great number of e-h pairs can be generated by one interaction, and therefore the signal fluctuation is low.

The benefits of semiconductor detectors, shown above, confirm that they are a good choice for α particle detection. At present, the most popular semiconductor materials for nuclear radiation detection are HPGe (high-purity germanium) and Si (silicon). HPGe can achieve an extraordinary energy resolution, better than 21 keV FWHM at 5486 keV [3]. However, HPGe cannot work normally unless the environment temperature is only a few kelvins. Silicon detectors have an astonishing energy resolution of 16 keV FWHM at 5486 keV [4]. Perovskite $CsPbBr_3$ is a member of new-generation semiconductor materials, able to be operated at room temperature but has an unsatisfying resolution of 15% FWHM at 5486 keV [5]. Gallium nitride (GaN) has good high-temperature stability and therefore functions well under extreme conditions. Its energy resolution is about 2.2% FWHM at 5486 keV [6].

CZT (CdZnTe, cadmium zinc telluride) is one of the second-generation semiconductor materials. In the past three decades, it has attracted the attention of many researchers because of its physical properties appropriate for radiation detection, which are listed in Table 1, together with those of other semiconductor materials. CZT has a wide forbidden band and high resistivity, which keeps the leaking current of CZT detectors at a low level and makes it capable of working at room temperature. The atomic number of CZT is large, which results in unignorable interaction cross-sections for X-rays, γ rays, and charged particles.

Table 1 Physical properties of different semiconductor materials [7]

Material	Atomic number	Density (g/cm^3)	Forbidden bandwidth (eV)	Ionization energy (eV)	Resistivity (Ω·cm)	$\mu_e\tau_e$[a] (cm^b/V)	$\mu_h\tau_h$[b] (cm^b/V)
Si	14	2.33	1.12	3.62	$>10^4$	>1	>1
Ge	32	5.33	0.67	2.96	50	~1	>1
GaAs	31,33	5.32	1.43	4.2	10^7	8×10^{-5}	4×10^{-6}
GaAs	31,33	5.32	1.43	4.2	10^7	10^{-5}	10^{-6}
InP	15,49	4.78	1.35	4.2	10^6	5×10^{-6}	$<2 \times 10^{-5}$
CdTe	48,52	6.20	1.44	4.43	10^9	10^{-3}	10^{-4}
$Cd_{0.9}Zn_{0.1}Te$	48,30,52	5.78	1.57	4.64	$10^{10} \sim 10^{11}$	$10^{-3} \sim 10^{-2}$	$10^{-6} \sim 10^{-5}$
$Cd_{0.8}Zn_{0.2}Te$	48,30,52	6.02	1.5 ~ 2.2	5.0	$10^{10} \sim 10^{11}$	10^{-3}	$10^{-6} \sim 10^{-5}$
HgI_2	80,53	6.40	2.13	4.2	10^{13}	10^{-4}	$10^{-5} \sim 10^{-4}$
PbI_2	82,53	6.20	2.3 ~ 2.6	4.9	10^{12}	10^{-6}	10^{-7}
TlBr	81,35	7.56	2.68	6.5	10^{12}	10^{-5}	10^{-6}

[a] $\mu_e\tau_e$ represents the mobility-lifetime product of electrons
[b] $\mu_h\tau_h$ represents the mobility-lifetime product of holes

CZT detectors are very popular in the field of X-ray and gamma-ray detection. Many Compton cameras, single-photon emission computed tomography (SPECT), and positron emission tomography (PET) systems have been developed based on CZT detectors to realize photon imaging. CZT detectors also play an important role in cosmic radiation detection, medical diagnosis, and nuclide identification. However, there is a lack of research on the performance of CZT detectors in alpha particle detection. In previous related works, alpha particle detection was mainly employed for the measurement of some transport properties of CZT detectors, like the trapping time and the mobility-time products of charge carriers [8–12]. The best energy resolution of CZT detectors for α particles is 1.18% FWHM at 5486 keV, which is better than most other semiconductor detectors [13].

In this chapter, an α-particle detection system based on a planar CZT detector is built. An α-particle radiation source made of Am-241 and Pu-239 is detected to test the performance of the system. The best energy resolution of the detector is measured as follows: 1.22% FWHM at 5.486 MeV and 1.17% FWHM at 5.157 MeV in vacuum (0.01 atm). Simultaneously, a spectrum simulation methodology is proposed. The simulation of the transport of charge carriers and the interaction between α particles and CZT have been made to construct a simulated spectrum, which is similar to that obtained from the experiment.

2 CZT α-Particle Detection System

The CZT α-particle detection system employed in this chapter can be divided into three modules: a CZT detector, a tailor-made electronics, and a data acquisition (DAQ) program. The scheme of the detection system is shown in Fig. 1. In this section, the function of each module in detection will be introduced.

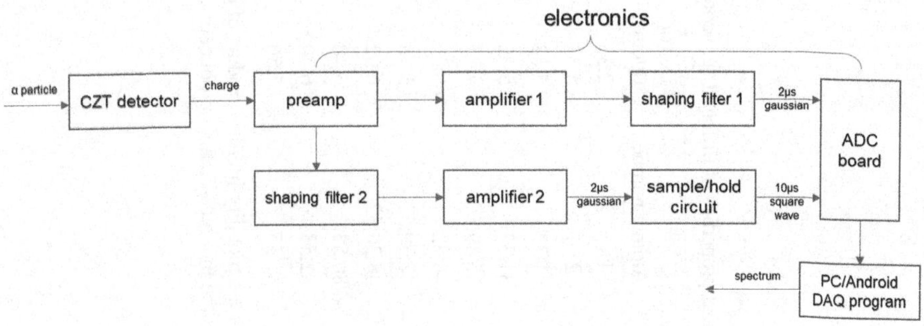

Fig. 1 The scheme of the detection system

2.1 CZT Detector

In a semiconductor detector, the electrical signal generated from the change of induced charge on electrodes is proportional to the energy deposition of the incident radiation. The change of the induced charge generated by an e-h pair, which is collected by the electronics of a planar semiconductor detector and marked as Q_i, can be calculated by the following Hecht formula [14]:

$$Q_\mathrm{i} = Q_\mathrm{e} + Q_\mathrm{h} = \frac{\mu_\mathrm{e}\tau_\mathrm{e}Ee}{d}\left(1 - \exp\left(-\frac{z}{\mu_\mathrm{e}\tau_\mathrm{e}E}\right)\right) + \frac{\mu_\mathrm{h}\tau_\mathrm{h}Ee}{d} \times \left(1 - \exp\left(-\frac{d-z}{\mu_\mathrm{h}\tau_\mathrm{h}E}\right)\right) \tag{1}$$

where Q_e is the induced charge caused by electrons, Q_h is the induced charge caused by holes, e is the elementary charge, E is the electric field, d is the thickness of the detector, and z is the distance from the position where the electron-hole pair is generated to the detector's anode. As a result of the limitation of the CZT crystal growth method, the thickness of a CZT detector is usually a few millimeters or centimeters. Therefore, the values of $z/\mu_e\tau_eE$ and $(d - z)/\mu_h\tau_hE$ are close to 0, resulting in a simplified formula of Eq. 1:

$$Q_\mathrm{i} = e\left(1 - \frac{1}{2dE}\left(\frac{z^2}{\mu_\mathrm{e}\tau_\mathrm{e}} + \frac{(d-z)^2}{\mu_\mathrm{h}\tau_\mathrm{h}}\right)\right) \tag{2}$$

It is obvious that the value of the induced charge relies on the mobility-lifetime products of holes and electrons. Since in a CZT detector the mobility-lifetime product of holes is always two orders of magnitude lower than that of electrons, the trapping of holes causes a wide fluctuation of induced charge, which deeply affects the energy resolution of the CZT detector. It is reported that in the γ-ray spectra acquired by a planar CZT detector, the 662 keV peak is unable to be recognized because of the trapping of holes, while the energy resolution at 59.6 keV is about 7% FWHM [15].

The contact formed by the electrodes and CZT was carefully chosen when we designed the detector. There are two types of metal-semiconductor contacts: Ohmic contact and Schottky contact. The characteristics of these two contacts are different, which could affect the performance of the CZT detector. According to the principle of metal-semiconductor contact, if the work function of metal W_m is apparently larger than the one of semiconductor W_s , their contact will be Schottky. For Schottky contact, if the potential of the semiconductor is higher than that of the metal, the leaking current will be quite low. Otherwise, the current will be high. If $W_m \approx W_s$, the contact will be Ohmic [16]. Hence, a proper electrode material should be chosen to form Ohmic contact. The work function of several potential materials is

Table 2 Work functions of different materials [18, 19]

Material	Al	Au	In	Pt	Cr	C	CZT
Work function (eV)	4.28	5.1	4.12	5.65	4.5	5.0	≈4.6

Fig. 2 Picture of the planar CZT detector. The crystal and cathode are beneath the square hollow made of Cr

listed in Table 2. It is reported that Au-CZT Ohmic contact has the lowest leaking current [17]. Thus, for our CZT detector, gold (Au) was chosen to be the material of the electrodes.

The detector utilized in our detection system is made from a 22 × 22 × 0.7 mm^3 CZT crystal. Two Au electrodes separately cover the two 22 × 22 mm^2 surfaces of the detector, as seen in Fig. 2. The whole cathode surface is faced with 100 nm thick Cr.

2.2 *Electronics*

In a CZT detection system, electronics are employed to transform the change of the induced charge on the anode into a digital signal. It can be separated into several parts, including a high-voltage module, a preamp, two amplifiers, two shaping filters, a trigger circuit, a sample/hold circuit, and an analog-to-digital converter (ADC). The high-voltage module is utilized to provide bias voltage. The preamp is a charge-sensitive preamplifier, where the change of induced charge collected by the anode is turned into a voltage signal. An MCU (microcontrol unit) and a Wi-Fi module are attached to the ADC. The digital signal created by the ADC could be delivered to a computer or a mobile phone through Wi-Fi or a USB data cable.

2.3 *DAQ and Data Processing Computer Program*

A DAQ (data acquisition) program is developed to receive and process data from the ADC board. The acquiring time is allowed to be set at a certain value. When acquiring a spectrum, the real-time count rate, acquisition time, and spectrum pattern are all shown in the user interface. The file format for saved spectra is ".chn", which can be read by GammaVision, a widely used spectrum analyzer.

3 α-Particle Spectrum Measurement

An Am-Pu alpha-particle radiation source is utilized to test the performance of our CZT detector. The source contains two nuclides: Am-241 and Pu-239. In this section, the linear response of the detector is tested first, then the electronic noise is also measured. Finally, the detector is employed in the detection of the alpha-particle source, the detection results are analyzed.

3.1 *Linear Response Test*

The energy deposition of an alpha particle in a CZT detector generates an electrical signal. The signal will then result in a count in the spectrum. Ideally, the channel of the count is always proportional to the energy deposition. However, many factors could render the relation between them nonlinear, such as polarization and signal saturation. Thus, before alpha particle detection, the linear response of the CZT detector has to be tested.

If there are a large number of charge carriers trapped in the detector, a relative electric field whose direction is opposite that of the electric field built by the bias voltage will be created. This phenomenon is called polarization. Since the directions of the two fields are opposite, the velocity of the charge carriers decreases, leading to longer drift time and a higher possibility of being trapped. As a consequence, the extra loss of charge carriers destroys the linear response. In our experiment, the activity of the radiation source we used was weak. The number of incident alpha particles per second for the detector was only a few tens. Therefore, the effect of polarization was negligible.

In the detection system, signal saturation can be caused due to the limitation of electronic components. If the amplitude of a signal exceeds the threshold of an amplifier, the signal will be distorted and saturated. The part of the signal higher than the threshold will disappear, and a platform flush with the threshold will take its place. Consequently, in the spectrum, the counts tend to accumulate at the channel, representing the amplifier's threshold rather than the amplitudes of original signals.

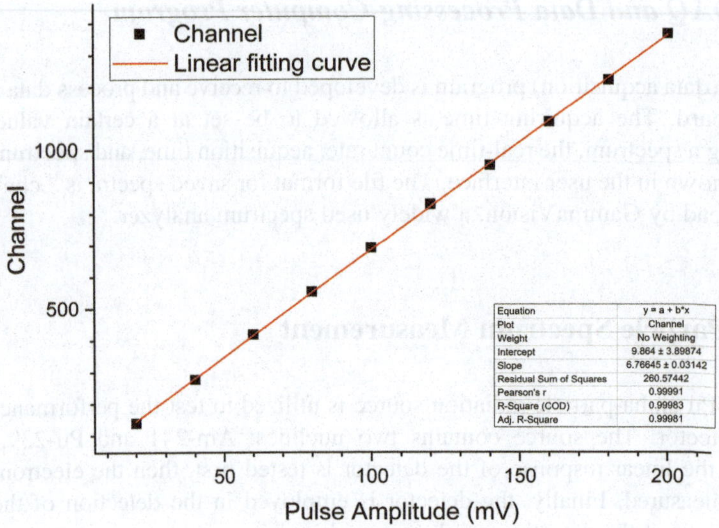

Fig. 3 The relationship between signal peak channel and pulse amplitude. The line is the linear fitting result

In our electronics, in order to avoid signal saturation, the amplification factor was set below 12, depending on the amplitude of the signal generated by the full energy deposition of alpha particles.

During the test of the linear response of the detection system, since alpha particle sources with different energy were unavailable, a reference pulser (Canberra model 1407) was used to produce an analog signal. The signal was transformed into charge by a 1 pF capacitor before entering the preamplifier. The anode was biased to 83 V, which ensured that the signal-noise ratio (SNR) reached the maximum. The relationship between a signal peak channel and pulse amplitude is shown in Fig. 3, confirming that the alpha detector has a good linear response.

The amplitudes of the analog signals corresponding to 5.486 MeV alpha particles from Am-241 and 5.157 MeV from Pu-239 are calculated. The results are 189 mV for 5.486 MeV and 177 mV for 5.157 MeV. They are in the linear response interval of the detector, which indicates that our detector is capable of detection.

3.2 *Electronic Noise Measurement*

Electronic noise was measured to evaluate its influence on the energy resolution of the CZT detector. The anode was biased to 83 V, while the cathode was connected to the ground. The test signal was also from the pulser Canberra model 1407. The FWHM (full width at half maxima) of the peak in the spectrum collected by the detector was measured to estimate the system's electronic noise. It was found that the FWHM is a constant and is equal to 5.63. What is more, the influence of electronic noise decreases when the signal amplitude increases.

3.3 Am-Pu α-Particle Spectrum Acquisition

The Am-Pu alpha particle source we detected is a wafer with a radius of 11.5 mm. The source was set parallel to the cathode of the CZT detector with a distance of 1.4 cm. The source and the CZT detector were put into an aluminum box, which served as an electromagnetic screen. The electronics was powered by a portable charger outside the box. In order to prevent the deposition of the alpha particle's energy into the air, the box was laid into a vacuum oven and air pressure was lowered to 1 kPa.

According to the spectra collected from the experiment, the energy resolution achieved by the detector system was 1.47% @ 5.486 MeV and 1.32% @ 5.157 MeV. A better one, 1.22% @ 5.486 MeV and 1.17% @ 5.157 MeV, was achieved after biasing the cathode to −92 V and grounding the anode. As a prototype, the detector has shown a quite satisfying energy resolution. It should be emphasized that biasing the cathode will increase the risk of electric shock due to the electroconductive Cr coating, and that is why this design was abandoned. The energy resolution of Am-241 is worse than that of Pu-239, whereas electronic noise should have less impact on the energy resolution of 5.486 MeV alpha particles. The reason for this abnormal result is the radiation source itself.

3.4 Electron Mobility-Lifetime Product Measurement

Mobility-lifetime product is a significant parameter that characterizes the quality of semiconductor radiation detectors. In Eq. 2, it is shown that there is a tight connection between CCE (charge collection efficiency) and mobility-lifetime product. CCE is equal to Q_i/Ne.

In this work, the ^{241}Am-^{239}Pu radiation source was aimed at the cathode of the CZT detector to measure $\mu_e\tau_e$, while the bias voltage of the anode was controlled by a high-voltage power supply (Canberra model 3105). The range of α particles in CZT is about 20 μm, and the detector is far more thick. Therefore, the contribution of the transport of the hole to CCE can be ignored. The electric field E is the quotient of the bias voltage U divided by the thickness d. Thus, Eq. 1 is simplified as follows:

$$\mathrm{CCE} = \frac{\mu_e\tau_e E}{d}\left(1 - \exp\left(-\frac{z}{\mu_e\tau_e E}\right)\right) \tag{3}$$

The peak channels of the alpha particles are proportional to CCE. The $\mu_e\tau_e$ is $2.74 \times 10^{-7}\ m^2 \cdot V^{-1}$ and $2.62 \times 10^{-7}\ m^2 \cdot V^{-1}$ from the fitting result of the bias voltage and the peak channels of the alpha-particle peaks of Am-241 and Pu-239, respectively.

4 Simulations of CZT Detectors' Spectral Response

A series of simulations were accomplished to construct the alpha particle spectrum of the planar CZT detector, consisting of the simulation of the interaction between α particles and CZT, the distribution of CCE in the CZT detector, and the electronic noise of the CZT detector. They will be introduced in this chapter.

4.1 *Simulation of α Source and α-CZT Interaction*

The interaction between alpha particles and CZT was simulated by a popular Monte Carlo simulation toolkit, Geant4 [20]. The configuration of the detector and the source in the simulation was identical to the one in the experiment. The directions of the emitted alpha particles were random. The energy of the emitted alpha particle in each event was selected based on the branch ratios and the count rates of Am-241 and Pu-239 measured in the experiment. We found that the number of steps in an incident event was always no more than three. Moreover, most of the energy of the alpha particle was deposited in the first step. Therefore, in order to simplify the data processing, only the deposited energy and a weighted energy deposition position were recorded for each event. The weighted energy depositing position is calculated by the following equation:

$$\boldsymbol{r} = \frac{\sum_{i=1}^{n} \boldsymbol{r}_i \cdot E_i}{\sum_{i=1}^{n} E_i} \tag{4}$$

where E is the energy deposition of the alpha particle in the step, $\boldsymbol{r}_i$ is the intermediate position between the PreStepPoint and the PostStepPoint, the subscript i is the symbol of the step, $E = \sum_{i=1}^{n} E_i$ is the total energy deposition, and n is the number of steps in an event. As all the deposited energy of the α particle is transformed to e-h pairs in CZT and most of the deposition concentrates on a short length compared to the entire incident range, $\boldsymbol{r}$ is appropriate to represent the initial position of the e-h pairs created by a single α particle. There are 500,000 events generated per simulation.

4.2 Simulation of Charge Carrier Transport

The trapping of holes influences the performance of CZT detectors in radiation detection. To estimate the influence, the CCE distribution in the planar CZT detector was simulated using finite element method (FEM) software, COMSOL Multiphysics 5.5. The differential equations of CCE employed in the simulation are as follows [21]:

$$\frac{dn^+}{dt} - \nabla \cdot (\mu_e n^+ \nabla \phi) - \nabla \cdot (D_e \nabla n^+) + \frac{n^+}{\tau_e} = \mu_e \nabla \phi \nabla \psi \tag{5}$$

$$\frac{dp^+}{dt} + \nabla \cdot (\mu_h p^+ \nabla \phi) - \nabla \cdot (D_h \nabla p^+) + \frac{p^+}{\tau_h} = \mu_h \nabla \phi \nabla \psi \tag{6}$$

$$D_{e,h} = \frac{k_B T}{e} \cdot \mu_{e,h} \tag{7}$$

where n^+ and p^+ are, respectively, the CCE contributed by electrons and holes, $D_{e,\,h}$ corresponds to the diffusion coefficient, ϕ is the electric potential, τ is the lifetime of charge carriers, ψ is the weighting potential for anodes, and k_B is the Boltzmann constant. The definition of weighting potential can be seen in references [22, 23].

Before the calculation of the CCE distribution, the distribution of the electric potential and anode weighting potential in the detector should be calculated first. Since the effect of the charge carriers generated in the interaction between alpha particles and CZT on the electric field is too small, it is assumed that the electric potential always obeys the following Laplace equation:

$$\nabla^2 \phi = 0 \tag{8}$$

Meanwhile, the electric potential of the anode and cathode is set at 83 and 0 V, respectively. All the incoming and outgoing fluxes through the other four detector surfaces are set to 0. The weighting potential can also be computed by a Laplace equation:

$$\nabla^2 \psi = 0 \tag{9}$$

The weighting potential at the anode is 1 and 0 at the cathode, and there are no fluxes on the other surfaces.

In the simulation, the distributions of the electric potential and the weighting potential were calculated first. Then the CCE distribution in the detector was simulated. It should be noted that in our work, the boundary condition of CCE was different from the one in Ref. [24]. Here, it is $\hat{n} \cdot \nabla D_h p^+ = -S_q p^+$ for the anode and $\hat{n} \cdot \nabla D_e n^+ = -S_n n^+$ for the cathode, where $S_n = 20\,\mathrm{m \cdot s^{-1}}$ and $S_p = 0.1\,\mathrm{m \cdot s^{-1}}$ are the surface recombination velocities. We changed the boundary condition into this because the original one, $p^+ = 0$ for the anode and $n^+ = 0$ for the cathode, will

Table 3 Physical properties utilized in the simulation

Property	Value
Temperature	300 K
Electron mobility μ_e	$2.315 \times 10^{-1}\ m^2 \cdot s^{-1} \cdot V^{-1}$
Hole mobility μ_h	$2 \times 10^{-3}\ m^2 \cdot s^{-1} \cdot V^{-1}$
Electron lifetime τ_e	1.175×10^{-6} s
Hole lifetime τ_h	1×10^{-6} s

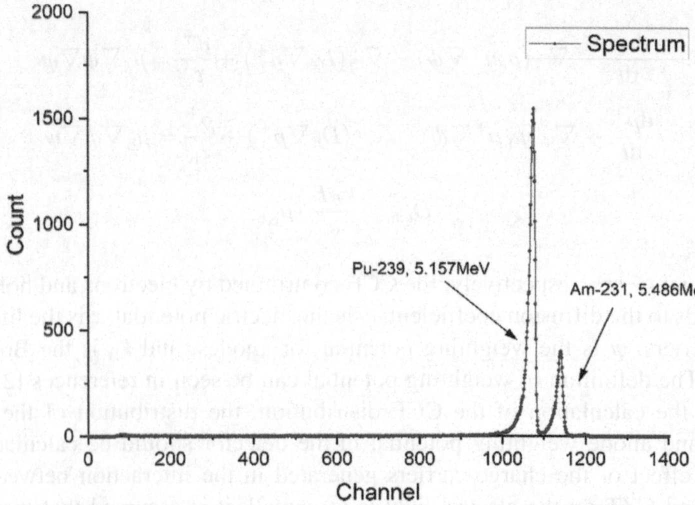

Fig. 4 Spectrum of Am-Pu α-particle source from the experiment, obtained when the anode was biased to 83 V and the voltage of the cathode was 0

cause a sharp drop of electron CCE near the anode and a steep cliff of hole CCE near the cathode, which does not fit the fact that electron CCE and hole CCE reach their maximum on the anode and cathode, respectively.

Our simulation results of the CCE distribution are drawn in Fig. 5. It can be found that the contribution of hole CCE to the total CCE near the cathode is so small that it can be ignored. The simulation result is the same as the one from the calculation, depending on the Hecht formula. The physical properties utilized in the simulation are summarized in Table 3. The total CCE becomes stable after 4000 ns, which means the charge carrier transport time will not be longer than 4 μs.

4.3 *Simulation of Electronic Noise*

The electronic noise of the detector has been estimated in the experiment, as seen in Sect. 3.2, and it is a constant. The relationship between the channel in the pulse height spectrum and the deposited energy of alpha particles is given by the peak channels of 5.486 and 5.157 MeV measured in the experiment (shown in Fig. 4), which can be expressed as:

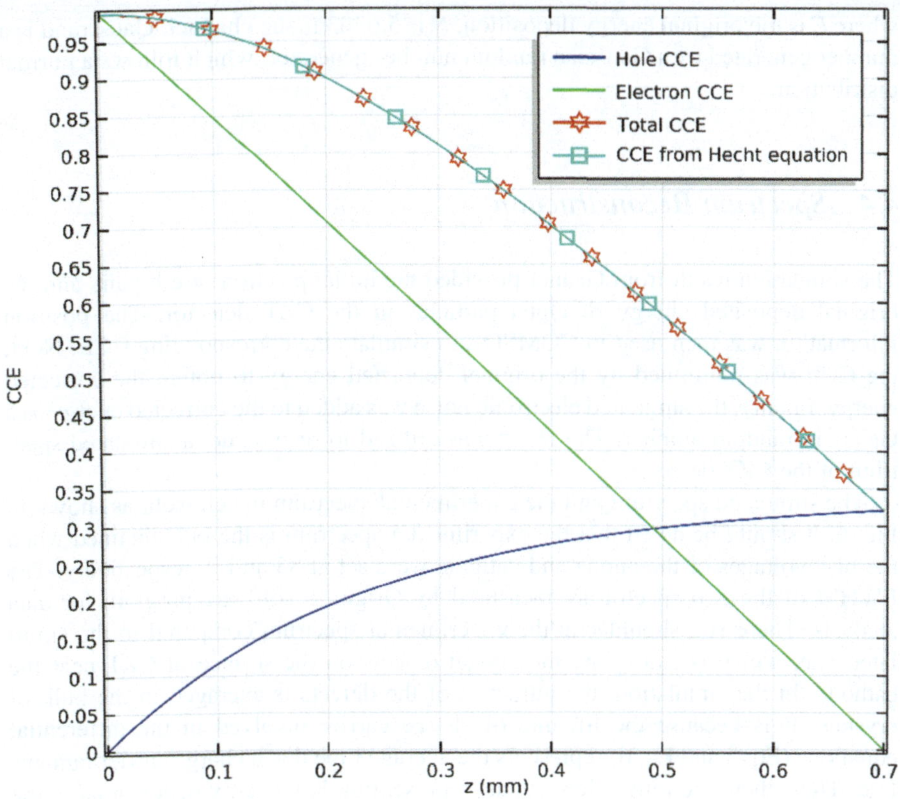

Fig. 5 CCE as a function of depth z. The dark blue curve is the hole CCE, the green curve is the electron CCE, the red curve is the sum of the electron and hole CCE, and the light blue curve is the calculation result of CCE, depending on the Hecht equation

$$E = bC \tag{10}$$

where E is energy expressed in keV, C is the channel, and $b = 4.796$ is a constant. The corresponding standard deviation σ of electronic noise N could be calculated by the following equation:

$$\sigma = \frac{\text{FWHM}}{2\sqrt{2\ln 2}} = \frac{\text{FWHM}}{2.355} = \frac{N}{2.355} \tag{11}$$

Finally, the energy E' concluding the influence brought by the electronic noise of the detector can be expressed as:

$$E' = E + b\sigma \cdot \text{Gaussrand} = E + \frac{bN}{2.355} \cdot \text{Gaussrand} \tag{12}$$

where E is the original energy deposition, N is 5.6303 in the channel, Gaussrand is a number generated by a Gaussian random number generator, which follows a normal distribution.

4.4 *Spectrum Reconstruction*

The simulation result from Geant4 provided the initial position of e-h pairs and the original deposited energy of alpha particles in the CZT detector. The position information was then used in COMSOL to simulate the corresponding CCE. Next, the CCE was multiplied by the original deposited energy to obtain the corrected energy. Finally, the simulated electronic noise was added to the corrected energy as a Gaussian random number. The result was utilized to reconstruct a simulated spectrum of the CZT detector.

The simulated spectrum and the experimental spectrum match well, as shown in Fig. 6. It should be noted that the experimental spectrum is the one obtained when the bias voltages of the anode and cathode were set at 83 and 0, respectively. The FWHM of the two spectra are measured by Origin, a software program for data analysis. There is a shoulder in the experimental spectrum compared to the simulated one, which is caused by the crystal defects on the surface of CZT near the cathode. In the simulation, the influence of the defects is averaged in the bulk of crystals. It is because the lifetime of charge carrier involved in the differential equations (Eq. 5 and Eq. 6) represents the average time that a charge carrier remains free. Thus, there are only a few counts representing low energy depositions in the simulated spectrum. In comparison, in experiment, the defects severely decrease the CCE in the regions surrounding them, and therefore many counts representing low energy depositions are created in the experimental spectrum. That's why the shoulder is shaped.

4.5 *Estimation of the Influence of Cr Coating on the Performance of CZT Detectors*

As introduced in Sect. 2.1, there is a 100 nm thick Cr coating on the whole surface of the CZT detector's cathode. Because of the hardness of Cr, the coating could protect the fragile Au electrode. Since alpha particles can be easily stopped in metal, the impact of Cr coating on the energy deposition of alpha particles in the CZT detector and the energy resolution of the detector need to be evaluated. A detector model without Cr coating was built, and a simulated alpha-particle spectrum of the detector was generated, which was compared with the simulated spectrum. It was found that

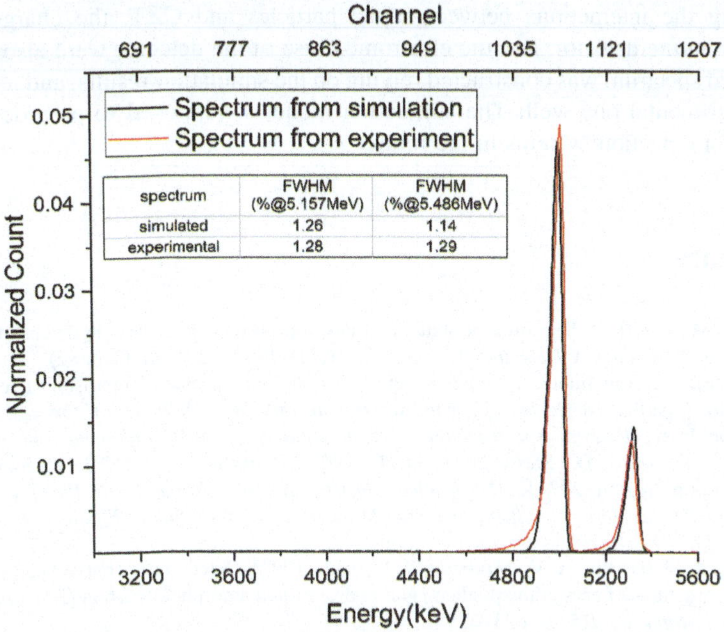

Fig. 6 The simulated spectrum and the experimental one. Black: the spectrum of the Am-Pu source from the simulation. Red: the spectrum of the Am-Pu source from the experiment. The bottom axis corresponds to the simulated spectrum, while the top axis is related to the experimental spectrum

the peak of the alpha particles from Pu-239 changed from 4984 keV toward 5024 keV after removing the coating, while the peak of the alpha particles from Am-241 changed from 5315 to 5354 keV. What is more, there was no apparent increase in the number of incident alpha particles, which lost their energy in the detector. The coating was too thin to stop the alpha particles. The energy resolution of the CZT detector for the alpha particles from Pu-239 and Am-241 improved: 1.02% @ 5157 and 0.93% @5486 keV, respectively.

Although in this work the coating did not apparently decrease the performance of the detector, it does not mean that the coating can be overlooked under any conditions. If the energy of the incident alpha particles is lower and the coating is thicker, the coating can severely decrease the detector's efficiency and cause a lower signal-noise ratio. So do the Au electrodes. Therefore, when manufacturing a detector, the coating and electrodes should be as thin as possible.

5 Conclusion

An α-particle detection system based on a planar CZT detector was built and tested in this work. The energy resolution achieved by the system was 1.47% @ 5.486 MeV and 1.32% @ 5.157 MeV. A series of corresponding simulations involving the interactions between alpha particles and CZT, the charge carrier transport in the detector, and the electronic noise of the detector were also done. A simulated spectrum was constructed relying on the simulation results, and it matched the experimental one well. The simulation model is supposed to provide a better design for detection systems in the future.

References

1. Yang, M. T. (2011). The present status and develop current of α particle detect instrument. *Nuclear Electronics & Detection Technology, 31*(11), 1198–1201. (in Chinese).
2. Morishita, Y., Yamamoto, S., Izaki, K., et al. (2014). Performance comparison of scintillators for alpha particle detectors [J]. *Nuclear Instruments and Methods in Physics Research Section A: Accelerators, Spectrometers, Detectors and Associated Equipment, 764*, 383–386.
3. Venos, D., Srnka, D., Slesinger, J., et al. (1995). Performance of HPGe detectors in the temperature region 277 K [J]. *Nuclear Instruments and Methods in Physics Research Section A: Accelerators, Spectrometers, Detectors and Associated Equipment, 365*(2–3), 419–423.
4. Kotina, I. M., Derbin, A. V., Morozov, V. F., et al. (1992). Detection of charged particles using heterostructures of crystalline p-silicon and hydrogenated amorphous carbon [J]. *Diamond and Related Materials, 1*(5–6), 623–625.
5. He, Y., Liu, Z., McCall, K. M., et al. (2019). Perovskite CsPbBr3 single crystal detector for alpha particle spectroscopy [J]. *Nuclear Instruments and Methods in Physics Research Section A: Accelerators, Spectrometers, Detectors and Associated Equipment, 922*, 217–221.
6. Xu, Q., Mulligan, P., Wang, J., et al. (2017). Bulk GaN alpha-particle detector with large depletion region and improved energy resolution [J]. *Nuclear Instruments and Methods in Physics Research Section A: Accelerators, Spectrometers, Detectors and Associated Equipment, 849*, 11–15.
7. Wang, X. (2013). *Principle and experimental characteristics of pixellated CdZnTe detector for nuclear radiation*. Ph.D. thesis, Chongqing University (in Chinese).
8. Eisen, Y., Shor, A., & Mardor, I. (1999). CdTe and CdZnTe gamma ray detectors for medical and industrial imaging systems [J]. *Nuclear Instruments and Methods in Physics Research Section A: Accelerators, Spectrometers, Detectors and Associated Equipment, 428*(1), 158–170.
9. Erickson, J. C., Yao, H. W., James, R. B., et al. (2000). *Alpha particle time of flight measurements on a new class of CdZnTe radiation detectors [R]* (Vol. 29, p. 699). Sandia National Labs.
10. Vizkelethy, G., Doyle, B. L., Walsh, D. S., et al. (2000). Nuclear microprobe studies of the electronic transport properties of cadmium zinc telluride (CZT) radiation detectors [C]. In *Hard X-ray, gamma-ray, and neutron detector physics II. SPIE* (Vol. 4141, pp. 178–185).
11. Ivanov, V., Alekseeva, L., Dorogov, P., et al. (2004). Correlation between quality of CZT crystals and spectrometric performance of hemispherical radiation detectors[C]. In *IEEE symposium conference record nuclear science 2004* (Vol. 7, pp. 4415–4419). IEEE.
12. Kim, K. O., Ahn, W. S., Kwon, T. J., et al. (2011). Sensitivity analysis to evaluate the transport properties of CdZnTe detectors using alpha particles and low-energy gamma-rays [J]. *Nuclear Engineering and Technology, 43*(6), 567–572.
13. Amman, M. S., Lee, J. S., & Luke, P. N. (2001). Alpha particle response characterization of CdZnTe[C]. Hard X-ray and gamma-ray detector physics III. *International Society for Optics and Photonics, 4507*, 1–11.
14. Hecht, K. (1932). Zum Mechanismus des lichtelektrischen Primrstromes in isolierendenKristallen [J]. *Zeitschrift fr Physik, 77*(3–4), 235–245. (in German).

15. 王莹,王凯,曲延涛,马吉增.平面型CZT探测器对中低能γ射线的响应[J].中国原子能科学研究院年报,2007(00):356. (in Chinese).
16. Liu, E. K., Zhu, B. S., & Luo, J. S. (2017). *The physics of semiconductors* (7th ed.). Publishing House of Electronics Industry.
17. Nemirovsky, Y., Ruzin, A., Asa, G., et al. (1997). Study of contacts to CdZnTe radiation detectors [J]. *Journal of Electronic Materials, 26*(6), 756–764.
18. Michaelson, H. B. (1977). The work function of the elements and its periodicity [J]. *Journal of Applied Physics, 48*(11), 4729–4733.
19. Liang, X. Y., Min, J. H., Chen, J., et al. (2012). Metal/semiconductor contacts for schottky and photoconductive CdZnTe detector [J]. *Physics Procedia, 32*, 545–550.
20. Agostinelli, S., Allison, J., Amako, K., et al. (2003). GEANT4—A simulation toolkit [J]. *Nuclear Instruments and Methods in Physics Research Section A: Accelerators, Spectrometers, Detectors and Associated Equipment, 506*(3), 250–303.
21. Kolstein, M., Ario, G., Chmeissani, M., et al. (2014). Simulation of charge transport in pixelated CdTe [J]. *Journal of Instrumentation, 9*(12), C12027.
22. Shockley, W. (1938). Currents to conductors induced by a moving point charge. *Journal of Applied Physics, 9*, 635.
23. Ramo, S. (1939, September). Currents induced by electron motion. *Proceedings of the I.R.E*, 584.
24. Prettyman, T. H. (1999). Method for mapping charge pulses in semiconductor radiation detectors [J]. *Nuclear Instruments and Methods in Physics Research Section A: Accelerators, Spectrometers, Detectors and Associated Equipment, 422*(1–3), 232–237.

Role of Selenium in CdZnTeSe as a Defect Engineering Agent

Utpal N. Roy, Joshua W. Kleppinger, and Ralph B. James

1 Introduction

Radiation detection technology, in various forms, is an integral part of our day-to-day life, often in a manner that is unbeknownst to consumers. Radiation detectors, especially nuclear radiation detectors used over a wide range of X-ray to high-energy gamma-ray energies, are crucial to medical imaging for diagnosis, surgery, astronomy, nonproliferation, the mining industry, environmental monitoring, health physics, etc. [1–8] Many of these applications need superior energy resolution to distinguish closely spaced gamma peaks and identify the source material with little or no false alarms. The best energy resolution for most applications is obtained with high-purity germanium (Ge) detectors [9]; however, these detectors need to be cooled to liquid-nitrogen temperature and are impractical for many field applications. To combat this issue, several semiconductors capable of operating at room temperature have been invented and have been broadly labeled as room-temperature semiconductor detectors (RTSDs). The basic criteria for ideal RTSD materials are that the material should have both high atomic number (Z) and high density as high-energy photoelectric absorption is proportional to Z^5. Additionally, the candidate material should have a wider bandgap than Ge to lower the bulk leakage current and thermal noise to provide an increased signal-to-noise ratio.

The invention of cadmium telluride (CdTe) and eventually cadmium zinc telluride (CdZnTe, CZT) has proven to be pivotal toward the development of instruments operating at room temperature for the detection of X- and gamma rays due to the materials' wide bandgap, high resistivity, excellent optoelectronic properties, and large attenuation of X- and gamma rays derived from the high atomic mass and density. Among other prospective semiconductors, CdTe/CZT has been regarded as

U. N. Roy (✉) · J. W. Kleppinger · R. B. James
Savannah River National Laboratory, Aiken, SC, USA
e-mail: Utpal.Roy@srnl.doe.gov

K. Iniewski (ed.), *CdTe and CdZnTe Materials*,
https://doi.org/10.1007/978-3-031-64521-1_7

the gold standard for RTSD materials, and the materials are the most common commercially available ones to address the growing needs in various fields. Despite its enormous commercial success as a room-temperature radiation detection material, CZT suffers from three major detrimental defects related to its material properties: [9–14] (1) a high concentration of secondary phases, (2) high concentrations of subgrain boundaries/dislocation walls, and (3) compositional inhomogeneity due to the nonunity segregation coefficient of zinc. The presence of high concentrations of subgrain boundary networks and Te inclusions remains a long-standing issue associated with the CZT material. In general, these subgrain boundaries and their network are often decorated with Te inclusions [9]. These extended defects act as charge trapping and recombination centers that can severely impact the spatial uniformity of the material's charge-transport characteristics, limiting the thickness and volume of high-performing detectors. Additionally, the compositional inhomogeneity severely hampers the overall yield of detector-grade material, and the compositional gradient imposes considerable strain on the CZT ingot [15, 16]. Finally, CdTe/CZT materials are prone to the formation of defects during growth and the subsequent cooling process due to their inherent poor thermophysical properties, such as low thermal conductivity, low critical resolved shear stress, and retrograde solubility of tellurium, among other relevant properties [17–19]. Thus, enhanced yield and, therefore, the reduced production cost of high-quality detector-grade materials demand the elimination or reduction of these long-standing material issues pertaining to CZT. Improved homogeneity in terms of both composition and defect distribution as well as minimized defect concentrations are critical to further enhance the performance of the detectors and increase the overall yield.

The partial replacement of tellurium with selenium (Se) has long been known to be effective in resolving some of the primary issues affecting CZT materials. This article will discuss the successful mitigation of inhomogeneity and defects in the CZT matrix by the addition of selenium.

2 Effect of Selenium on Compositional Homogeneity

CZT heavily suffers from axial and radial compositional inhomogeneity due to the nonunity segregation of Zn in the CdTe matrix [13, 14]. Zhang et al. [14] simulated the segregation of Zn using a transient, coupled model with the help of CrysMAS and Cats2D software and then validated the model by growing a 45-mm-diameter $Cd_{0.9}Zn_{0.1}Te$ ingot via the electrodynamic gradient freeze (EDG) technique. Figure 1a (left) shows the photoluminescence (PL) map of the wafer cut along the axis of the CZT ingot grown through the EDG technique as well as the contours of the model prediction. Figure 1a shows that the model and experimental results match reasonably well. It should be noted that the Zn concentrations show an upward bowing near the central part of the ingot and a prominent dip near the ampoule wall. Figure 1b displays the experimental and simulated results for the corresponding Zn concentration along the center line of the axial cross-section. Once again, the

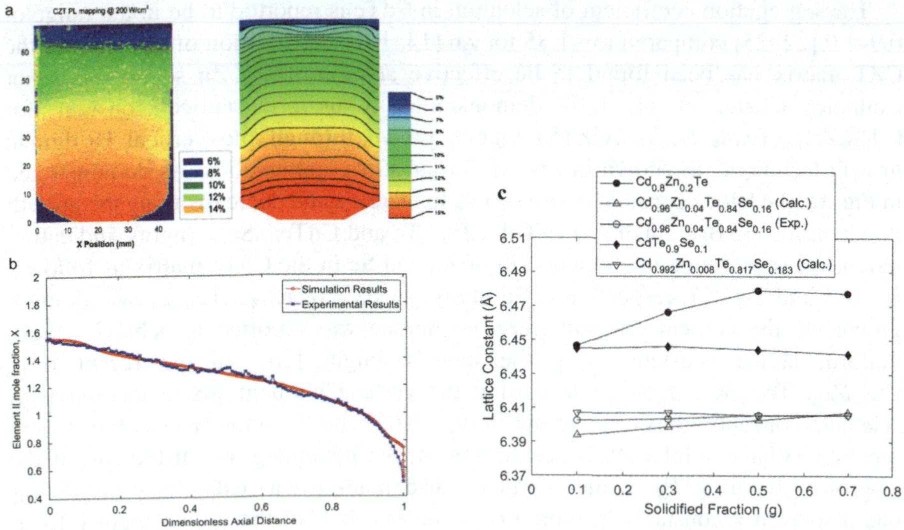

Fig. 1 Prediction of the zinc composition by the coupled model compares well with that measured in growth experiments. (**a**) A Zn composition along a lengthwise axial cut of the crystal is indicated for the experiment using PL mapping (left) and by color and contours in the model computation (right). (**b**) Comparison of the axial concentration of zinc along the centerline. (Taken from N. Zhang et al. [14]), and (**c**) Lattice constant versus the solidified fraction of the ingots of CdTe alloyed with Zn and Se. (Taken from C. Chang et al. [20])

simulated results agree well with the experimental data, except for the last-to-freeze region [14]. The zinc concentration is observed to decrease axially as the ingot grows. Some nonuniformity of the Zn concentration is also observed along the diameter of the grown ingot.

The compositional inhomogeneity profoundly hampers the overall yield of detector-grade materials, particularly for thick detectors, and enacts considerable strain on the CZT ingot [15, 16]. The segregation coefficient of zinc is reported to be ~1.35 [14], leading to a monotonic axial concentration distribution of Zn that steadily decreases along the growth direction. There is, however, a large dispersion of the reported segregation coefficient value for Zn in the CdTe matrix as the segregation coefficient is dependent on various growth parameters, such as temperature gradient, growth speed, etc. Since the bandgap is directly dependent on the concentration of Zn, the bandgap decreases along the direction of growth. Consequently, the resistivity and number of generated electron-hole pairs as a function of the absorbed gamma-ray energy also decrease along the same direction [13, 21], deteriorating the detector's performance. It is thus crucial that the composition has a high degree of uniformity to enhance the overall detector yield and performance, as well as for substrate applications for night-vision cameras. Several different options have been proposed to minimize Zn segregation and boost compositional uniformity [15], and, so far, selenium incorporation into the CZT matrix has been established to be the most viable and cost-effective approach.

The segregation coefficient of selenium in CdTe is reported to be in the range of 0.9–1.0 [22–25] compared to ~1.35 for Zn [13, 14]. The addition of selenium in the CZT matrix has been found to be effective at minimizing Zn segregation. For example, Chang et al. [20] demonstrated a uniform lattice constant for $Cd_{0.96}Zn_{0.04}Te_{0.84}Se_{0.16}$ (CZTS) ingots, grown through the vertical Bridgman growth technique, as shown in Fig. 1c for substrate applications. As demonstrated in Fig. 1c, the lattice constants were shown to increase and decrease along the growth direction for the Bridgman-grown $Cd_{0.8}Zn_{0.2}Te$ and $CdTe_{0.9}Se_{0.1}$ ingots, indicating individual segregation coefficients (*k*) of Zn and Se in the CdTe matrix as follows: $k_{Zn} > 1$ and $k_{Se} \leqslant 1$, respectively. Similarly, a $Cd_{0.9904}Zn_{0.0096}Te_{0.964}Se_{0.036}$ ingot grown via the vertical gradient freeze technique was reported to achieve a very uniform lattice constant along the ingot's length [26]. In the recent past, $Cd_{0.9}Zn_{0.1}Te_{1-y}Se_y$ ingots grown using the vertical gradient freeze method with selenium concentrations of 2 atomic % ($y = 0.02$) and 7 atomic % ($y = 0.07$) were reported to have similar Zn concentrations at the beginning and at the end of the as-grown ingot [27], while vertical Bridgman-grown $Cd_{0.9}Zn_{0.1}Te_{0.98}Se_{0.02}$ had displayed a constant Zn profile over the first half of the grown ingot [28]. It should be noted that unlike selenium, the segregation of Zn is determined by the segregation of sulfur (S) for S-containing quaternary compounds (e.g., CdZnTeS) [20].

The traveling heater method (THM) [29–34] is conceivably the key to the vast success of high-quality detector-grade CdTe/CZT production. The THM technique offers several advantages over conventional melt growth techniques. THM growth is facilitated by a Te-rich melt, which drastically lowers the growth temperature compared to melt growth, well below the melting point of the material. A lower-growth temperature reduces thermal stress, resulting in a lowering of the concentration of defects in the grown ingots, less impurity migration from the crucible, and a lower probability of ampoule explosion from the Cd vapor pressure. Furthermore, the technique offers self-purification of the grown material as Te is known to have an efficient extraction effect on impurities [29–33]. The THM technique is also reported to produce better macroscopic compositional homogeneity [34]. Owing to all these advantages, the THM technique, in due course, has emerged to be the most promising technique for the industrial production of large-volume high-quality detector-grade CdTe and CZT. The first commercial production of detector-grade CZT was based on the high-pressure vertical Bridgman growth technique in the early 1990s. However, due to its multipronged advantages, producers eventually switched to the THM technique [35, 36]. At present, the major producers of detector-grade CdTe and CZT rely primarily on the THM technique [36–39]. As opposed to melt growth techniques, better compositional homogeneity can be achieved with the THM technique for CZT ingots; however, the segregation of Zn is still dependent on the growth parameters [12, 40]. Figure 2a shows an optical photograph of the axial cross-section of a THM-grown 2-inch-diameter CZT ingot. The ingot was grown from a Te-rich solution at a lowering speed of 3–5 mm/day with a temperature gradient of 10–15 °C/cm near the growth interface [41, 42]. Figure 2b shows the room-temperature PL mapping of the corresponding wafer, up to the white line, as

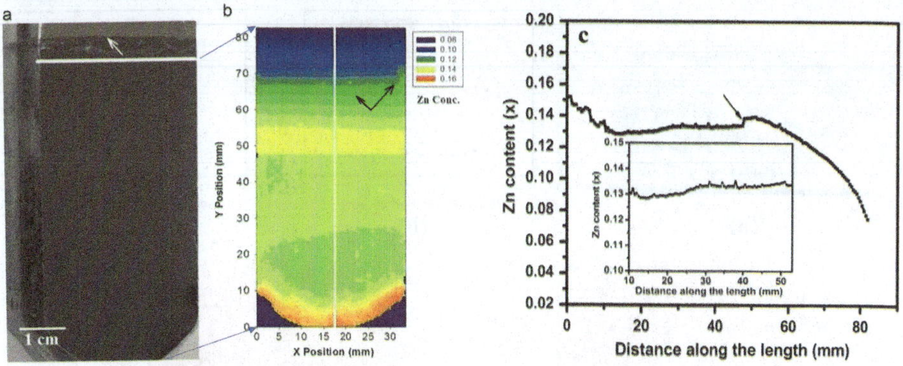

Fig. 2 (**a**) THM-grown CZT ingot cut along the length, (**b**) Zn concentration mapping of the as-grown ingot by THM, and (**c**) Zn concentration along the length of the ingot. The inset shows Zn concentration along the length in the central part of the ingot. (Taken from U. N. Roy et al. [41])

shown in Fig. 2a, avoiding the cracked portion near the Te-rich/CZT interface. It is evident from Fig. 2b that both axial and radial compositional homogeneity are superior to the melt growth technique, as illustrated in Fig. 1a. The striations are visible near the upper part of the ingot, as indicated by the arrow in Fig. 2b, and the growth interfaces are flat macroscopically, which also corroborates the interface shown in Fig. 2a (shown by a white arrow). Figure 2c illustrates the axial Zn composition profile of the ingot along the center line of the axial cross-section shown in Fig. 2b, indicated by the white line. The Zn concentration profile of the selected portion (~1.05 cm and 5.25 cm from the tip of the ingot) with uniform concentration is shown in the inset of Fig. 2c [41]. The Zn concentration variation in this little over 4 cm region has been reported to be within ±0.3%. Thus, overall, the composition remains equitably uniform over more that 33% of the total length of the THM-grown CZT ingot.

As was observed for melt-grown CdZnTeSe, a reduced Zn segregation has also been observed for THM-grown CZTS [43, 44]. Figure 3a shows a THM-grown 2-inch diameter CZTS ingot with 2 atomic % Se, employing the same growth parameters for the previously discussed THM-grown CZT ingot (shown in Fig. 2). Figure 3b, c illustrates the variation of Zn and Se and the calculated bandgap along the central line of the axial wafer (shown in the inset of Fig. 3b). The bandgap of the CZTS ingot was found to be uniform for ~90% of the total length of the ingot (Fig. 3c). However, considerable depletion of Zn was observed within 1 cm of the Te-rich/CZTS interface. It should be noted that inevitable cracks occur for all THM-grown CdTe-based ingots near the interface. The cracks are associated with the large thermal expansion coefficient difference between the Te-rich CZT/CZTS alloy and are most extensively found in the unusable part of the ingot. The higher compositional uniformity along the length of the CZTS ingot compared to THM-grown CZT ensures a higher yield, thus potentially lowering the cost of production for relatively large-volume detectors. The axial compositional uniformity of THM-grown CZTS has been reported to be further enhanced with increased

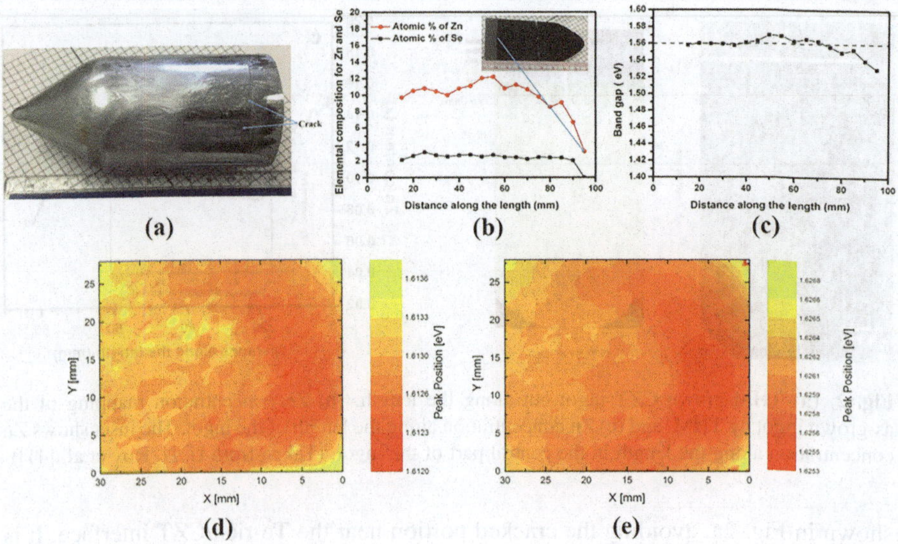

Fig. 3 (**a**) As-grown $Cd_{0.9}Zn_{0.1}Te_{0.98}Se_{0.02}$ ingot with a 52-mm diameter, grown via THM and (**b**) variation of Zn and Se (atomic %) along the length of the ingot. The inset shows the wafer cut along the length of the as-grown $Cd_{0.9}Zn_{0.1}Te_{0.98}Se_{0.02}$ ingot. (**c**) The calculated bandgap along the length of the ingot. The map of PL peak energy positions at 7 K for the as-grown $Cd_{0.9}Zn_{0.1}Te_{0.98}Se_{0.02}$ wafer cut perpendicular to the growth axis. (**d**) (A°, X) peak mapping and (**e**) (D°, X) peak mapping. The mapping area is ~2.8×3.0 cm^2. (Taken from U. N. Roy et al. [43])

selenium in the matrix [44]. Thus, the incorporation of selenium is effective in modifying the segregation coefficient of Zn in THM-grown CZTS ingots as well. The spatial variation of the bandgap of the material highly reflects the compositional inhomogeneity. The PL peak energy positions are very sensitive to the bandgap, hence the composition of the material. The radial compositional uniformity has been studied by the same group by performing low-temperature (7 K) PL mapping on a THM as-grown $Cd_{0.9}Zn_{0.1}Te_{0.98}Se_{0.02}$ wafer cut perpendicular to the ingot axis [43]. Figure 3d, e illustrates the peak energy mapping for two dominant acceptor-bound excitons (A°, X) and donor-bound excitons (D°, X), respectively, over a ~ 2.8×3 cm^2 area. For both the (A°, X) and (D°, X) peaks, the spatial peak energy variations (ΔE) are reported to be within 0.0015 eV over the scanned area of ~2.8×3.0 cm^2, confirming excellent radial compositional uniformity for the as-grown (THM-grown) quaternary compound $Cd_{0.9}Zn_{0.1}Te_{0.98}Se_{0.02}$. It is thus evident that selenium is effective in modifying the segregation coefficient of Zn, resulting in higher compositional homogeneity radially and along the ingot length. Thus, CZTS has remarkable potential to increase the overall production yield of high-quality uniform materials and allow for the manufacturing of detector blanks at a lower cost.

3 Effect of Selenium on Subgrain Boundaries and Their Networks

Unlike conventional elemental semiconductors such as Si and Ge, various structural defects are present in compound semiconductors [45], such as CZT [9], which adversely affect the charge-transport characteristics of the material and the performance of fabricated devices. In general, dislocation cell structures or subgrain boundaries and their network consist of dislocation walls with high dislocation density present in the grown materials, especially in compound semiconductors. The subgrain boundaries and their network cannot be removed by postgrowth annealing, so the minimization of subgrain boundaries and their networks is required in the as-grown material. The generation and propagation of such dislocation walls during growth and subsequent cooling sequence originate from thermo-mechanical stress fields. For undoped compound semiconductors, it is generally difficult to avoid the formation and propagation of the dislocation walls due to the material's inherent poor thermo-physical properties. Because of the high ionicity of CdTe at 0.55 [15], it is not surprising that CdTe-based ingots show the presence of subgrain boundaries and their networks as well as residual strain since the creation energy of dislocations and stacking faults are inversely proportional to the ionicity [15]. This makes the highly ionic II–VI crystal lattices very sensitive to any strain and prone to stress-related defects. Furthermore, the subgrain boundaries and their networks form charge-trapping and recombination centers that severely hinder charge transport, resulting in the stark degradation of detector performance. The subgrain boundary networks in general are distributed randomly inside the CdTe/CZT matrix and induce spatial nonuniformity of the charge-transport properties, which is perhaps the most dominant reason for the degradation of the device's performance. These subgrain boundaries have also been known to be decorated with performance-limiting Te inclusions [9]. In addition, they severely hinder the yield of high-quality detector-grade materials due to their random distribution. Bolotnikov et al. [9] demonstrated the effect of a subgrain boundary network on device performance for a Frisch grid geometry. These subgrain boundaries and their networks can be observed after the preferential etching of polished surfaces and also appear in X-ray topographic images, preferably in the reflection mode. Figure 4 shows an X-ray diffraction topographic image taken in the reflection mode for two different Frisch grid detectors with dimensions of $6 \times 6 \times 15$ mm^3. The presence of subgrain boundary networks is evident as black and white lines in both the detectors, as shown in Fig. 4. Figure 4a shows a sample that contains few subgrain boundaries and their network, while Fig. 4b depicts the presence of well-resolved subgrain boundaries and a higher density of subgrain boundary network, as compared to the detector represented in Fig. 4a. As expected, the as-measured energy resolution at 662 keV was reported to be better (~ 2.3%) for the less defective detector (Fig. 4a), while the detector represented by Fig. 4b revealed an enhanced degradation of detector response with an energy resolution at 662 keV of 4.2% [9]. The subgrain boundaries

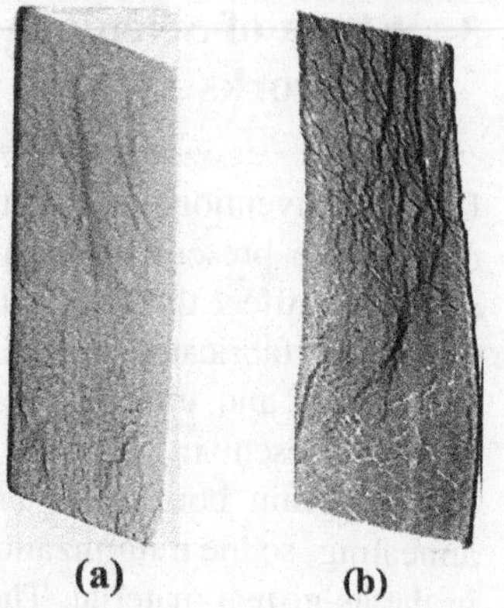

Fig. 4 X-ray diffraction topographic images for two CZT Frisch grid detectors with dimensions of $6 \times 6 \times 15$ mm^3. (Taken from A. E. Bolotnikov et al. [9])

and their network are also known to impose adverse effects on medical imaging applications [46]. It is thus crucial to develop detector materials free from subgrain boundary networks for both homeland security and medical imaging applications.

It is known that the formation and propagation of such dislocation walls/subgrain boundaries can be mitigated by adding selective dopants, such as indium (In) in gallium arsenide (GaAs) [45] and Se in CdTe [19, 47–49]. At the early stage of CZT development, Fiederle et al. [50] reported enhanced charge-transport properties in $CdTe_{0.9}Se_{0.1}$ compared to CZT. Selenium has also been found to be an effective lattice-hardening agent in the $Cd_{1-x}Zn_xTe_{1-y}Se_y$ matrix, producing quaternary compounds free from subgrain boundary networks for a wide range of Se concentrations, ranging from 1.5 atomic % to 7 atomic %; however, the occasional presence of subgrain boundaries has been observed [43, 44, 47, 51–53]. Recently, Naydenov and Pritula [54] predicted, via thermodynamic modeling, an optimum selenium concentration of ~1.8 atomic % for effective solid solution hardening in the CZTS matrix and the minimization of extended defects, validating earlier experimental findings.

Figure 5a, b shows X-ray diffraction topographic images in the reflection mode for two Frisch grid detectors with dimensions of (a) $5 \times 5 \times 12.3$ mm^3 and (b) $4.5 \times 4.5 \times 10.8$ mm^3, fabricated from a THM as-grown $Cd_{0.9}Zn_{0.1}Te_{0.98}Se_{0.02}$ ingot. For the detector sample shown in Fig. 5a, a prominent subgrain boundary is evident, as marked by the black arrows, while no subgrain boundary was observed for the detector shown in Fig. 5b. Both detector samples were free from subgrain boundary networks as opposed to the CZT samples shown in Fig. 4. The blue arrows indicated in Fig. 5 represent scratches that occurred during polishing and handling. Selenium has thus been found to be very effective as a solid solution hardening agent to arrest the formation of subgrain boundary networks. The absence of subgrain boundary networks in the as-grown CZTS indicates superior crystalline quality compared to present-day detector-grade CZT. CZTS materials that are free from subgrain boundary networks offer a substantial advancement for the RTSD field and their applications. The absence of any subgrain boundary network in as-grown CZTS detectors is expected to lead to high-resolution detection properties. The

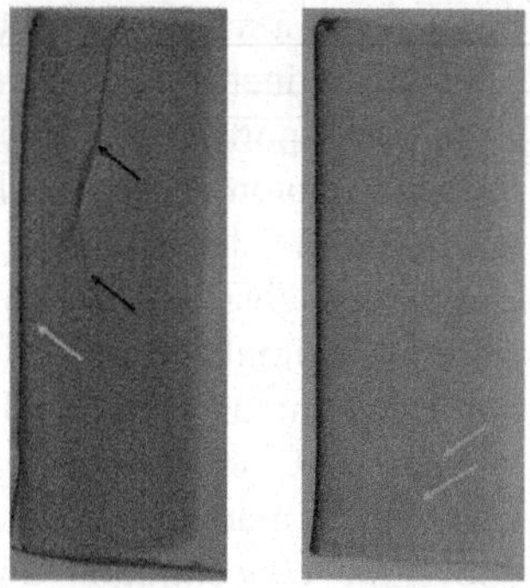

Fig. 5 X-ray diffraction topographic images for two Frisch grid detectors with dimensions of (**a**) $5 \times 5 \times 12.3\ mm^3$ and (**b**) $4.5 \times 4.5 \times 10.8\ mm^3$. (Taken from U. N. Roy et al. [55])

energy resolutions at 662 keV of the detectors presented in Fig. 5a, b were 1.08% and 0.9%, respectively. The detector presented in Fig. 5a had a slightly worse resolution due to the presence of a prominent subgrain boundary compared to that in Fig. 5b.

4 Effect of Selenium on Secondary Phases (Te Inclusions)

Secondary phases, commonly Te inclusions, embedded within the CZT matrix is another performance-limiting defect that compromises CZT detector performance. Secondary phases are grouped into two categories—viz., inclusions and precipitates—depending on how they are formed during growth and the subsequent cooling process [19]. The Te-rich phases generated due to morphological instability at the growth interface are generally known as inclusions, and the size of the Te inclusions are generally in the range of 1–20 μm [19, 56–58]; however, much larger inclusions have been observed, especially in THM-grown CdTe/CZT samples as they are grown from Te-rich solutions. On the other hand, Te precipitates form during the cooling process of the grown ingot due to the retrograde solubility of Te [19]. The sizes of Te precipitates are substantially smaller than Te inclusions and are reported to be in the range of 10–50 nm [19]. The retrograde solubility of Te is increased with Zn incorporation, and its concentration in the CdTe matrix, as described by the bulge on the Te side of the solidus line, becomes more pronounced with increasing ZnTe content in the solid solution [59]. The Te inclusions are opaque in the infrared (IR) region and are easily recognized in the IR transmission microscopic studies of CdTe/CZT.

Te inclusions act as charge-trapping centers and severely hinder charge transport beyond the actual volume of the inclusions [60], and the effect is cumulative with increased thickness of the detector, which is eventually responsible for the degradation of detector performance, especially as the concentration and size of the inclusions increase. The mobility lifetime product for electrons ($(\mu\tau)_e$) has been reported to vary over a wide range, from 0.2×10^{-3} to 20×10^{-3} cm^2/V for regions with and without Te inclusions, respectively [10]. This significant spatial variation of the charge-transport properties produces different pulse heights in detector response for different photon absorption locations and broadens the resulting photo peak for the entire volume, degrading the spectroscopic performance of the detector [10, 60]. The cumulative adverse effects of Te inclusions on the energy resolution of the detector have been simulated for different sizes and concentrations of Te inclusions [61]. It has been demonstrated that Te inclusions of a size ~1 μm or less with concentrations as high as 10^7 cm^{-3} do not impose any degradation on the detector energy resolution, while an adverse effect is noticeable for Te inclusions of a size ~3 μm or less above the concentration threshold of 10^6–10^7 cm^{-3} for 15-mm-thick detectors [61]. The extent of performance degradation is more prominent for larger Te inclusions with lower concentrations. Significant research efforts have been dedicated toward eliminating these inclusions. Postgrowth annealing in a Cd atmosphere is effective in annihilating them, and it is regularly implemented in industrial manufacturing processes. However, annealing in Cd vapor produces star-like defects with sizes 50 to 100 times larger than the Te inclusions [62, 63] that are invisible in the IR transmission mode, which also act as substantial charge-trapping defects. Thus, the key to increasing the yield of high-quality detectors is to produce materials with reduced Te inclusions in the as-grown ingots.

The incorporation of selenium in the CZT matrix has been found to be effective in substantially reducing the concentrations of Te inclusions irrespective of the growth techniques [28, 43, 44, 52, 55, 64, 65]. The reduction of Te inclusions in CZTS has been observed over a wide range of selenium concentrations, from 2 atomic % to 7 atomic %, while CZTS with a selenium concentration of 1.5 atomic % has not been found to be effective in reducing Te inclusion concentration [52]. Figure 6a shows an IR transmission image for a typical Frisch grid detector sample with dimensions of $3.5 \times 3.5 \times 9.15$ mm^3. A higher magnification IR transmission image is shown in Fig. 6b, illustrating the presence of reduced concentrations of Te inclusions.

A selenium addition in the CZT matrix has been found to be effective in resolving many of the decade-long issues suffered by CZT. The material quality requirements for semiconductor materials in X- and gamma-ray detector applications are stringent compared to other applications, requiring high compositional homogeneity with fewer defects. The material properties of CZTS have been found to be superior to present-day CZT. Compositional homogeneity, the absence of subgrain boundary networks, and reduced concentrations of Te inclusions allow for higher spatial uniformity of the charge-transport properties, making CZTS an attractive alternative to CZT at a lower cost. The reduced defects and better compositional homogeneity are expected to offer a heightened yield of high-quality detector-grade material that can readily achieve excellent energy resolution [43, 51, 55]. Reduced concentrations of electrically active defects have also been observed in CZTS [66, 67], Additionally, the device performance of CZTS under high flux X-ray irradiation has been reported to surpass conventional CZT [67]. Considering its exceptional material properties, CZTS is evolving as the potential next-generation low-cost detector material for homeland security and medical imaging applications.

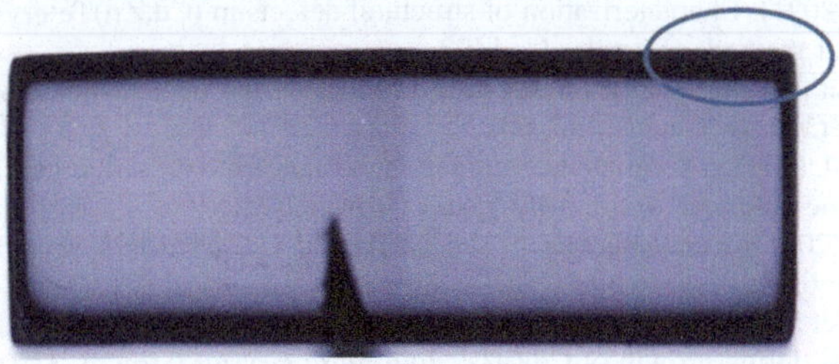

(a)

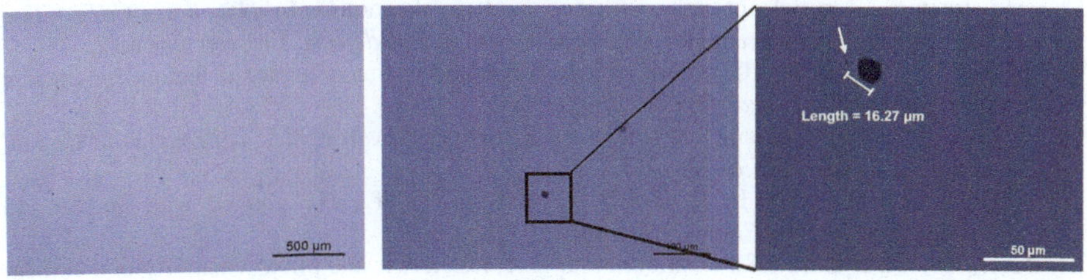

(b)

Fig. 6 (**a**) IR transmission image for a Frisch grid detector sample with dimensions of $3.5 \times 3.5 \times 9.15$ mm^3 and (**b**) high-magnification IR transmission images with increased magnification from left to right. (Taken from U.N. Roy et al. [51])

References

1. Schlesinger, T. E., et al. (2001). Cadmium zinc telluride and its use as a nuclear radiation detector material. *Materials Science and Engineering R, 32*, 103.
2. Yang, G., & James, R. B. (2009). Applications of CdTe, CdZnTe, and CdMnTe radiation detectors. In *Physics, defects, hetero- and nano-structures, crystal growth, surfaces and applications part II* (EDAX, Triboulet R. et al.) (p. 214). Elsevier.
3. Scheiber, C., & Giakos, G. C. (2001). Medical applications of CdTe and CdZnTe detectors. *Nucl Instrum Methods Phys Res A, 458*, 12.
4. Li, Z., et al. (2024). Research on the technological Progress of CZT Array detectors. *Sensors, 24*, 725.
5. Harrison, F. A., et al. (2013). The nuclear spectroscopic telescope array (NuSTAR) high-energy X-ray mission. *The Astrophysical Journal, 770*, 103.
6. Krawczynski, H. S., et al. (2016). X-ray polarimetry with the polarization spectroscopic telescope array (PolSTAR). *Astroparticle Physics, 75*, 8.
7. Slomka, P. J., et al. (2019). Solid-state detector SPECT myocardial perfusion imaging. *Journal of Nuclear Medicine, 60*, 1194.
8. Wahl, C. G., et al. (2015). The Polaris-H imaging spectrometer. *Nucl Instrum Methods A, 784*, 377.
9. Bolotnikov, A. E., et al. (2013). Characterization and evaluation of extended defects in CZT crystals for gamma-ray detectors. *Journal of Crystal Growth, 379*, 46.
10. Carini, G. A., et al. (2007). High-resolution X-ray mapping of CdZnTe detectors. *Nucl Instrum Methods A, 579*, 120.
11. Hossain, A., et al. (2010). Extended defects in CdZnTe crystals: Effects on device performance. *Journal of Crystal Growth, 312*, 1795.

12. Zou, J., et al. (2021). Characterization of structural defects in (Cd,Zn)Te crystals grown by the travelling heater method. *Crystals, 11*, 1402.
13. Fougeres, P., et al. (1999). Zinc segregation in HPB grown nuclear detector grade Cd_{1-x} Zn_x Te. *Journal of Crystal Growth, 197*, 641.
14. Zhang, N., et al. (2011). Anomalous segregation during electrodynamic gradient freeze growth of cadmium zinc telluride. *Journal of Crystal Growth, 325*, 10.
15. Triboulet, R. (2005). Fundamentals of the CdTe and CdZnTe bulk growth. *Physica Status Solidi C, 5*, 1556.
16. Perfeniuk, C., et al. (1992). Measured critical resolved shear stress and calculated temperature and stress fields during growth of CdZnTe. *Journal of Crystal Growth, 119*, 261.
17. Triboulet, R., & Siffert, P. (2009). CdTe and related compounds; physics, defects, hetero- and nano-structures, crystal growth, surfaces and applications. In *Physics, CdTe-based nanostructures, CdTe-based semimagnetic semiconductors, defects*. Elsevier Science.
18. Johns, P. M., & Nino, J. C. (2019). Room temperature semiconductor detectors for nuclear security. *Journal of Applied Physics, 126*, 040902.
19. Rudolph, P. (1994). Fundamental studies on Bridgman growth of CdTe. *Progress in Crystal Growth and Characterization, 29*, 275.
20. Chang, C. Y., & Tseng, B. H. (1997). Crystal growth of CdTe alloyed with Zn, Se and S. *Materials Science and Engineering, B49*, 1.
21. Babentsov, V., et al. (2008). Distribution of zinc, resistivity, and photosensitivity in a vertical Bridgman grown $Cd_{1-x}Zn_xTe$ ingot. *Journal of Crystal Growth, 310*, 3482.
22. Strauss, A. J., & Steininger, J. (1970). Phase diagram of the CdTe – CdSe pseudobinary system. *Journal of the Electrochemical Society, 117*, 1420.
23. Johnson, C. J. (1989). Recent progress in lattice matched substrates for HgCdTe epitaxy. *SPIE, 1106*, 56.
24. Kim, K. H., et al. (2008). Electrical properties of semi-insulating CdTe0.9Se0.1:Cl crystal and its surface preparation. *Journal of Crystal Growth, 310*, 91.
25. Roy, U. N., et al. (2015). High compositional homogeneity of $CdTe_xSe_{1-x}$ crystals grown by the Bridgman method. *Applied Physics Letters Materials, 3*, 026102.
26. Tanaka, A., et al. (1989). Zinc and selenium co-doped CdTe substrates lattice matched to HgCdTe. *Journal of Crystal Growth, 94*, 166.
27. Herraiz, L. M., et al. (2021). Vertical gradient freeze growth of two inches $Cd_{1-x}Zn_xTe_{1-y}Se_y$ ingots with different Se content. *Journal of Crystal Growth, 537*, 126291.
28. Hwang, S., et al. (2019). Anomalous Te inclusion size and distribution in CdZnTeSe. *IEEE Transactions on Nuclear Science, 66*, 2329.
29. Triboulet, R., et al. (1974). Undoped high-resistivity cadmium telluride for nuclear radiation detectors. *Journal of Applied Physics, 45*, 2759.
30. Zanio, K. (1974). Purification of CdTe from tellurium-rich solutions. *Journal of Electronic Materials, 3*, 327.
31. Triboulet, R., & Marfaing, Y. (1981). CdTe growth by multipass THM and sublimation THM. *Journal of Crystal Growth, 51*, 89.
32. Mochizuki, K., & Masumoto, K. (1986). Growth of CdTe single crystals by THM (travelling heater method) and its repetition effect. *Materials Letters, 4*, 298.
33. Triboulet, R., et al. (1990). "Cold travelling heater method", a novel technique of synthesis, purification and growth of CdTe and ZnTe. *Journal of Crystal Growth, 101*, 216.
34. El Morki, A., et al. (1994). Growth of large, high purity, low cost, uniform CdZnTe crystals by the "cold travelling heater method". *Journal of Crystal Growth, 138*, 168.
35. Szeles, C., et al. (2002). Advances in the crystal growth of semi-insulating CdZnTe for radiation detector applications. *IEEE Transactions on Nuclear Science, 49*, 2535.
36. Prokesh, M., et al. (2018). CdZnTe detectors for gamma spectroscopy and x-ray photon counting at 250×10^6 photons/(mm^2 s). *Journal of Applied Physics, 124*, 044503.
37. Amman, M., et al. (2009). Evaluation of THM-grown CdZnTe material for large-volume gamma-ray detector applications. *IEEE Transactions on Nuclear Science, 56*, 795.

38. Shiraki, H., et al. (2009). THM growth and characterization of 100-mm diameter CdTe single crystals. *IEEE Transactions on Nuclear Science, 56*, 1717.
39. Chen, H., et al. (2008). Characterization of large cadmium zinc telluride crystals grown by traveling heater method. *Journal of Applied Physics, 103*, 014903.
40. Zou, J., et al. (2021). Analysis of Te inclusion striations in (Cd,Zn)Te crystals grown by traveling heater method. *Crystals, 11*, 649.
41. Roy, U. N., et al. (2012). Zinc mapping in THM grown detector grade CZT. *Journal of Crystal Growth, 347*, 53.
42. Roy, U. N., et al. (2009). Growth of spectroscopic grade Cd0.9Zn0.1Te:In by THM technique. *Journal of Crystal Growth, 312*, 33.
43. Roy, U. N., et al. (2019). Evaluation of CdZnTeSe as a high-quality gamma-ray spectroscopic material with better compositional homogeneity and reduced defects. *Scientific Reports, 9*, 7303.
44. Roy, U. N., et al. (2019). Role of selenium addition to CdZnTe matrix for room-temperature radiation detector applications. *Scientific Reports, 9*, 1620.
45. Rudolph, P. (2005). Dislocation cell structures in melt-grown semiconductor compound crystals. *Crystal Research and Technology, 40*, 7.
46. Buis, C., et al. (2013). Effects of dislocation walls on image quality when using cadmium telluride X-ray detectors. *IEEE Transactions on Nuclear Science, 60*, 199.
47. Ono, H. (1988). Dislocation reactions and lineage formation in liquid encapsulated Czochralski grown GaAs crystals. *Journal of Crystal Growth, 89*, 209.
48. Hähnert, I., et al. (1994). Study of the detect structure of CdTe-rich II–VI single crystals. *Journal of Crystal Growth, 142*, 310.
49. Roy, U. N., et al. (2014). Evaluation of $CdTe_xSe_{1-x}$ crystals grown from a Te-rich solution. *Journal of Crystal Growth, 389*, 99.
50. Fiederle, M., et al. (1994). Comparison of CdTe, $Cd_{0.9}Zn_{0.1}Te$ and $CdTe_{0.9}Se_{0.1}$ crystals: Application for γ- and X-ray detectors. *Journal of Crystal Growth, 138*, 529.
51. Roy, U. N., et al. (2021). Impact of selenium addition to the cadmium-zinc-telluride matrix for producing high energy resolution X-and gamma-ray detectors. *Scientific Reports, 11*, 10338.
52. Roy, U. N., et al. (2021). Optimization of selenium in CdZnTeSe quaternary compound for radiation detector applications. *Applied Physics Letters, 118*, 152101.
53. Roy, U. N., et al. (2020). X-ray topographic study of Bridgman-grown CdZnTeSe. *Journal of Crystal Growth, 546*, 125753.
54. Naydenov, S. V., & Pritula, I. M. (2023). Hardening of multi-component CdZnTeSe solid solutions: A theoretical approach. *Applied Physics A: Materials Science & Processing, 129*, 812.
55. Roy, U. N., et al. (2019). High-resolution virtual Frisch grid gamma-ray detectors based on as-grown CdZnTeSe with reduced defects. *Applied Physics Letters, 114*, 232107.
56. Datta, A., et al. (2011). Experimental studies on control of growth interface in MVB grown CdZnTe and its consequences. In *IEEE nuclear science symposium conference record* (p. 4720).
57. Zhou, B., et al. (2018). Modification of growth interface of CdZnTe crystals in THM process by ACRT. *Journal of Crystal Growth, 483*, 281.
58. Roy, U. N., et al. (2010). Growth and interface study of 2 in diameter CdZnTe by THM technique. *Journal of Crystal Growth, 312*, 2840.
59. Guskov, V. N., et al. (2004). Vapour pressure investigation of CdZnTe. *Journal of Alloys and Compounds, 371*, 118.
60. Amman, M., et al. (2002). Electron trapping nonuniformity in high-pressure-Bridgman-grown CdZnTe. *Journal of Applied Physics, 92*, 3198.
61. Bolotnikov, A. E., et al. (2007). Performance-limiting defects in CdZnTe detectors. *IEEE Transactions on Nuclear Science, 54*, 821.

62. Yang, G., et al. (2009). Internal electric field investigations of a cadmium zinc telluride detector using synchrotron X-ray mapping and Pockels effect measurements. *Journal of Electronic Materials, 38*, 1563.
63. Yang, G., et al. (2013). Post-growth thermal annealing study of CdZnTe for developing room-temperature X-ray and gamma-ray detectors. *Journal of Crystal Growth, 379*, 16.
64. Kleppinger, J. W., et al. (2021). Growth of $Cd_{0.9}Zn_{0.1}Te_{1-y}Se_y$ single crystals for room-temperature gamma ray detection. *IEEE Transactions on Nuclear Science, 68*, 2429.
65. Chaudhuri, S. K., et al. (2021). Quaternary semiconductor $Cd_{1-x}Zn_xTe_{1-y}Se_y$ for high-resolution, room-temperature gamma-ray detection. *Crystals, 11*, 827.
66. Gul, R., et al. (2017). A comparison of point defects in $Cd_{1-x}Zn_xTe_{1-y}Se_y$ crystals grown by Bridgman and traveling heater methods. *Journal of Applied Physics, 121*, 125705.
67. Yakimov, A., et al. (2019). Growth and characterization of detector-grade CdZnTeSe by horizontal Bridgman technique. *Proceedings of SPIE, 11114*, 111141N.

Advancements in CdZnTe Detectors: Overcoming Challenges Through Physical and Digital Correction Techniques

Sandeep K. Chaudhuri and Krishna C. Mandal

1 Introduction

Over the last three decades, significant research efforts have been devoted to high-atomic number (high-Z) semiconductor materials, leading to the development of the well-established wide-bandgap ternary compound semiconductor $Cd_{1-x}Zn_xTe$ (CZT) for room-temperature gamma-photon detection [1–13]. Prior to the emergence of high-resolution cadmium zinc telluride (CZT) detectors, the stringent low-temperature requirements for operating conventional high-resolution HPGe (high-purity germanium) detectors in detecting high-energy gamma rays compelled the use of lower-resolution scintillation detectors or indirect readout detection systems, particularly in imaging applications. The quality of radiological images relies heavily on the energy resolution of the detectors as low-resolution and poor-contrast images can result in both over- and underdiagnosis, posing a significant challenge in gaining public trust in the application of digital radiography for diseases like breast cancer [14].

Indirect conversion imaging systems, such as those employing scintillator detectors, rely on multiple photomultipliers to detect photons. The summation of signals from all photomultiplier tubes (PMTs) forms the energy signal, while the weighted sum provides the spatial location of the event in the scintillator array. However, this process may lead to miscalculations if photons are lost due to self-absorption within the crystal or at the coupling window (scintillator-PMT interface) [15].

Semiconductor detectors like CZT function as direct readout devices, where photons interact with the detector, generating charge pairs (electrons and holes) collected by pixels (anode) and the cathode. The resulting electric current is directly converted into an electrically measurable signal by front-end electronics. Besides

S. K. Chaudhuri · K. C. Mandal (✉)
Department of Electrical Engineering, University of South Carolina, Columbia, SC, USA
e-mail: mandalk@cec.sc.edu

K. Iniewski (ed.), *CdTe and CdZnTe Materials*,
https://doi.org/10.1007/978-3-031-64521-1_8

direct conversion, semiconductor detectors offer advantages such as high material density, efficient absorption of high-energy neutral particles like gamma photons, and improved detection efficiency and resolution due to high-Z constituents like those in CZT. Additionally, compound semiconductors'bandgap can be engineered to optimize absorption efficiency and leakage current, enhancing electrical resistivity and, consequently, reducing electronic noise for improved detector resolution. Despite these advantages, the success of CZT is hindered by the lack of large-volume detector-grade single crystals and presence of defects [8, 10, 11, 16–19].

CZT, a ternary compound composed of Cd, Zn, and Te, fulfills all the essential criteria for a room-temperature, compact, high-Z semiconductor x−/γ-ray detector. Its constituent elements ($Z_{Cd} = 48$, $Z_{Zn} = 30$, $Z_{Te} = 52$) have high atomic numbers, and CZT has a material density of 5.78 g/cm^3. Specific stoichiometries, such as $Cd_{0.9}Zn_{0.1}Te$, exhibit a bandgap of ~1.6 eV and high bulk electrical resistivity (up to 10^{11} Ω-cm).

Despite CZT's tremendous success, its performance is constrained by poor charge carrier (particularly hole) transport properties, challenges in crystal growth yielding low-quality crystals, and the necessity for operation at elevated temperatures. In most CZT growth processes, the concentration of impurities in the solidified phase is higher than that in the melt phase due to thermodynamic equilibrium factors. This causes a decrease in the concentration of impurities along the growth axis, leading to compositional inhomogeneity. Apart from that, CZT crystals contain many extended as well as point defects that lead to charge trapping and recombination.

To address these limitations, this chapter focuses on overcoming the challenges associated with large-volume CZT detectors. The emphasis lies in the fabrication of x−/γ-ray detectors in a single-polarity configuration, circumventing the impact of poor hole transport. Additionally, digital correction techniques are discussed that help further enhance the spectral response quality.

2 Defects and Effects of Charge Trapping

The demonstrated superior charge transport property in CZT single crystals is, however, localized and short range. CZT single crystals suffer from compositional inhomogeneity, making the crystal growth yield less than 33% [20]. Crystal growth yield refers to the part of the usable (detector-grade) crystal in a fully grown ingot. A lower growth yield implies that only single crystals with limited volume can be harvested from a grown ingot. Large-volume crystals are required when higher detection efficiency for penetrating radiations, such as intermediate to high-energy gamma rays, is needed. Harvesting large-volume crystals inevitably results in the inclusion of inhomogeneities, extended defects, and an increased concentration of charge-trapping defects. These factors affect the charge transport through the trapping and/or recombination of radiation-induced electron-hole pairs. Hole transport, in particular, is severely affected by trapping and recombination in CZT crystals and poses as the primary challenge in attaining the expected energy resolution and charge collection efficiency [21].

2.1 *Extended Defects*

There are several types of extended defects that limit the drift distance uniformity in CZT detectors [3]. Among the most studied are cracks, grain boundaries, subgrain boundaries, twin boundaries, tilt boundaries, dislocations, and secondary phases such as tellurium inclusion. Cracks are fissures in the grown ingot that propagate through the grain and twin boundaries during the cooldown period and lower the growth yield severely [22]. Twin boundaries are the interface of two adjacent crystal regions that have different crystallographic orientations but are related through a specific symmetry operation, generally reflection. Tilt boundaries are similar interfaces, except for the fact that the different crystallite regions are related through rotational symmetry. A subgrain boundary is an interface within a grain separating two crystallographic regions having slightly different orientations [3]. Tellurium inclusions are secondary phases that are basically small regions within the CZT crystal rich in elemental tellurium [18, 23, 24]. Tellurium inclusions form during CZT growth and typically have diameters between 1 and 30 μm.

The twin, subgrain, and tilt boundaries may be considered walls of edge dislocation where dangling bonds are likely to exist in abundance. The dangling bonds accumulate space charge to form potential barriers, which in turn impede the motion of drifting carriers [25]. The diffusion of unreacted precursors across the boundaries may lead to the accumulation of secondary phases, which impedes carrier transport. Dislocation cores facilitate electrical conduction, which alters the local electric field created by the external bias for detector operation. Also, charged dislocations attract impurities that participate in charge trapping. Te inclusions themselves participate in severe charge trapping [18, 24, 26, 27]. Overall, the extended defects result in significant fluctuations in charge transport, which deteriorate charge collection efficiency as well as the energy resolution of the detector [25, 28, 29].

2.2 *Point Defects*

Point defects in CZT include tellurium precipitates, impurities, vacancies, interstitials, antisites, and their complexes. Most impurities, especially elements from groups I, III, IV, V, and VII, generally form shallow-level (close to either the valence band maxima or the conduction band minima) defects in CZT. Shallow-level defects are less harmful as far as detector operation is concerned. The shallow levels release any trapped charge, also known as detrapping, even at room temperature. Some impurities may also cause significant lattice relaxations, also referred to as DX centers, impacting the recombination and lifetime of charge carriers. Elements such as aluminum (Al), indium (In), bromine (Br), and iodine (I) may form DX centers in CZT. Transition-metal impurities form relatively deeper defect levels. Tellurium precipitates have been reported to form a deep-level trap center in CZT at 1.1 eV above the valence band maximum [3].

Among the widely seen deep-level defects are A centers that are donor-vacancy complexes. Group III elements at substitutional Cd cite in the CZT lattice behave as donors, especially In, which is deliberately added to CZT crystals during growth to increase its resistivity. These donors can form complexes with vacancy-type defects, such as cadmium vacancy, to form A centers, denoted by $\left(\mathrm{In}_{\mathrm{Cd}}^{+} V_{\mathrm{Cd}}^{2-}\right)^{-}$, which is located at 0.18 eV above the valence band maximum [30]. Chlorine (Cl)-related A centers have hole trap levels at 0.13 eV [31].

Purely native defects include cadmium vacancies (V_{Cd}), tellurium vacancies (V_{Te}), cadmium interstitials (Cd_{I}), tellurium interstitials (Te_{I}), and tellurium-substituted cadmium antisites, such as $\mathrm{Te}_{\mathrm{Cd}}$. While V_{Cd} mostly acts as an acceptor level, the rest of them behave generally as donor levels. A commonly observed vacancy-type defect is the doubly negatively ionized cadmium vacancy V_{Cd}^{2-}, located at 0.47 eV above the valence band. These defects exhibit various ionization energies depending on their charge states as well as doping concentration [3, 32].

2.3 *Effects of Charge Trapping on Detection Performance*

Radiation detectors work on the principle of collection of electron-hole charge pairs within the detector generated by incident radiation. In its simplest form, a planar detector comprises a resistive block of semiconductors with metal contacts deposited on two opposite sides, as shown in Fig. 1a. When a potential difference (bias) is applied across the detector, an electric field is established that drives the electron-hole pairs in opposite directions—toward the electrodes with favorable bias polarities. Detector-grade CZT is generally highly resistive (10^{10} Ω.cm) and exhibits very low leakage (dark) currents at room temperature. Leakage current adds to the electronic noise of the detector signal [33].

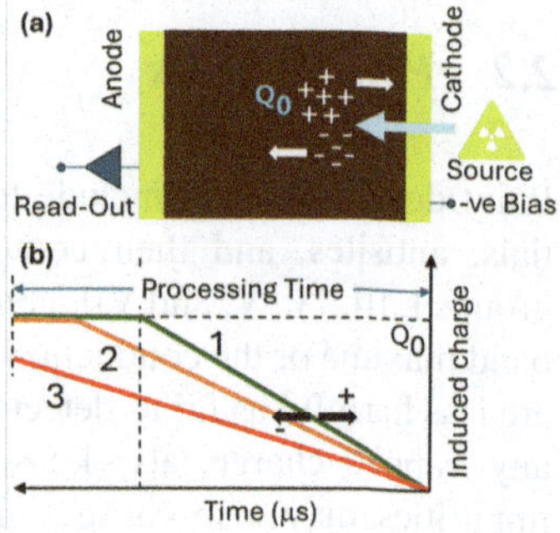

Fig. 1 (**a**) Schematic of a planar radiation detector. (**b**) Time evolution of the induced charge in a planar detector in the cases of (1) no trapping, (2) detrapping followed by trapping, and (3) recombination followed by trapping. Q_0 is the total charge induced in a single interaction

Usually, one of the electrodes, referred to as the collecting electrode, is used to connect to the front-end electronics, such as a preamplifier that produces a measurable voltage signal. When a charge pair is generated at a certain location inside the detector, the electrons start to drift toward the anode (positive potential) and the holes toward the cathode (negative potential). The charges start inducing a charge on the collecting electrode as soon as they start drifting. The induced charge is proportional to the proximity of the charge carrier to the collecting electrode. When the generated charges reach the collecting electrode physically, the induced charge becomes equal to the total charge generated by the incident radiation. The time evolution of the detector signal obtained from a planar detector connected to a charge-sensitive preamplifier is depicted in Fig. 1b.

Obviously, in the case of charge trapping, the drifting of the charge carriers would be impeded. In the case of the repeated trapping of charge carriers followed by detrapping, the charge pairs eventually reach the collecting electrode, however with a longer drift time. Such repeated trapping and detrapping lower the drift mobility of the charge carriers. Drift mobility is defined as the rate of increase in drift velocity as a function of an electric field. The detector signals in such a scenario get slower, as shown in Fig. 1b. For very high trapping concentrations, the pulses become so slow that the pulse processing time in conventional spectrometers becomes inadequate for processing the entire signal, resulting in partial pulse height determination. In the case of trapping followed by recombination in deep-level defects, the charge carriers do not reach the electrodes physically, which hence induces a fraction of the total charge generated on the collecting electrode. Hence, recombination lowers the lifetime of the charge carriers. The lifetime of charge carriers is defined as the average time between their generation and recombination. The partial deposition of charge due to the loss of charge carriers during transit is referred to as ballistic deficit. The effect on the detector signal of both the above phenomena of charge trapping and recombination has been depicted in Fig. 1b. Drift mobility and lifetime are the general measures of the charge transport properties of a semiconductor crystal.

The height of the detector pulses (pulse height) plays a significant role in spectroscopic measurements as the pulse height is directly proportional to the incident energy. The pulse heights are determined after filtering the preamplifier signal using a shaping amplifier. The time constant of the shaping amplifier determines the pulse processing time. The shaped pulse heights are plotted as a histogram to form a pulse height spectrum (PHS). The PHS comprises peaks corresponding to different incident energies. The width of a peak depicts the distribution of the measured pulse heights corresponding to that incident energy. This distribution is generally measured as the full width at half maximum (FWHM) calculated followed by a Gaussian fitting of the peak. The peak FWHM, commonly referred to as energy resolution, depends on several factors such as the electronic noise of the spectrometer, which also includes the detector leakage current and capacitance, the shaping time of the amplifier, and the applied bias. The energy resolution is often expressed as percentage resolution, defined as the percentage of the FWHM to the incident energy. Fluctuations in the detector output signal due to the random trapping of charge carriers cause fluctuations in the pulse height which lead to the deterioration of energy resolution [34].

Apart from energy resolution, another parameter that defines detector performance is charge collection efficiency (CCE). CCE is defined as the ratio of charge collected at the detector output to the charge generated by incident radiation after a single interaction. One hundred percent CCE is obtained when all radiation-induced charge pairs are collected by the acquisition setup, even in the presence of charge-trapping centers, provided adequate bias is applied. However, in the case of substantial recombination, 100% CCE cannot be achieved. Other spectral features include peak-to-valley (P-V) and peak-to-Compton (P-C) ratios. The P-V ratio is calculated as the ratio of full-energy peak height to the average height at a distance σ (standard deviation of the Gaussian peak fit) from the peak centroid, and the P-C ratio is calculated as the ratio of full-energy peak height to the Compton height measured approximately 100 keV below the Compton edge.

The fluctuations in the detector pulse height make spectroscopic measurements even more complicated in the case of gamma-ray detection. Unlike charged particle radiation, gamma rays do not interact at a specific depth within the detector. The random location of interactions adds to the fluctuation in the induced charge at the collecting electrode. For instance, two electrons created at different proximity to the collecting electrode by gamma photons having the same energy will induce different amounts of charge, even though they drift for the same distance and recombine before reaching the collecting electrode physically. This leads to severe broadening of the peak width, especially in the case of thicker detectors (>2 mm). Moreover, when charge transport properties drastically differ between electrons and holes, such as in CZT (hole transport is severely impeded), the fluctuations in the detector output signals become much more pronounced as holes are highly likely to be trapped in defect centers compared to electrons.

While advances in crystal growth methods aim to improve the transport properties of charge carriers, some of the above fluctuations can be prevented by adopting innovative detector geometries, which essentially make the devices single polarity charge sensitive. Such device structures electrostatically shield the contribution of the drifting charge carriers with poor charge transport to the detector signal. Further, digital correction schemes can compensate for the discrepancies arising from poor charge transport properties in the detector crystal. The following two sections describe how these preventive and corrective measures work.

3 Single-Polarity Charge Sensing

In single-polarity charge-sensing geometry in CZT detectors, the output signal is electrostatically shielded from the effect of hole transit to a great extent, as shown in Fig. 2. This substantially improves the fluctuations in the pulse-height measurements

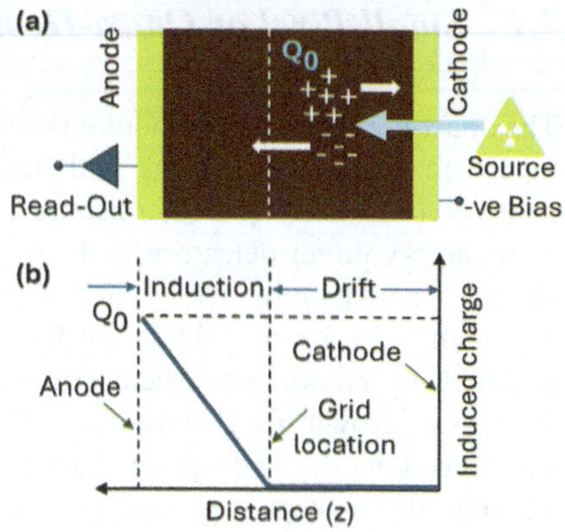

Fig. 2 (**a**) Schematic of a Frisch grid detector. The dotted line shows the location of the grid. (**b**) Variation of the induced charge as a function of depth measured from the cathode side

and hence the energy resolution. Single-polarity charge sensing may be achieved through various electrode geometries. These geometries are collectively called virtual Frisch grids (VFGs) in contrast to actual Frisch grids, which are physically present inside detectors such as gas-filled detectors. In gas-filled detectors, the Frisch grid, typically a metallic mesh kept at ground potential, prevents slow-moving heavy ions from inducing a charge on the collecting anode. The anode senses the movement of electrons only, and that too when they transit the drift region and appear at the induction region. The drift region is the interaction region, and the induction region begins from where the transiting charge starts to induce a charge on the collecting electrode. Similar effects are achieved with innovative electrode geometries in CZT detectors without including any actual grid/electrode inside the detector.

CZT detectors are generally rectangular parallelopiped in shape. The most common electrode designs that render such CZT blocks as VFG detectors are quasi-hemispherical (small single pixel) [35] and capacitive Frisch grid geometries [36]. Other single-polarity geometries are parallel-strip Frisch grid (the earlier generation of capacitive Frisch grid detectors) [37] and two-terminal detectors [38]. The parallel-strip Frisch grid employs two parallel strips of conducting material deposited on two parallelly opposite surfaces (transverse to the cathode and anode). The two-terminal detector design is achieved when the cathode metallization extends to the sides of the detector block up to a certain distance while the anode is a full contact and deposited on one side only. Besides rectangular parallelopiped geometry, a VFG effect has been reported to be obtained in geometrically weighted trapezoid-prism-shaped detectors as well [39]. The most used VFG configurations, viz., small-pixel and capacitive Frisch grid detectors, will be discussed in greater detail in the following subsections. A different class of VFG detectors, called coplanar grid [40], which works on the principle of electronically subtracting the contribution of the hole transit from the detector signal rather than electrostatic shielding of the hole signal, will also be discussed.

3.1 Small-Pixel or Quasi-Hemispherical Detectors

The quasi-hemispherical structure is achieved by depositing a single anode, which is much smaller in dimensions with respect to the cathode [35]. Among all VFG configurations, only quasi-hemispherical detectors enable efficient charge collection over large-volume detectors (≈4 cm^3) with a single readout channel. Figure 3a shows a picture of a small-pixel detector fabricated on an 11.3 mm × 11.3 mm × 11.1 mm high-resistive CZT ($Cd_{0.9}Zn_{0.1}Te$) single crystal grown using a modified vertical Bridgman technique in our laboratory. The details of the crystal growth may be found in Ref. [41]. The circular dot at the center is the small-pixel anode. The square-shaped metallization around the pixel is a guard ring structure that is kept at ground potential to minimize any current flow between the anode and the cathode through the detector surface. The pixel diameter is 3.2 mm, which is smaller than the square-shaped cathode, with an edge length of 11.3 mm. The pixel size in a quasi-hemispherical detector plays a crucial role in the detector's performance and needs to be optimized. The smaller is the pixel size relative to the cathode, the more effective is the hole screening [42]. However, a small pixel size also weakens the electric field, leading to inhomogeneous charge collection. Figure 3b shows a pulse height spectrum obtained using a benchtop spectrometer, with the detector exposed to a ^{137}Cs source emitting 662 keV gamma photons. The detector demonstrated an energy resolution of 3.7% when biased at 3500 V. Recent studies have enabled the identification of an optimum electrode configuration through first-principle simulations [21]. CZT detectors with dimensions of 15 mm × 15 mm × 10 mm and a small-pixel diameter of 2 mm optimized using the above simulation studies have been reported to demonstrate an energy resolution of 1.7% for 662-keV gamma photons when the detector was biased at 2750 V.

3.2 Coplanar Grid Detectors

A coplanar grid (CPG) geometry, as mentioned above, does not electrostatically shield the hole-transit-induced current, unlike other VFGs. In a CPG, instead of a continuous metallic anode, a pair of interdigitated electrodes (grids) with identical dimensions are deposited on one of the surfaces, and a continuous metallic cathode is

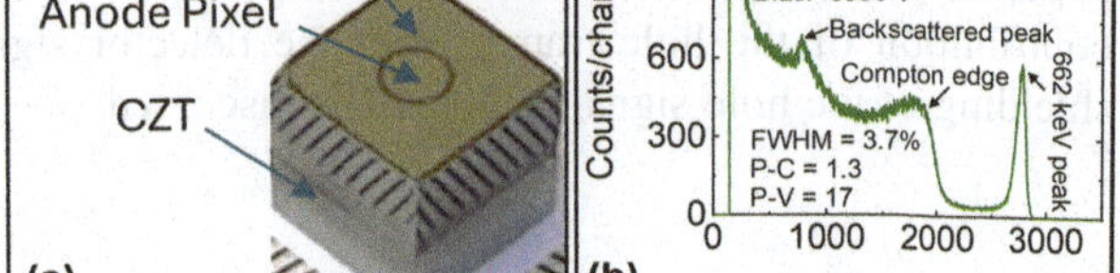

Fig. 3 (**a**) Photograph of a small-pixel CZT detector equipped with a guard ring. (**b**) Pulse height spectrum obtained using the small-pixel CZT detector when exposed to a ^{137}Cs gamma source [41]

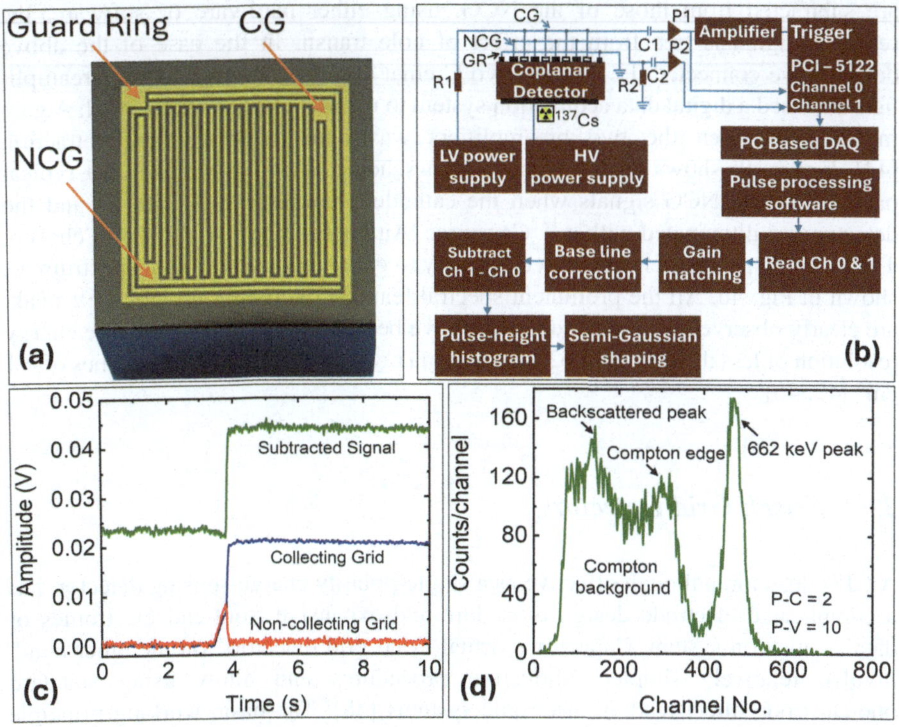

Fig. 4 (**a**) Photograph of a coplanar grid CZT detector equipped with a guard ring. (**b**) Flowchart depicting the data acquisition and processing scheme. (**c**) Preamplifier outputs connected to the collecting and noncollecting grids and the subtracted signal. The subtracted signal has been vertically offset. (**d**) Pulse height spectrum obtained using the coplanar grid detector when exposed to a ^{137}Cs gamma source [44]

deposited on the opposite surface [40, 43]. Figure 4a shows a CPG detector fabricated on an 11.1 mm × 11.1 mm × 10.0 mm high-resistive CZT ($Cd_{0.9}Zn_{0.1}Te$) single crystal grown using a modified vertical Bridgman technique in our laboratory [41]. It also has a guard ring structure connected to the ground that helps lower the flow of surface current.

For spectroscopic measurements, both electrodes are connected to two separate charge-sensitive preamplifiers, and bias is applied to the cathode through a noise-filtering biasing circuit. The bias is usually of negative polarity with respect to one of the grids kept at ground potential, called the collecting grid (CG), to facilitate the collection of electrons. The other grid, called the noncollecting grid (NCG), is biased at a higher positive potential (intergrid bias) with respect to the collecting grid. Under such a biasing scheme, the charge induced at the CG is due to both the transiting electrons and holes, while the charge induced at the NCG is due to hole transit only. When the signals obtained from the preamplifier connected to the CG are subtracted from those of the NCG, using either hardware or software, the resulting signal is free from the effect of hole transit. In the case of the above detector, we connected the CPG to two Cremat CR110 charge-sensitive preamplifiers and used a digital data acquisition system to record the subtracted signal. A gain matching between the two preamplifiers was done prior to the subtraction [44]. Figure 4b shows the data acquisition scheme, and Fig. 4c shows a typical pair of CG and NCG signals when the cathode was biased at −2500 V and the detector was illuminated with a ^{137}Cs source. An intergrid bias of 50 V was chosen. The recorded pulses were processed digitally to generate a pulse height spectrum, as shown in Fig. 4d. All the prominent spectral features, including the 662-keV peak, are clearly observed [44]. CPG detectors have been demonstrated to provide energy resolution of less than 2% for 662-keV gamma rays for detectors with volumes of ≈1 cm^3 [45, 46].

3.3 Frisch Grid Detectors

A CPG detector, although effective as a single-polarity charge-sensing detector, has a complicated electrode design procedure and expensive front-end electronics or data acquisition system. Capacitive virtual Frich grid detectors, on the other hand, involve relatively simpler fabrication procedures and allow using standard one-electrode direct readout electronic systems [38]. The basic working principle of capacitive Frisch grid designs is similar to that of parallel-strip Frisch grid and two-terminal detectors but with a significant improvement in terms of detector leakage current and hence energy resolution. The capacitive Frisch grid employs a conducting sheath kept equipotential with the cathode, which acts as an extended cathode. However, unlike a two-terminal or parallel-strip Frisch grid detector where the resistive path length between the anode and the cathode decreases, the capacitive Frisch grid is electrically isolated from the other surfaces of the detector using a thin layer of dielectrics such as Kapton. This isolation restores the leakage path between the anode and the extended cathode to the width of the detector [36, 39].

Figure 5a shows a photograph of a capacitive Frisch grid detector fabricated on an 11.1 mm × 11.1 mm × 10.0 mm high-resistive CZT ($Cd_{0.9}Zn_{0.1}Te$) single crystal grown using a modified vertical Bridgman technique in our laboratory [41]. Figure 5b shows a pulse height spectrum obtained using the detector biased at 1300 V and illuminated with a ^{137}Cs source. The length of the Frisch grid measured from the cathode side of the above detector is 9.0 mm. The length of the Frisch grid is an important part of this design. It is determined according to the detector aspect ratio, and it plays an important role in the device's performance when gamma photons of varied energies need to be detected [26, 47, 48].

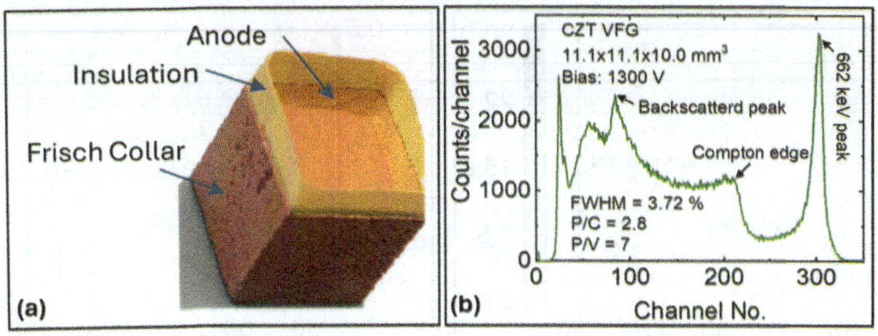

Fig. 5 (**a**) Photograph of a capacitive Frisch grid CZT detector. (**b**) Pulse height spectrum obtained using the capacitive Frisch grid detector when exposed to a ^{137}Cs gamma source [41]

4 Digital Correction Schemes

In their pioneering study, Richter and Siffert observed in thin CdTe planar detectors that the detector output pulses have much longer risetimes when the gamma photons interact close to the cathode and the corresponding pulse heights become lower [49]. In a planar detector, while the pulse height of the output signal indicates the amount of charge generated by an interaction event, its risetime suggests the depth of the interaction within the detector. Events occurring close to the anode imply a longer transit of the holes compared to those occurring close to the cathode. Pulse rejection may be used to reject the pulses with longer lifetimes as these events lead to poor resolution, however at the cost of loss of efficiency [50]. Additionally, Richter and Siffert found that the observed reduction in pulse height and the corresponding increase in pulse risetimes for various interactions in their detector had a linear relationship [49]. This type of dependence is referred to as biparametric correlation. Based on the linear relationship, they designed a compensating circuit that greatly restored the peak efficiency and enhanced the energy resolution. The compensating circuit used was very sophisticated and contained many pulse processing modules integrated into a single printed circuit board (PCB) card. Auricchio et al. reported a different approach wherein the detector signal obtained after the preamplifier was parallelly processed by two shaping amplifiers with different shaping time constants—one slow and the other fast [51]. The ratio of the amplitude of the two-shaped signals gives an indirect measure of the depth of interaction. The recovery was achieved offline through a correction algorithm.

A fast and efficient digital data acquisition system enables the acquisition of a large number of pulses with evenly distributed depths of interactions and calculating their risetime and pulse heights [52]. Modern-day digital spectrometers even allow one to visualize the correlation between risetimes and pulse heights in real time. Digital correction schemes, often referred to as biparametric corrections, are then applied to rectify the events affected by charge trapping [53].

Biparametric correction schemes can also be applied to VFG detectors should any correlated trapped events be noticed due to inefficient electrostatic shielding by the grid [1, 54]. Figure 6a shows a biparametric plot obtained corresponding to the

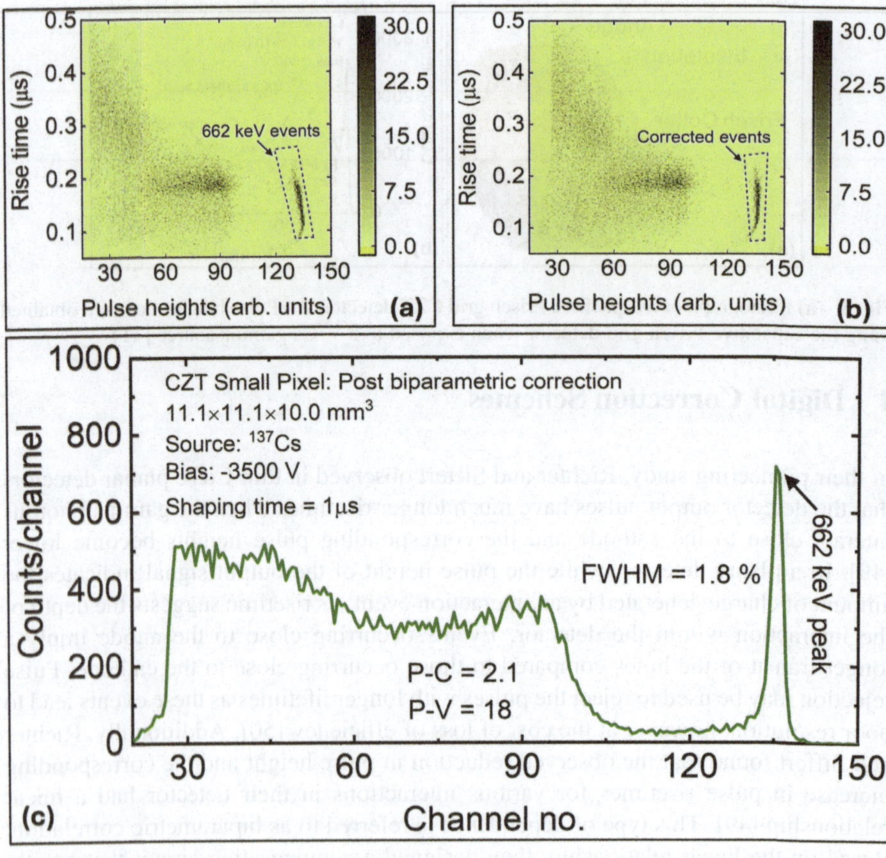

Fig. 6 Biparametric correlation plots obtained for the small-pixel detector described in Sect. 3.1, exposed to a ^{137}Cs source before (**a**) and after (**b**) the biparametric correction. (**c**) Pulse height spectrum generated from the corrected biparametric plot [41]

small-pixel detector described in Sect. 3.1. The events enclosed within the dotted line are the events corresponding to the interactions for the maximum energy deposited from the 662-keV events. The rest of the events in the biparametric plot are primarily from the Compton interactions. Contributions from the backscattered peak and the electronic noise are also embedded. The inclination of the 662-keV events with higher risetimes toward the lower pulse heights shows the effect of trapping. The affected events can be seen to be corrected in Fig. 6b which shows the biparametric plot after a correction algorithm was applied. Figure 6c shows the corrected pulse height spectrum regenerated from the BP plot. The substantial improvement in the energy resolution (1.8% for 662 keV gamma photons), peak-to-Compton (P-C) ratio, and peak-to-valley (P-V) ratio over the as-seen pulse height spectrum shown in Fig. 2b indicates the importance of biparametric corrections for VFG detectors. Biparametric corrections are applied to CZT-based multipixel imaging detectors as well [55, 56].

5 Conclusion

In conclusion, the chapter provides a comprehensive overview of the challenges and advancements in high-atomic number semiconductor materials, particularly focusing on $Cd_{1-x}Zn_xTe$ (CZT), for room-temperature gamma-photon detection. CZT has emerged as a promising candidate due to its wide bandgap and high atomic number constituents, offering advantages such as improved detection efficiency and resolution over conventional detectors. However, the performance of CZT detectors is hindered by various factors, including poor charge carrier transport properties, crystal growth limitations, and the presence of defects such as extended defects and point defects. These defects significantly impact charge transport and can deteriorate detector performance, particularly affecting energy resolution and charge collection efficiency.

To address these limitations, the chapter emphasizes the development of large-volume CZT detectors and discusses strategies such as single-polarity charge sensing and digital correction techniques. Single-polarity charge-sensing configurations, such as virtual Frisch grid detectors, effectively mitigate the effects of poor hole transport, leading to improved energy resolution. Additionally, digital correction schemes enable the rectification of events affected by charge trapping, enhancing overall detector performance. Furthermore, the chapter highlights various electrode geometries and innovative detector designs, such as quasi-hemispherical detectors, coplanar grid detectors, and capacitive Frisch grid detectors, each offering unique advantages in terms of charge sensing and energy resolution.

Overall, the research presented in this chapter underscores the ongoing efforts to overcome the challenges associated with CZT detectors, aiming to enhance their performance for applications in gamma-ray spectroscopy and medical imaging. By addressing these challenges and implementing advanced techniques, CZT detectors hold great potential for revolutionizing the field of nuclear radiation detection and the medical imaging field.

Acknowledgments As a principal investigator (PI), K.C.M acknowledges funding from the DOE Office of Nuclear Energy's Nuclear Energy University Programs (NEUP), Grant No. DE-NE0008662; South Carolina (SC) National Aeronautics and Space Administration (NASA) Established Program to Stimulate Competitive Research (EPSCoR) Program, Grant No. 10011935; SC NASA EPSCoR Research and Education Awards Program (REAP), Grant No. 10013007; SC Center of Research Excellence (CORE) Innovation, Grant No. 10013350; and Los Alamos National Laboratory/DOE, Grant No. 143479.

References

1. Chaudhuri, S. K., & Mandal, K. (2022). In K. Iniewski (Ed.), *Advanced materials for radiation detection* (pp. 211–234). Springer.
2. Veale, M. C., Bell, S., Cline, B. D., Church, I., Cross, S., Day, C., French, M., Gardiner, T., Ghorbanian, N., Hart, M. D., Jones, L. L., Lipp, J., Nicholls, T., Nobes, J., Prydderch, M., Schneider, A., Seller, P., Sole, D., Wilson, M. D., Dhamgaye, V., Fox, O., & Sawhney,

K. (2023). Preliminary characterisation of the HEXITECMHz spectroscopic X-ray imaging detector. *Journal of Instrumentation, 18*, P07048.

3. Schlesinger, T. E., Toney, J. E., Yoon, H., Lee, E. Y., Brunett, B. A., Franks, L., & James, R. B. (2001). Cadmium zinc telluride and its use as a nuclear radiation detector material. *Materials Science and Engineering, 32*, 103–189.
4. Bell, S. J., Baker, M. A., Duarte, D. D., Schneider, A., Seller, P., Sellin, P. J., Veale, M. C., & Wilson, M. D. (2017). Performance comparison of small-pixel CdZnTe radiation detectors with gold contacts formed by sputter and electroless deposition. *Journal of Instrumentation, 12*, 06015.
5. Szeles, C. (2004). CdZnTe and CdTe materials for X-ray and gamma ray radiation detector applications. *Physica Status Solidi B, 241*, 783–790.
6. Del Sordo, S., Abbene, L., Caroli, E., Mancini, A. M., Zappettini, A., & Ubertini, P. (2009). Progress in the development of CdTe and CdZnTe semiconductor radiation detectors for astrophysical and medical applications. *Sensors, 9*, 3491–3526.
7. Roy, U. N., Weiler, S., Stein, J., Cui, Y., Groza, M., Buliga, V., & Burger, A. (2012). Zinc mapping in THM grown detector grade CZT. *Journal of Crystal Growth, 347*, 53–55.
8. Mandal, K. C., Krishna, R. M., Muzykov, P. G., & Hayes, T. C. (2012). Fabrication and characterization of high barrier $Cd_{0.9}Zn_{0.1}Te$ Schottky diodes for high resolution nuclear radiation detectors. *IEEE Transactions on Nuclear Science, 59*, 1504–1509.
9. Bolotnikov, A. E., Ackley, K., Camarda, G. S., Cui, Y., Eger, J. F., De Geronimo, G., Finfrock, C., Fried, J., Hossain, A., Lee, W., Prokesch, M., Petryk, M., Reiber, J. L., Roy, U. N., Vernon, E., Yang, G., & James, R. B. (2005). High-efficiency CdZnTe gamma-ray detectors. *IEEE Transactions on Nuclear Science, 62*, 3193–3198.
10. Mandal, K. C., Kang, S. H., Choi, M., Bello, J., Zheng, L., Zhang, H., Groza, M., Roy, U. N., Burger, A., Jellison, G. E., Holcomb, D. E., Wright, G. W., & Williams, J. A. (2006). Simulation, modeling, and crystal growth of $Cd_{0.9}Zn_{0.1}Te$ for nuclear spectrometers. *Journal of Electronic Materials, 35*, 1251–1256.
11. Mandal, K. C., Kang, S. H., Choi, M., Kargar, A., Harrison, M. J., McGregor, D. S., Bolotnikov, A. E., Carini, G. A., Camarda, G. C., & James, R. B. (2007). Characterization of low-defect $Cd_{0.9}Zn_{0.1}Te$ and CdTe crystals for high-performance Frisch collar detectors. *IEEE Transactions on Nuclear Science, 54*, 802–806.
12. Fougères, P., Chibani, L., Hageali, M., Koebel, J. M., Hennard, G., Zumbiehl, A., Siffert, P., & Benkaddour, M. (1999). Zinc segregation in HPB grown nuclear detector grade $Cd_{1-x}Zn_xTe$. *Journal of Crystal Growth, 197*, 641–645.
13. Rodríguez, M. E., Gutiérrez, A., Zelaya-Angel, O., Vázquez, C., & Giraldo, J. (2001). Influence of crystalline quality on the thermal, optical and structural properties of $Cd_{1-x}Zn_xTe$ for low zinc concentration. *Journal of Crystal Growth, 233*, 275–281.
14. Esserman, L. (2020). We need more evidence to answer questions about screening. *Nature, 579*, S5.
15. Iniewski, K. (2014). CZT detector technology for medical imaging. *Journal of Instrumentation, 9*, C11001.
16. Pak, R. O., & Mandal, K. (2015). C.: Defect levels in nuclear detector grade $Cd_{0.9}Zn_{0.1}Te$ crystals. *ECS J. Solid State Science and Technology, 5*, P3037–P3040.
17. Krishna, R. M., Muzykov, P. G., & Mandal, K. C. (2013). Electron beam induced current imaging of dislocations in $Cd_{0.9}Zn_{0.1}Te$ crystal. *Journal of Physics and Chemistry of Solids, 74*, 170–173.
18. Bolotnikov, A. E., Camarda, G. S., Carini, G. A., Cui, Y., Li, L., & James, R. B. (2007). Cumulative effects of Te precipitates in CdZnTe radiation detectors. *Nuclear Instruments and Methods in Physics Research, 571*, 687–698.
19. Cavallini, A., Fraboni, B., Castaldini, A., Marchini, L., Zambelli, N., Benassi, G., & Zappettini, A. (2013). Defect characterization in fully encapsulated CdZnTe. *IEEE Transactions on Nuclear Science, 60*, 2870–2871.
20. Roy, U. N., Camarda, G. S., Cui, Y., Gul, R., Yang, G., Zazvorka, J., Dedic, V., Franc, J., & James, R. B. (2019). Evaluation of CdZnTeSe as a high-quality gamma-ray spectroscopic material with better compositional homogeneity and reduced defects. *Scientific Reports, 9*, 7303.

21. Vicini, V., Zanettini, S., Amadè, N. S., Grill, R., Zambelli, N., Calestani, D., Zappettini, A., Abbene, L., & Bettelli, M. (2023). Optimization of quasi-hemispherical CdZnTe detectors by means of first principles simulation. *Scientific Reports, 13*, 3212.
22. James, R. B., Brunett, B., Heffelfinger, J., Van Scyoc, J., Lund, J., Doty, F. P., Lingren, C. L., Olsen, R., Cross, E., Hermon, H., Yoon, H., Hilton, N., Schieber, M., Lee, E. Y., Toney, J., Schlesinger, T. E., Goorsky, M., Yao, W., Chen, H., & Burger, A. (1998). Material properties of large-volume cadmium zinc telluride crystals and their relationship to nuclear detector performance. *Journal of Electronic Materials, 27*, 788–799.
23. Bolotnikov, A. E., Camarda, G. S., Carini, G. A., Cui, Y., Kohman, K. T., Li, L., Salomon, M. B., & James, R. B. (2007). Performance-limiting defects in CdZnTe detectors. *IEEE Transactions on Nuclear Science, 54*, 821–827.
24. Bolotnikov, A. E., Abdul-Jaber, N. M., Babalola, O. S., Camarda, G. S., Cui, Y., Hossain, A. M., Jackson, E. M., Jackson, H. C., James, J. A., Kohman, K. T., Luryi, A. L., & James, R. B. (2008). Effects of Te inclusions on the performance of CdZnTe radiation detectors. *IEEE Transactions on Nuclear Science, 55*, 2757–2764.
25. Bolotnikov, A. E., Babalola, S. O., Camarda, G. S., Chen, H., Awadalla, S., Cui, Y., Egarievwe, S. U., Fochuk, P. M., Hawrami, R., Hossain, A., James, J. R., Nakonechnyj, I. J., Mackenzie, J., Yang, G., Xu, C., & James, R. B. (2009). Extended defects in CdZnTe radiation detectors. *IEEE Transactions on Nuclear Science, 56*, 1775–1783.
26. Bolotnikov, A. E., Camarda, G. C., Carini, G. A., Fiederle, M., Li, L., McGregor, D. S., McNeil, W., Wright, G. W., & James, R. B. (2006). Performance characteristics of Frisch-ring CdZnTe detectors. *IEEE Transactions on Nuclear Science, 53*, 607–614.
27. Rudolph, P. (2003). Non-stoichiometry related defects at the melt growth of semiconductor compound crystals—A review. *Crystal Research and Technology, 38*, 542–554.
28. Szeles, C., & Eissler, E. E. (1998). Current issues of high-pressure Bridgman growth of semi-insulating CdZnTe. In R. B. James, T. E. Schlesinger, P. Siffert, W. Dusi, M. R. Squillante, M. O'Connell, & M. Cuzin (Eds.), *Semiconductors for room temperature radiation detector applications II* (pp. 3–12). Materials Research Society.
29. James, R. B., Schlesinger, T. E., Lund, J. C., & Schieber, M. (1995). Cadmium zinc telluride spectrometers for gamma and X-ray applications. In R. B. James & T. E. Schlesinger (Eds.), *Semiconductors for room temperature nuclear detector applications* (p. 334). Academic.
30. Gul, R., Roy, U. N., Camarda, G. S., Hossain, A., Yang, G., Vanier, P., Lordi, V., Varley, J., & James, R. B. (2017). A comparison of point defects in $Cd_{1-x}Zn_xTe_{1-y}Se_y$ crystals grown by Bridgman and traveling heater methods. *Journal of Applied Physics, 121*, 125705.
31. Toney, J. E., Brunett, B. A., Schlesinger, T. E., & James, R. B. (1997). Extracting trap parameters from PICTS spectra in cadmium zinc telluride radiation detector material. *MRS Proceedings, 487*, 109.
32. Kleppinger, J. W., Chaudhuri, S. K., Roy, U. N., James, R. B., & Mandal, K. C. (2021). Growth of $Cd_{0.9}Zn_{0.1}Te_{1-y}Se_y$ single crystals for room temperature gamma-ray detection. *IEEE Transactions on Nuclear Science, 68*, 2429–2434.
33. Krishna, R. M., Chaudhuri, S. K., Zavalla, K. J., & Mandal, K. (2013). C.: Characterization of $Cd_{0.9}Zn_{0.1}Te$ based virtual Frisch grid detectors for high energy gamma ray detection. *Nuclear Instruments and Methods in Physics Research A, 701*, 208–213.
34. Knoll, G. F. (2000). *Radiation detection and measurements* (3rd ed.). Wiley.
35. Szeles, C., Bale, D., Grosholz, J., Jr., Smilth, G. L., Blostein, M., & Eger, J. (2006). Fabrication of high performance CdZnTe quasi-hemispherical gamma-ray CAPture™ plus detectors. *Proceedings of SPIE, 6319*, 191.
36. Montemont, G., Arques, M., Verger, L., & Rustique, J. (2001). A capacitive Frisch grid structure for CdZnTe detectors. *IEEE Transactions on Nuclear Science, 48*, 278–281.
37. McGregor, D. S., He, Z., Seifert, H. A., Rojeski, R. A., & Wehe, D. K. (1998). CdZnTe semiconductor parallel strip Frisch grid radiation detectors. *IEEE Transactions on Nuclear Science, 45*, 443–449.
38. Bolotnikov, A. E., Abdul-Jabbar, N. M., Babalola, S., Camarda, G. S., Cui, Y., Hossain, A., Jackson, E., Jackson, H., James, J. R., Luryi, A. L., & James, R. B. (2007). Optimization of

virtual Frisch-grid CdZnTe detector designs for imaging and spectroscopy of gamma rays. *Proceedings of SPIE, 6706*, 670603.
39. McGregor, D. S., & Rojeski, R. A. (2001). High-resolution ionization detector and array of such detectors. US Patent 6,175,120 (16 01 2001).
40. Luke, P. N. (1998). Single-polarity charge sensing in ionization detectors using coplanar electrodes. *Applied Physics Letters, 65*, 2884–2886.
41. Chaudhuri, S. K., Krishna, R. M., Zavalla, K. J., Matei, L., Buliga, V., Groza, M., Burger, A., & Mandal, K. C. (2013). Cd Zn Te crystal growth and fabrication of large volume single-polarity charge sensing detectors. *IEEE Transactions on Nuclear Science, 60*, 2853–2858.
42. Berrett, H. H., Eskin, J. D., & Barber, H. B. (1995). Charge transport in arrays of semiconductor gamma-ray detectors. *Physical Review Letters, 75*, 156–159.
43. Luke, P. N. (1995). Unipolar charge sensing with coplanar electrodes – Application to semiconductor detectors. *IEEE Transactions on Nuclear Science, 42*, 207–213.
44. Chaudhuri, S. K., Sajjad, M., Kleppinger, J. W., & Mandal, K. C. (2020). Real-Time Pulse Height Spectroscopy Using $Cd_{0.9}Zn_{0.1}Te$ Coplanar Grid Digital Spectrometer. In *2020 IEEE nuclear science symposium and medical imaging conference (NSS/MIC)*, Boston, MA, USA.
45. He, Z., Knoll, G. F., Wehe, D. K., & Miyamoto, W. J. (1997). Position-sensitive single carrier CdZnTe detectors. *Nuclear Instruments and Methods in Physics Research A, 388*, 180–185.
46. Carini, G. A., Bolotnikov, A. E., Camarda, G. S., Wright, G. W., De Geronimo, G., Siddons, D. P., & James, R. B. (2004). Synchrotron radiation response characterization of coplanar grid CZT detectors. In *IEEE symposium conference record nuclear science 2004*, Rome, Italy.
47. Kargar, A., Jones, A. M., McNeil, W. J., & McGregor, D. S. (2006). CdZnTe Frisch ring detectors for low-energy gamma ray spectroscopy. *Nuclear Instruments and Methods in Physics Research A, 558*, 487–503.
48. Polack, J. K., Hirt, M., Sturgess, J., Sferrazza, N. D., Bolotnikov, A. E., Babalola, S., Camarda, G. S., Cui, Y., Egarievwe, S. U., Fochuk, P. M., Gul, R., Hossain, A., Kim, K., Kopach, O. V., Marchini, L., Yang, G., Xu, L., & James, R. B. (2010). Variation of electric shielding on virtual Frisch-grid detectors. *Nuclear Instruments and Methods in Physics Research A, 621*, 424–430.
49. Richter, M., & Siffert, P. (1992). High resolution gamma ray spectroscopy with CdTe detector system. *Nuclear Instruments and Methods in Physics Research A, 322*, 529–537.
50. Squillante, M. R., Entine, G., Frederick, E., Cirignano, L., & Hazlett, T. (1989). Development of two new M-π-n CdTe sensors. *Nuclear Instruments and Methods in Physics Research A, 283*, 323–329.
51. Auricchio, N., Caroli, E., Donati, A., Roggio, A., Sephen, J. B., & Ventura, G. (2007). Application of a biparametric method using a twin shaping filter on CZT detectors. In *2007 IEEE nuclear science symposium conference record*, Honolulu, HI, USA.
52. Jordanov, V. T., & Knoll, G. F. (1994). Digital synthesis of pulse shapes in real time for high resolution radiation spectroscopy. *Nuclear Instruments and Methods in Physics Research A, 345*, 337–345.
53. Chaudhuri, S. K., Lohstroh, A., Nakhostin, M., & Sellin, P. J. (2012). Digital pulse height correction in HgI_2 γ-ray detectors. *Journal of Instrumentation, 7*, T04002.
54. Bolotnikov, A. E., Camarda, G. S., Cui, Y., Egarievwe, S. U., Fochuk, P. M., Fuerstnau, M., Gul, R., Hossain, A., Jones, F., Kim, K., Kopach, O. V., Taggart, R., Yang, G., Ye, Z., Xu, L., & James, R. B. (2010). Array of virtual Frisch-grid CZT detectors with common cathode readout and pulse-height correction. *Proceedings of SPIE, 7805*, 780504.
55. Verger, L., Gentet, M. C., Gerfault, L., Guillemaud, R., Mestais, C., Monnet, O., Montemont, G., Petroz, G., Rostaing, J. P., & Rustique, J. (2003, October). Performance and perspectives of a gamma camera based on CdZnTe for medical imaging. In *2003 IEEE nuclear science symposium. Conference record*, Portland, OR, USA.
56. Abbene, L., Gerardi, G., Principato, F., Bettelli, M., Seller, P., Veale, M. C., Fox, O., Sawhney, K., Zambelli, N., Benassi, G., & Zappettini, A. (2018). Dual-polarity pulse processing and analysis for charge-loss correction in cadmium–zinc–telluride pixel detectors. *Journal of Synchrotron Radiation, 25*, 1078–1092.

High-Resolution 3D CdZnTe Drift Strip Detectors for Astrophysical and Medical Applications

Manuele Bettelli, Leonardo Abbene, and Ezio Caroli

1 Introduction

Over the past 20 years, significant progress has been made in the development of three-dimensional (3D) cadmium-zinc-telluride (CZT) detectors capable of operating at room temperature for gamma-ray spectroscopic imaging. The high demand for gamma-ray spectroscopic imagers, able to offer sub-millimetre spatial resolution in 3D and energy resolution close to cooled high-purity germanium (HPGe) detectors (0.3% FWHM at 662 keV) [1], has led to extensive research into the development of 3D CZT detectors. The measurement of photon energy, timing, and 3D positioning up to the MeV range is mandatory for several applications in astrophysics, medical imaging, and nuclear security. Since the first spectroscopic-grade detector was manufactured [2], CZT has become the preferred detector material over high-impedance (high-Z) and wide-band-gap compound semiconductors. The success of CZT detectors can be attributed to significant improvements in both crystal growth and device fabrication technologies, as well as their appealing physical properties (such as high atomic number, wide band gap, and high density).

Recent developments have been made in the fabrication of imaging devices that use cadmium-zinc-telluride (CZT) detectors to perform X-ray and gamma-ray spectroscopy at room temperature. These advanced imagers offer incredibly precise

M. Bettelli (✉)
IMEM-CNR, Parma, Italy
e-mail: manuele.bettelli@imem.cnr.it

L. Abbene
DiFC, University of Palermo, Palermo, Italy
e-mail: leonardo.abbene@unipa.it

E. Caroli
INAF/OAS, Bologna, Italy
e-mail: ezio.caroli@inaf.it

K. Iniewski (ed.), *CdTe and CdZnTe Materials*,
https://doi.org/10.1007/978-3-031-64521-1_9

sub-millimetre spatial resolution in 3D and can provide high-energy resolution up to 1 MeV [3–6]. With potential applications in various fields, such as astrophysics, medical imaging, and nuclear security, these systems have the ability to deliver highly accurate measurements of energy, timing, and the 3D positioning of X-rays and gamma rays. CZT is well suited for room-temperature measurements because of its unique properties [7–10].

Typically, three different types of 3D CZT imagers have been proposed and developed. These include pixelated thick detectors [6], virtual Frisch-grid detectors [5], and orthogonal drift-strip detectors [11]. Pixelated thick detectors, typically used for gamma cameras, are based on large and thick detectors, which result in high-energy resolution. However, they are impacted by strong charge sharing and limited spatial resolution (typically 1.5–2 mm) and require a high number of read-out channels per unit volume.

Virtual Frisch-grid detectors are created using CZT blocks shaped like bars, with electrodes deposited on two opposing faces and a single isolated pad placed on each of the remaining faces close to the anode. Four pads are used for read-out to enable precise sub-voxel position sensing. This geometry results in a satisfactory energy resolution and enables the reading of a large volume with a small number of read-out channels. However, the spatial resolution is typically limited to 2–3 mm [12].

On the other hand, the orthogonal drift-strip detector [11] is based on two sets of orthogonal strips: one on the anode and the other on the cathode. This electrode configuration enables 3D detection with sub-millimetre resolution while using only a few read-out channels. Moreover, the energy resolution is aligned with virtual Frisch-grid detectors. These detectors have moderate leakage currents between the drift strips, allowing for high drift bias voltages and resulting in improved charge collection efficiency and energy resolution. Unfortunately, fabricating these detectors is challenging due to the presence of strips on both the anode and cathode and the need to mitigate surface leakage currents.

Detector geometry	Low-energy threshold	FWHM at ^{137}Cs	Spatial resolution in 3D	Volume per channel	Fabrication complexity
Pixelated	25 keV [6]	0.5% [6]	1.7 mm [6]	49 mm^3/ch	Low
Virtual Frisch grid	30 keV [5]	0.6% [5]	2.8 mm [5]	341 mm^3/ch	Medium
Orthogonal drift strips	10 keV [11, 13]	0.8% [11, 13]	<0.4 mm [14]	112 mm^3/ch	High

Orthogonal drift-strip detectors are used in planar transverse field (PTF) geometry and offer 2D positioning through cross-strip electrode patterns on the cathode (C)/ anode (A) electrodes, with the third coordinate provided by the C/A ratio.

Creating read-out electronics for 3D CZT detectors is a challenging task because of the complex processing required to analyse the various pulse shapes produced by these detectors. Due to the specific electrode patterns and various mechanisms in pulse formation (such as drift, induction, charge trapping, charge sharing, and crosstalk), a wide range of different charge pulses is typically generated by 3D CZT detectors. The digital pulse processing (DPP) approach, which analyses digitised detector pulses, is more effective than traditional analogue electronics, thus providing better results [15, 16].

A novel orthogonal strip 3D detector concept was developed in the framework of a collaboration between Italian research groups. In this chapter, the performance of new high-resolution CZT drift-strip detectors is reported. These detectors (19.4 mm × 19.4 mm × 6 mm) are equipped with collecting anode strips (with a pitch of 1.6 mm) and drift strips (with a pitch of 0.4 mm) that are negatively biased to optimise electron charge collection. The cathode is divided into strips orthogonal to the anode strips with a pitch of 2 mm. Using custom digital read-out electronics, a dedicated pulse processing analysis was performed on a wide range of collected and induced charge pulse shapes. The detectors were found to achieve excellent room-temperature energy resolution (1.3% FWHM at 662 keV) without any spectral corrections, and further improvements (0.8% FWHM at 662 keV) were made using a novel correction technique based on the analysis of collected/induced charge pulses from anode and drift strips.

These developments are part of two Italian research projects aimed at developing spectroscopic gamma-ray imagers (10–1000 keV) for astrophysical and medical applications, the 3DCaTS and 3CaTM projects, funded by INFN and ASI, respectively.

The 3CaTS project aims to develop a new single-photon emission computed tomography (SPECT) system for real-time therapeutic dose monitoring in boron neutron capture therapy (BNCT). BNCT is a hadron therapy that uses the nuclear reaction $^{10}B(n,\alpha)^{7}Li$, selectively depositing energy inside ^{10}B-enriched cells while sparing surrounding tissues. The INFN 3CaTS project aims to develop a BNCT-dedicated SPECT using innovative 3D strip-cell CdZnTe (CZT) photon detectors. To evaluate the feasibility of this technology in BNCT applications, the response of CZT detectors to a mixed $(n + \gamma)$ radiation field, particularly the 558 keV signal induced by thermal neutrons on ^{113}Cd, needs to be examined.

The 3DCaTM project aims to develop detectors that can provide simultaneous spectroscopy (10–1000 keV), imaging, timing, and polarimetric scattering as both focal plane detectors on a space telescope implementing new high-energy optics (such as broadband Laue lens) and small wide-field instruments used on clusters of micro-satellites. Additionally, in 2019, we proposed a stratospheric balloon flight of a small prototype (BADG3R), based on the aforementioned detector, in the framework of the H2020/HEMERA programme. This flight, which took place in September 2022 from the SSC ESRANGE base in Kiruna (Sweden), was crucial in assessing the reliability of the new technological solutions we have implemented in the 3D-CZT spectroscopic-imager module. Furthermore, this flight has demonstrated the flexibility and reliability of the a signals readout digital approach in a noisy pseudo-space environment.

2 The Drift Strip Concept in CZT Detectors

Gatti and Rehak were the first to introduce the concept of drift or steering electrodes for silicon radiation detectors in 1984 [17]. By combining this approach with electron-charge-sensing electrode geometries such as pixels, strips, and circular rings, CZT detectors have achieved significant improvements in both charge collection efficiency and energy resolution. This has been demonstrated in a number of studies, including Refs. [18–22].

The detectors presented in this chapter are designed to fit a standard Redlen crystal ($19.4 \times 19.4 \times 6\ mm^3$). The anode layout features gold-strip electrodes with a pitch of 0.4 mm (0.15 mm strip width and 0.25 mm inter-strip gap). The cathode side is divided into ten gold strips orthogonal to the anode strips with a pitch of 2 mm (1.9 mm strip width and 0.1 mm inter-strip gap).

To enhance the efficiency of electron charge collection on the green collecting strips (shown in Fig. 1), the neighbouring strips, called drift strips, are negatively biased. As displayed in Fig. 1a, the drift strips closest to the collecting strips (purple) are biased at 100 V, and the central drift strips (magenta) are biased at 200 V. The cathodes are biased at 350 V. This bias configuration was obtained by simulation (Fig. 1, right) and confirmed by real measurement. This values allow to optimize the charge collection from the whole detector volume keeping down the leakage current among strips. The signal read-out of the strips is organised in the following manner (Fig. 1b): the collecting anode (green) and the cathode (light blue) strips are read out individually, while the drift strips are combined in two groups of three strips. The right-side drift (RD) strips (pink) and the left-side drift (LD) strips (mustard) of each collecting strip are combined separately.

The detectors are divided into 12 functional strip groups, known as drift cells. Each drift cell contains a collection anode, the RD and LD strips, and 10 cathode strips. Each drift cell can provide both the energy and the 3D interaction position of the photons, as demonstrated in the literature [11, 23]. The pulses from the collecting

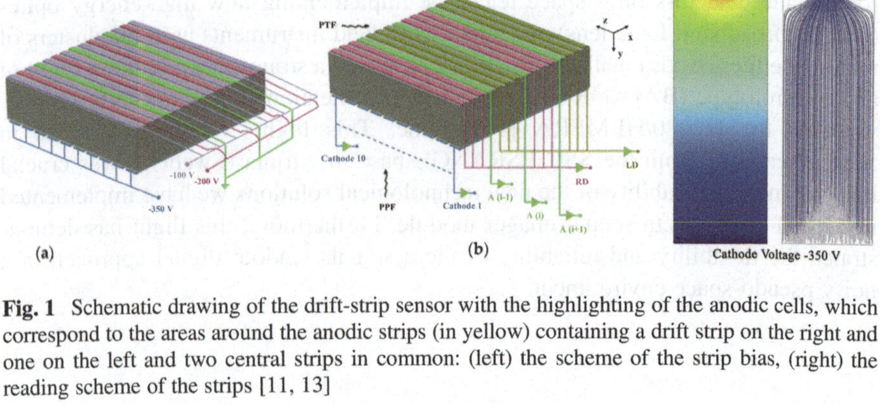

Fig. 1 Schematic drawing of the drift-strip sensor with the highlighting of the anodic cells, which correspond to the areas around the anodic strips (in yellow) containing a drift strip on the right and one on the left and two central strips in common: (left) the scheme of the strip bias, (right) the reading scheme of the strips [11, 13]

anodes give the energy of the photons, the x position is determined by the collecting anodes related to the RD/LD anode pulses, the y position is determined by the relation between the cathode and anode pulses, and the z position is determined by the cathode pulses. Despite the 1.6 mm pitch of the drift cells and the 2 mm pitch of the cathode strips, it is possible to improve the spatial resolution by analysing the pulses from the drift strips [11, 23].

3 Detectors Fabrication and Response Uniformity

Three-dimensional orthogonal drift-strip detectors were and are extensively studied by a large Italian joint collaboration composed of IMEM-CNR (Parma), INAF-OAS (Bologna), Due2lab (Reggio Emilia), UNIPA (Palermo), and UNIPV (Pavia). IMEM-CNR of Parma and Due2lab collaborated to fabricate the new CZT drift-strip detectors. The detectors were realised using CZT crystals (19.4 mm × 19.4 mm × 6 mm) grown through the travelling heater method (THM) technique. Starting from Redlen CZT pixel detectors, these prototypes were fabricated by removing the electrodes and treating the surface. Redlen is well known for producing CZT crystals with exceptional charge transport properties (mobility-lifetime products of electrons $\mu_e\tau_e > 10^{-2}$ cm^2/V) [24, 25]. The primary goal in developing CZT drift-strip detectors is to enable the application of high-bias voltages between the drift strips to optimise electron charge collection while keeping the leakage currents (i.e. the electronic noise) to a minimum. To achieve this, a significant effort was devoted to developing new detector-processing techniques.

Reducing inter-strip leakage current is crucial for the proper functioning of CZT sensors as the operational bias between anode strip pairs can range up to several tens of volts. High surface leakage current can significantly impact the performance of the CZT sensor for both spectroscopy and spatial resolution, so we consider a reasonable compliance value to be <5 nA at 50 V. Reducing leakage current is challenging since the anode strips are 20 mm long and very close together.

Over the years, IMEM-CNR has employed various techniques to achieve low surface leakage current. Initially, wet passivation (using 10% H_2O_2 in H_2O) was used, but it resulted in an unacceptably high leakage current for the application requirements. The second version of detectors, developed by Zanettini et al. in 2018 [26], used an innovative deposition technique and Al_2O_3 thin layer passivation to reduce leakage current. However, only one of the three detectors had compliant currents, and the technique requires a precise alignment between the photolithographic mask and the gold strips, making it a risky and complex process.

For the third batch of 3D samples, IMEM-CNR and Due2Lab s.r.l. developed an innovative surface passivation procedure involving a two-phase process [11]: (i) the deposition of a negative photoresist with anode patterns on the as-polished CZT surface and (ii) the deposition of Au strips using a wet-chemical electroless technique from a methanol solution [27]. This new passivation process reduced surface dark leakage current by a factor of 10 compared to standard passivation techniques

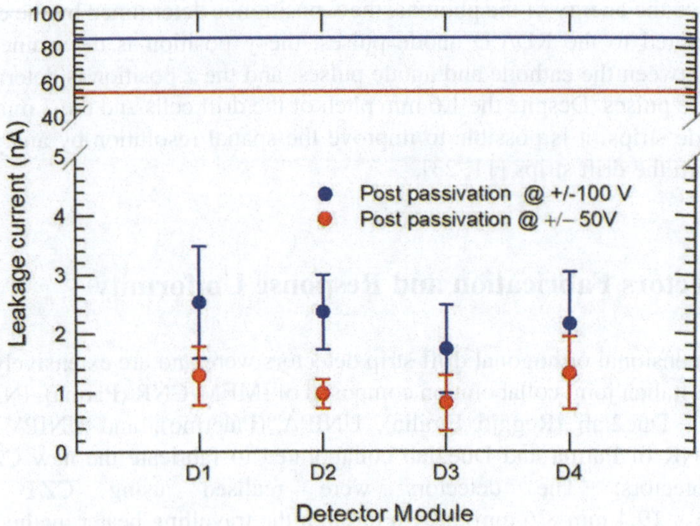

Fig. 2 Average values and standard deviation of the surface leakage currents between pairs of anode side strips at two different voltages measured on the realised set of 3D CZT sensors. The two upper solid lines represent the average value of the same currents before passivation for biases of 50 V (red line) and 100 V (blue line), respectively

and by a factor of 30 compared to unprocessed surfaces (Fig. 2). Additionally, the process showed a high level of reproducibility across different CZT crystals and good uniformity across the entire set of anode strips. Figure 2 shows the measured surface leakage current distribution between adjacent strip pairs across the entire set of anode strips of four 3D CZT sensors at $\Delta V = 50$ V before bonding the sensor to its support.

The CZT crystals underwent a multi-step preparation process before being introduced to the clean room environment. They were first lapped with abrasive silicon carbide (SiC) paper, then polished and thoroughly cleaned. The cathode strips were patterned using wet-chemical electroless gold deposition, standard photolithography, and subsequent bromine (Br)-based etching in ethylene glycol. On the other hand, a dedicated surface passivation procedure in three steps was performed on the anode side. First, a negative photoresist with the anode pattern was deposited on the as-polished CZT surface. Then a short Br-based etching for surface oxide removal was performed, and finally, tripe deposition was carried out using the wet-chemical electroless technique from a methanol solution. This lithographic process for the anode side was particularly challenging as it required the use of an alcoholic solution for the Au electroless deposition, which provides better mechanical adhesion than water-based deposition [27]. The patterning of the electrodes before Au deposition also avoided a detrimental etching step and ensured lower leakage currents than common wet passivation techniques. The detectors achieved very low leakage currents between the drift strips (<1 nA at 50 V and twice this value at 100 V) as well as uniformity among them.

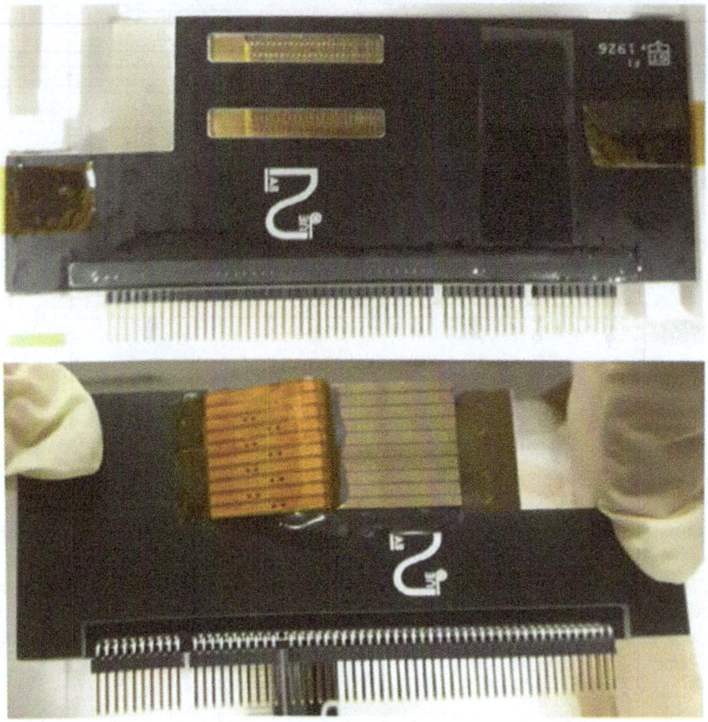

Fig. 3 A fabricated detector bonded on a custom PCB

The final packaging of the detectors was highlighted in Fig. 3, where the sensor unit support and electrical interface board were made up of five superimposed layers of different materials (Roger, Kapton, and FR4), rolled together. Each metallic strip of the 3D CZT sensor was electrically connected with a conductive glue to Au tracks insulated by Kapton film through a manual bonding procedure.

The sensors in their final configuration were tested with a standard electronics spectroscopy chain using calibration radioactive sources, with the main objective of further validating the construction process and its reproducibility by verifying their response uniformity in terms of linearity and spectroscopic resolution as a function of energy in both planar parallel field (PPF) irradiation and planar transverse field (PTF) configurations [3]. These measurements have been performed using, for all the sensor channels, the same preamplifier, with an equivalent noise charge of 100 electrons (RMS) and a decay constant of 150 ns, to make the measurements independent from the electronic read-out chain.

As regards the proportionality of the generated signal as a function of the deposited energy, the data confirmed that the difference at each energy between the calibration line and the measured value (ΔPeak) remains within $\pm 1\%$ for all 3D CZT sensors (Fig. 4, upper data point). The uniform response of the sensor set is further confirmed by the evaluation of the spectroscopic resolution (FWHM) as a

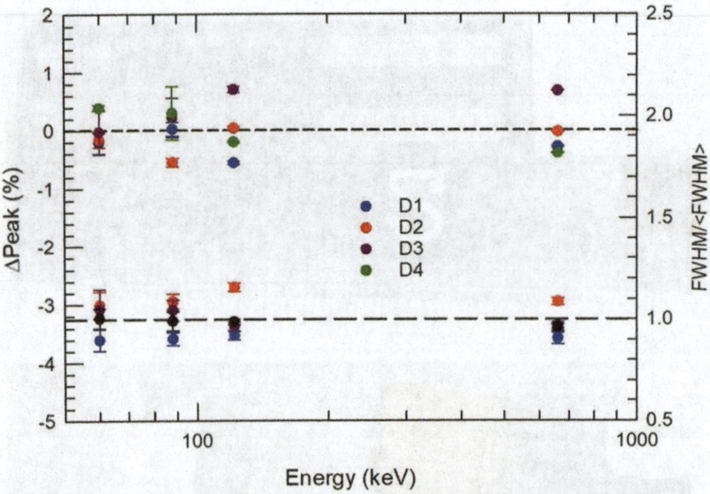

Fig. 4 Summary of the results on the response uniformity of 3D CZT sensors. The graph, for the complete set of 3D CZT sensors, shows the deviation from linearity (Δpeak(%)) in the upper part and, in the lower part, the ratio between the spectroscopic resolution values and the average value over the set (FWHM/<FWHM>) as a function of the source emission line energy. The two dashed lines represent the respective reference values: at zero for the ΔPeak axis and at one for the FWHM/<FWHM> one

function of the energy. The ratio between the spectroscopic resolutions (FWHM) measured at each energy and the average value (<FWHM>) calculated on the four 3D CZT sensors is overall uniform, remaining confined within ±1.2 (Fig. 4, lower data point). These data therefore confirm the quality and repeatability of the techniques developed and used for the realisation of this type of 3D CZT spectrometer.

4 Front-End and Digital Pulse Processing Electronics

The strips from both anode and cathode sides are *A.C.* coupled to hybrid charge-sensitive preamplifiers (CSPs) and processed by 32-channel digital electronics. Both the CSPs and the digital electronics are developed at DiFC of the University of Palermo (Italy). The CSPs are characterised by an equivalent noise charge (ENC) of about 100 electrons (with no detector) and equipped with a resistive feedback circuit with a decay time constant of 150 μs. The digital electronics consists of eight digitizers (DT5724, 16 bit, 100 MS/s, CAEN S.p.A., Italy; http://www.caen.it) driven by an original firmware [1, 11, 28–30]. The data from each channel are transmitted through USB to PCs, where the user can control both acquisition and analysis. Due to the complex analysis required by the different pulse shapes, we performed, at this step, a mixed online/offline pulse processing. For each CSP output channel, the digital system performs online pulse detection and time-tag triggering and provides the PCs with a sequence of selected pulses with the related arrival times (snapshot waveform); therefore, these pulses are analysed offline for dedicated pulse shape and height analysis. The details of the online operations and outputs from each CSP output channel are described below:

(a) Pulse detection and arrival time estimation: the CSP output waveforms are shaped using the classical single-delay line (SDL) shaping technique (Knoll, 2000), acting as the classical differentiation; the trigger time is generated and time-stamped through the ARC (amplitude and rise time compensation) timing marker (at the leading edge of the SDL pulses), able to reduce the effects of time jitters and walks.
(b) Snapshot waveforms (SWs): a sequence of CSP output pulses with the related arrival times are provided to PCs for the offline pulse shape and height analysis; each CSP pulse, selected through a pile-up rejection (PUR), is centred on a time window, the duration of which is termed 'snapshot time' (ST); a pulse is accepted if it is not preceded and not followed by another pulse in the ST/2 time windows.

The SWs are processed offline with a dedicated analysis, taking into account the different features of the pulses. The height (i.e. the photon energy) and the peaking time (i.e. the pulse shape) of the pulses are estimated offline after SDL and trapezoidal shaping.

5 Collecting Induced Charge Pulses and Modelling

The complex geometry of strip electrodes in CZT detectors, along with the various physical processes taking place within them (such as charge sharing and trapping), results in the generation of a multitude of charge pulses with distinct features and shapes. This section provides an overview of the charge pulses that arise in CZT drift-strip detectors, with the aim of comprehending the nature of these pulses and their impact on detector performance through simulations and measurements. This understanding is crucial for optimising pulse processing analysis and devising novel techniques to enhance spectroscopic and spatial performance. Typically, the generation of pulses in CZT detectors can be explained using the Shockley-Ramo theorem, along with the concept of weighting potential, as described by various studies [31–33]. The charge produced on an electrode is related to the variation in weighting potentials between the points of charge generation and collection. Monotonic and non-monotonic weighting potentials give rise to pulses of varying shapes, which depend heavily on the position and trajectory of the charge carriers. Detector pulses can be coarsely classified as either 'collected charge' or 'induced charge' pulses. Collected charge pulses are generated by charge carriers that are finally collected by an electrode, while induced charge pulses are generated on an electrode by charge carriers that are finally collected by another electrode. The movement of charge carriers near the collecting electrode can induce a small charge on the surrounding non-collecting electrodes, known as 'weighting potential crosstalk' [34, 35]; in the literature, these induced charge pulses are also called transient pulses [36, 37]. Ideally, induced charge pulses should have a bipolar amplitude and saturate to a zero-level baseline, indicating that no charge is collected by the electrode. But due to the presence of charge sharing and trapping, induced charge pulses in real detectors often saturate to a non-zero level. Additionally, pulse shapes can contain components arising from both charge collection and weighting potential crosstalk, as described by Bolotnikov et al. [16].

In this section, measured pulses and simulated pulses resulting from a photoelectric interaction at 662 keV photons are compared. The simulation tool developed by Bettelli et al. [38] was used to achieve the simulated pulses. The tool is derived from first-principle simulations and is based on three steps: (i) the computation of radiation-semiconductor interaction using Monte Carlo methods (Geant4), (ii) electric and weighting field calculation through the finite element method (FEM) using COMSOL Multiphysics, and (iii) the calculation of the charge carrier transport and pulse formation in a MATLAB environment. The well-known charge carrier transport properties of Redlen CZT crystals were used for the calculation of carrier drift: μ_e of 1100 $cm^2\ V^{-1}\ s^{-1}$ and τ_e of 11 ms for electrons and μ_h of 88 $cm^2\ V^{-1}\ s^{-1}$ and τ_h of 0.2 ms for holes.

In Fig. 4, an overview of the simulated and measured pulses obtained from anode, cathode, and drift strips is presented. The respective electrode voltages used were 350 for the cathodes, 200 for the central drift strips, and 100 for the adjacent drift strips. The differences in signals coming from photon interactions that occurred near the cathode (top) and the anode (central) are presented; the black circular points indicate the positions of these interactions. The measured pulses shown in the figure were detected in temporal coincidence for each photon interaction. It is important to underline that both the simulations and measurements produced similar results on the key features of the pulses.

In a study by Abbene et al. [11], the various collected and induced charge pulses observed in CZT drift-strip detectors were classified into several pulse classes. These pulse classes include:

- *Collected charge signals*: anode and cathode strips are responsible for generating positive and negative collected charge pulses, respectively. The anode strips' weighting potential, like in pixel detectors, is reduced near the anode to decrease the holes' contribution to the charge pulses. However, single-electron charge sensing is also reduced for photon interactions near the anode, as seen in pixel detectors. The shape of the collecting anode pulses heavily depends on the charge generation point (photon interaction). For interactions near the cathode, the anode pulses have a leading edge with slow and fast slopes, following the anode weighting potential's behaviour. However, for interactions near the anode, only a fast slope features the leading edge, which poses a critical issue in estimating the events' arrival time. Due to the cathode strips' coarse segmentation, their weighting potential spans between the cathode and anode electrodes, producing charge pulses more sensitive to the hole contribution. In contrast, the collected

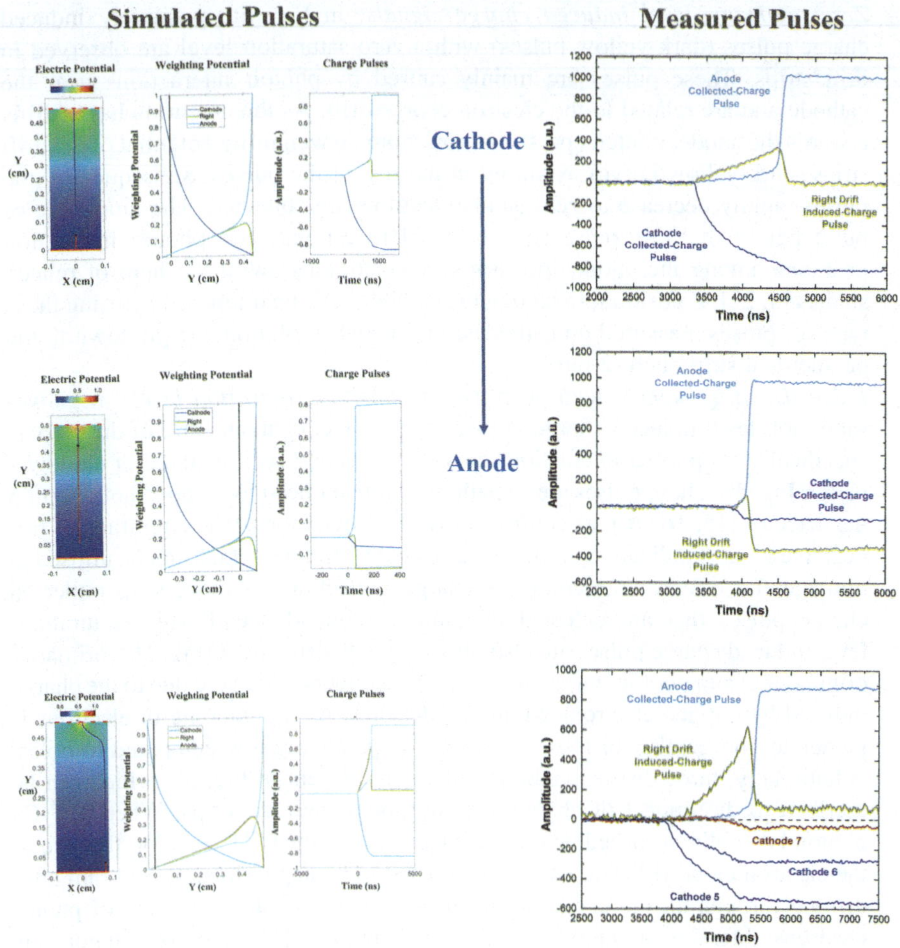

Fig. 5 Comparison of simulated and measured charge pulses achieved by all the electrodes and generated by photons interacting in different detector regions. (**a**) Measured and simulated pulses from anode, cathode, and drift strips. Calculated electric and weighting potential profiles are also reported. Collected and induced charge pulses are clearly visible, related to photon interaction near the cathode (top) and near the anode (bottom). (**b**) Simulated and measured pulses from the anode, cathode, and drift strips. Induced charge pulses with positive saturation levels can be attributed to electron trapping. Mixed induced/collected charge pulses (cathode 7) are also observed due to both charge sharing and weighting potential crosstalk [11]

charge pulses from cathode strips have a well-defined slope of the leading edge (negative blue pulses in Fig. 5a), which benefits arrival time estimation. The peaking times of the pulses can reach maximum values of 2 ms and 3 ms for the anode and cathode, respectively. This necessitates a long shaping time constant of the shaper amplifier to avoid ballistic deficit distortions.

- *Zero-saturation-level induced charge signals*: in Fig. 5a, positively induced charge pulses (dark yellow pulses) with a zero-saturation level are observed in drift strips. These pulses are mainly caused by photon interactions near the cathode and are related to the electron charge drift. As the electron charge drifts towards the anode, it intercepts the non-monotonic weighting potential of the drift strips (dark yellow curve), resulting in the rise of the pulses. Subsequently, the pulses rapidly decrease to zero as the electrons are collected. The amplitude or pulse height of these pulses can be used to estimate the photon interaction positions among the anode/drift strips (x positioning) with the help of related studies [4, 23]. Z positioning among the cathodes can be improved by using these types of pulses measured on cathodes; the spatial resolution can go down to the geometrical strip pitch (2 mm).
- *Induced charge signals with negative and positive saturation levels*: negative-saturation-level induced charge pulses are a distinct characteristic of drift strips, specifically for photon interactions close to the anode, as illustrated in the lower part of Fig. 5a. These pulses are a result of charge induced by trapped holes within the detector [15, 16, 36], as confirmed by our simulations. The amplitude of this negative level is indicative of the degree of hole trapping, which can be utilised to mitigate the effects of incomplete charge collection in anode-strip collected charge pulses that are detected in temporal coincidence. Positive-saturation-level induced charge pulses are also observed with drift strips (Fig. 5b), primarily originating from photon interactions near the cathode, which are due to the charge induced by trapped electrons within the detector. As the trapping of electrons is moderate, the number of these pulses is comparatively low compared to holes. Additionally, three distinct cathode pulses are present in Fig. 5b, which can be ascribed to the impact of charge sharing among cathode strips. When a ^{137}Cs photon is wholly absorbed near the cathode side of the CZT detector, the size of the electron cloud (FWHM) can be estimated to be approximately 350 mm after drifting throughout the crystal's thickness, which includes the range of photo-electrons, diffusion, and repulsion [37–39]. The cathode 7 pulse (brown pulse) is an example of mixed induced/collected charge pulses, which arise from both charge sharing and weighting potential crosstalk. Fortunately, these pulses are absent from collecting anode strips, thanks to the focusing effects of drift strips. In pixel detectors, mixed induced/collected charge pulses are common [16, 36] and pose a critical issue in determining pulse heights (photon energies).

6 Spectroscopic Response of the Detectors

The spectroscopic response of the detectors was investigated by using X-ray and gamma-ray calibration sources (main gamma lines: ^{241}Am, 59.5 and 26.3 keV; ^{57}Co, 122.1 and 136.5 keV; ^{137}Cs: 661.7 keV). The 3D working configuration of the detectors foresees the planar transverse field (PTF) irradiation geometry (X-Y plane), exploiting the 20 mm thickness of CZT material in the Z direction. All measurements were performed at $T = 25$ °C. The detectors show interesting room-

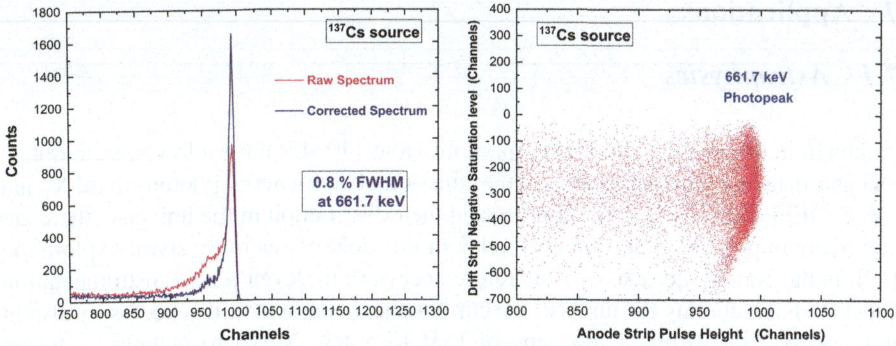

Fig. 6 (Right) Two-dimensional (2D) scatter plot of the negative saturation level of the pulses from a drift strip versus the height of pulses from an anode collecting strip. The curvature highlights the presence of pulses with incomplete charge collection related to high values of the negative saturation level. (Left) Measured ^{137}Cs energy spectra after spectral correction (violet line). The events in the tailing of the photopeak, i.e. characterised by incomplete charge collection, are well detected and corrected

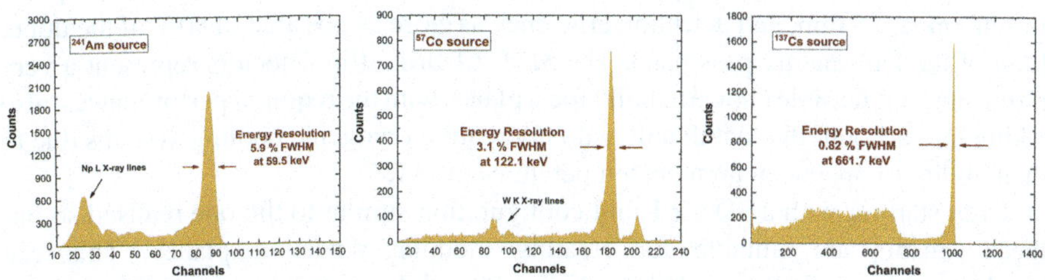

Fig. 7 Uncollimated ^{241}Am (left), ^{57}Co (centre), and ^{137}Cs (right) energy spectra measured with CZT drift-strip detectors. The excellent energy resolution was obtained after the application of a novel correction technique using negative saturation levels from the drift strips [11]

temperature spectroscopic performance even without the application of spectral correction techniques; in particular, the raw energy spectra (i.e. with no spectral correction) are characterised by energy resolution FWHM of 5.0% and 1.3% at 122.1 keV and 661.7 keV, respectively. It is possible to improve the energy resolution through the relation between the height (i.e. the energy) of the pulses of the investigated collecting anode strips and some features of the pulses from the cathode and drift strips. For example, it is possible to use the height or the peaking time of the cathode pulses, the peaking time of the collecting anode pulses, and the negative heights (i.e. the negative saturation levels) of the pulses from the drift strips. The best results are obtained by using negative saturation levels from the drift-strip pulses, as shown in Fig. 6. Figure 7 shows the energy spectra from uncollimated radiation sources (^{241}Am, ^{57}Co, ^{137}Cs), measured through a collecting anode after spectral correction using negative saturation levels from the drift strips. The detectors are characterised by excellent energy resolution at room temperature (5.9% FWHM at 59.5 keV; 3.1% FWHM at 122.1 keV; 0.82% FWHM at 661.7 keV).

7 Applications

7.1 Astrophysics

Although in recent decades various space missions [40–42] have allowed us to obtain valuable data on the characteristics of the emission of high-energy photons (hard X- and soft #xDEFE-rays) by celestial sources and their distribution in the universe, there are still many important open problems today in this field of scientific space exploration [43]. In the coming decades, it is therefore necessary to develop new instrumentation capable of overcoming the limits of the current one; in particular, it is necessary also in this energy band (from a few tens of keV to a few MeV) to achieve a drastic improvement in sensitivity and angular resolution, i.e. several orders of magnitude in sensitivity and angular resolution, and, further, to attain high polarimetric capability. High-energy focusing techniques, in particular broadband Laue lenses, are a challenging answer to these requirements. Even if these optics can be applied to energies up to several MeV, a particularly interesting band is that of hard and soft gamma X-rays between 50 and 600 keV [44]. However, this new type of 'telescope' requires focal plane detectors with very high performance: $\Delta E/E$ (FWHM) $< 1\%$ 500 keV, 3D spatial resolution < 0.5 mm, and detection efficiency as close as possible to 80% of the upper limit of the lens energy pass band. The 3D CZT drift-strip detectors represent a very promising and feasible choice to build focal planes with the required performances, also taking into account the fact that they do not require cryogenic cooling systems due to their ability to operate at ambient temperatures (300 K).

Even starting with a 3D CZT unit configuration similar to the one realised so far, these sensors are suitable for creating modular detection planes. This can be obtained by packaging side by side linear modules obtained assembling two 3D CZT units facing each other, as shown in Fig. 8 (left). These planes can then be stacked as presented in Fig. 8 (right) to achieve the desired thicknesses to ensure the required detection efficiency at the upper energy band limit. Such a configuration is the baseline of an ongoing study on a space mission concept with a telescope based on a Laue lens with a 50–600 keV bandwidth and a 20 m focal length [45].

The study carried out using Monte Carlo simulations based on GEANT4 and the MEGALib package [46] has allowed us to establish that by adopting a configuration with a focal plane consisting of a stack of four detection planes (as in Fig. 8, right), it almost satisfies the efficiency requirement (Fig. 9, left), and the entire telescope can guarantee a continuum sensitivity of at least a factor of 100, better than the instrumentation currently operating in the same band, and an angular resolution of 10 arcsec, with the advantage of allowing simultaneous measurements in spectroscopy, imaging, and polarimetry [47].

Thanks to the expected performance, this type of instrumentation for space astrophysics would allow polarimetry to become a standard mode of observation, allowing it to be extended at higher energies than those (up to 10 keV) where NASA's IXPE mission [48, 49] is currently achieving outstanding scientific results and in an energy band (hard X and soft gamma) where the theoretical models predict polarised emission from many classes of celestial sources.

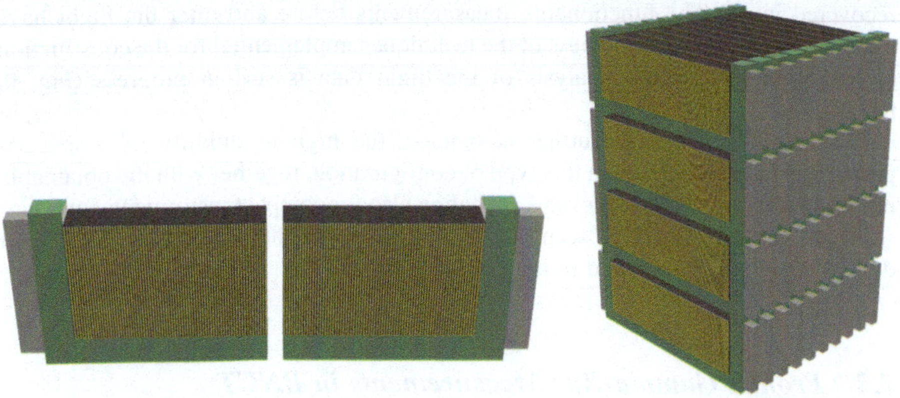

Fig. 8 A high-energy Laue lens focal plane schematic made of 3D CZT sensors based on $30 \times 20 \times 6$ mm^3 crystals: (left) the basic linear module of two 3D CZT sensor units facing each other on the free structure side; (right) the schematic of final 3D CZT detector assembly made of a stack of planes built packaging side by side the linear modules. Each plane of the stack is a mosaic of 2×10 3D CZT units. By the extension of the electrode configuration implemented in the realised 3D CZT sensors, each sensitive unit will need 31 read-out channels, for a total of only 2480 in overall stack

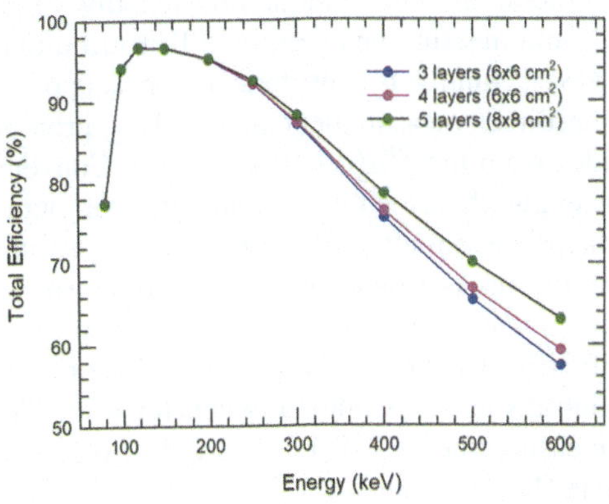

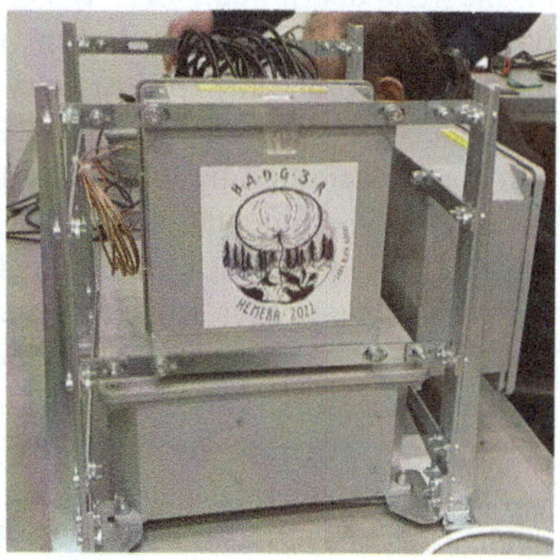

Fig. 9 (Left) The total detection efficiency evaluated by Monte Carlo simulation using a 3D CZT focal plane model, as depicted in the previous figure for different numbers of layers in the stack and different lateral sizes; (right) the BADG3R (balloon detector for gamma rays with 3D resolution) payload in the flight configuration before its integration in the balloon gondola at the SSC ESRANGE base in Kiruna (Sweden). In the foreground is the detection box containing the 3D CZT sensor and its preamplifier read-out electronics

In view of the use of 3D CZT sensors in space, and with the aim of verifying their reliability and the versatility of a digital approach to reading signals in a pseudo-space environment, a small payload was built for a stratospheric balloon mission based on a single 3D CZT unit. The prototype (BADG3R) was successfully flown from the SSC/ESRANGE base in Kiruna (Sweden) in autumn 2022 and was recovered intact. The functionality measurements before and after the flight have already confirmed the robustness of the techniques implemented for the construction of the sensor, while the analysis of the flight data is still in progress (Fig. 9, right) [50].

To conclude and as a further advantage, the high modularity of a spectro-polarimetric imager based on this type of configuration, together with the obtainable performances, makes these devices appealing also for compact instruments for nano-satellites in cluster imaging scenarios, which is one of the sectors that have rapid development also in the field of high energies [51].

7.2 Prompt Gamma-Ray Measurements in BNCT

Boron neutron capture therapy (BNCT) is a radiation therapy exploiting the high probability of ^{10}B for capturing thermal neutrons through the nuclear reaction ^{10}B(n, a)^{7}Li. The products of this reaction (^{7}Li nuclides and alpha particles) are characterised by high linear energy transfer (LET) of about 150 keV/μm for alpha particles and 175 keV/μm for ^{7}Li nuclei, with important potentialities in destroying cancer cells. The optimisation of the treatment effects in BNCT requires knowledge of the real distribution of ^{10}B concentration in both tumour and healthy cells, possibly during irradiation. For the real-time measurements of the ^{10}B distribution, the prompt gamma-ray spectroscopy (PGS) technique is considered the most appealing procedure. The key operation is the direct measurement of 478 keV prompt gamma rays emitted by 94% ^{7}Li nuclides from the ^{10}B(n, a)^{7}Li reaction. Conventional PGS instrumentation is based on single-photon emission computed tomography (SPECT) systems, equipped with the proper energy resolution to well detect the 478 keV gamma rays from the gamma-ray background created by various (n, g) reactions and annihilation phenomena. In the framework of the 3CaTS project (funded by INFN, Italy), 3D CZT drift-strip detectors were proposed for prompt gamma-ray spectroscopy in BNCT. High-resolution gamma-ray spectra from ^{10}B samples under thermal neutrons were measured at the T.R.I.G.A. Mark II research nuclear reactor of the University of Pavia (Italy), as shown in Fig. 10.

The 478 keV and 558 keV (gamma rays from ^{113}Cd (n, g)^{114}Cd neutron capture reaction) peaks are well detected and resolved, with important spectral improvements from the previous results obtained using other CZT detector prototypes [52]. The high energy resolution of the detectors also allows them to determine the expected different shapes between the 478 keV and the 558 keV peaks, based on the Doppler effect, since the prompt gamma rays are emitted in flight by ^{7}Li. In fact, the flat top of the 478 keV peak clearly highlights the effects of the Doppler broadening. To our knowledge, this is the first measurement performed with room-temperature CdTe/CZT detectors that clearly highlights this effect; generally, these different peak shapes are masked by the poor energy resolution of the detectors, and they are typically measured with high-resolution cryogenic cooled germanium

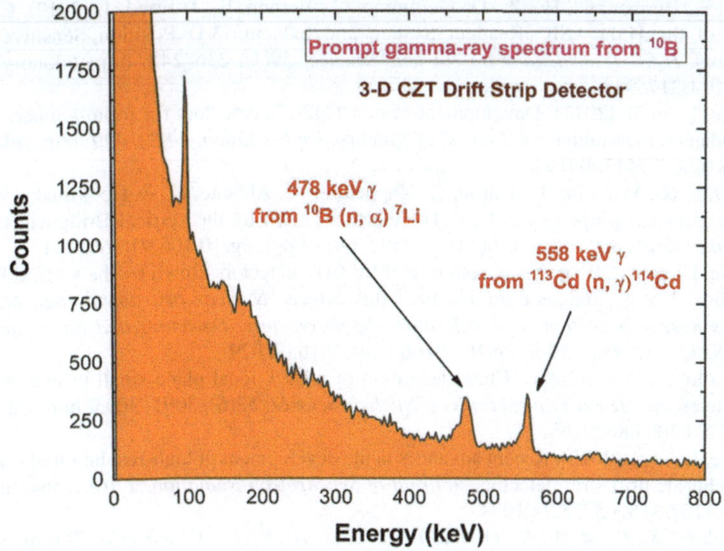

Fig. 10 Gamma-ray energy spectrum measured with CZT drift-strip detectors in a BNCT environment. The flat top of the 478 keV peak highlights the Doppler broadening effect

detectors. These measurements confirm the capabilities of these new CZT detectors to discriminate properly the ^{10}B signal in BNCT despite the close proximity of the activation peak of ^{113}Cd.

References

1. Abbene, L., Gerardi, G., Raso, G., Basile, S., Brai, M., & Principato, F. (2013). Energy resolution and throughput of a new real time digital pulse processing system for x-ray and gamma ray semiconductor detectors. *Journal of Instrumentation, 8*(07), P07019. https://doi.org/10.1088/1748-0221/8/07/P07019
2. Butler, J. F., Lingren, C. L., & Doty, F. P. (1992). Cd $_{1-x}$Zn$_x$Te gamma ray detectors. *IEEE Transactions on Nuclear Science, 39*(4), 605–609. https://doi.org/10.1109/23.159673
3. Caroli, E., et al. (2010). Development of a 3D CZT detector prototype for Laue lens telescope. In *High energy, optical, and infrared detectors for astronomy IV*. SPIE Proc. Vol. 7742, 77420V. https://doi.org/10.1117/12.858120
4. Budtz-Jørgensen, C., & Kuvvetli, I. (2017). New position algorithms for the 3-D CZT drift detector. *IEEE Transactions on Nuclear Science, 64*(6), 1611–1618. https://doi.org/10.1109/TNS.2017.2696574
5. Bolotnikov, A. E., et al. (2020). A 4×4 array module of position-sensitive virtual Frisch-grid CdZnTe detectors for gamma-ray imaging spectrometers. *Nuclear Instruments and Methods in Physics Research Section A: Accelerators, Spectrometers, Detectors and Associated Equipment, 954*, 161036. https://doi.org/10.1016/j.nima.2018.07.090

6. Zhang, F., Herman, C., He, Z., De Geronimo, G., Vernon, E., & Fried, J. (2012). Characterization of the H3D ASIC Readout System and 6.0 cm^3 3-D Position Sensitive CdZnTe Detectors. *IEEE Transactions on Nuclear Science, 59*(1), 236–242. https://doi.org/10.1109/TNS.2011.2175948
7. Abbene, L., et al. (2017). Development of new CdZnTe detectors for room-temperature high-flux radiation measurements. *Journal of Synchrotron Radiation, 24*(2), 429. https://doi.org/10.1107/S1600577517000194
8. Auricchio, N., Marchini, L., Caroli, E., Zappettini, A., Abbene, L., & Honkimaki, V. (2011). Charge transport properties in CdZnTe detectors grown by the vertical Bridgman technique. *Journal of Applied Physics, 110*(12), 124502. https://doi.org/10.1063/1.3667201
9. Abbene, L., et al. (2016). X-ray response of CdZnTe detectors grown by the vertical Bridgman technique: Energy, temperature and high flux effects. *Nuclear Instruments and Methods in Physics Research Section A: Accelerators, Spectrometers, Detectors and Associated Equipment, 835*, 1–12. https://doi.org/10.1016/j.nima.2016.08.029
10. Del Sordo, S., et al. (2005). Characterization of a CZT focal plane small prototype for hard X-ray telescope. *IEEE Transactions on Nuclear Science, 52*(6), 3091–3095. https://doi.org/10.1109/TNS.2005.862916
11. Abbene, L., et al. (2020). Recent advances in the development of high-resolution 3D cadmium–zinc–telluride drift strip detectors. *Journal of Synchrotron Radiation, 27*(6), 1564. https://doi.org/10.1107/S1600577520010747
12. Bolotnikov, A., Geist, B., & Jeffcoat, R. (2020). *24.1.3.1.1 – Frisch grid CZT spectrometer*. Brookhaven National Lab. (BNL), BNL-219869-2020-INRE. https://doi.org/10.2172/1763316
13. Abbene, L., et al. (2022). Potentialities of high-resolution 3-D CZT drift strip detectors for prompt gamma-ray measurements in BNCT. *Sensors, 22*(4), 1502. https://doi.org/10.3390/s22041502
14. Auricchio, N., et al. (2023). A CZT 3D imaging spectrometer prototype with digital readout for high energy astronomy. *Nuclear Instruments and Methods in Physics Research Section A: Accelerators, Spectrometers, Detectors and Associated Equipment, 1047*, 167869. https://doi.org/10.1016/j.nima.2022.167869
15. Zhu, Y., Anderson, S. E., & He, Z. (2011). Sub-pixel position sensing for pixelated, 3-D position sensitive, wide band-gap, semiconductor, gamma-ray detectors. *IEEE Transactions on Nuclear Science, 58*(3), 1400–1409. https://doi.org/10.1109/TNS.2011.2132738
16. Bolotnikov, A. E., et al. (2014). Comparison of analog and digital readout electronics on the processing of charge-sharing events in pixelated CdZnTe detectors. *IEEE Transactions on Nuclear Science, 61*(2), 787–792. https://doi.org/10.1109/TNS.2014.2299493
17. Gatti, E., & Rehak, P. (1984). Semiconductor drift chamber—An application of a novel charge transport scheme. *Nuclear Instruments and Methods in Physics Research, 225*(3), 608–614. https://doi.org/10.1016/0167-5087(84)90113-3
18. Kuvvetli, I., Budtz-Jørgensen, C., Caroli, E., & Auricchio, N. (2010). CZT drift strip detectors for high energy astrophysics. *Nuclear Instruments and Methods in Physics Research Section A: Accelerators, Spectrometers, Detectors and Associated Equipment, 624*(2), 486–491.
19. Kalemci, E., & Matteson, J. L. (2002). Investigation of charge sharing among electrode strips for a CdZnTe detector. *Nuclear Instruments and Methods in Physics Research Section A: Accelerators, Spectrometers, Detectors and Associated Equipment, 478*(3), 527–537. https://doi.org/10.1016/S0168-9002(01)00892-0
20. Lingren, C. L., et al. (1998). Cadmium-zinc-telluride, multiple-electrode detectors achieve good energy resolution with high sensitivity at room-temperature. *IEEE Transactions on Nuclear Science, 45*(3), 433–437. https://doi.org/10.1109/23.682422
21. Alruhaili, A., et al. (2014). Performance characteristics of CdTe drift ring detector. *Journal of Instrumentation, 9*(3), C03029. https://doi.org/10.1088/1748-0221/9/03/C03029
22. Abbene, L., et al. (2007). Spectroscopic response of a CdZnTe multiple electrode detector. *Nuclear Instruments and Methods in Physics Research Section A: Accelerators, Spectrometers, Detectors and Associated Equipment, 583*(2), 324–331. https://doi.org/10.1016/j.nima.2007.09.015

23. Owe, S. H., Kuvvetli, I., Budtz-JØrgensen, C., & Zoglauer, A. (2019). Evaluation of a Compton camera concept using the 3D CdZnTe drift strip detectors. *Journal of Instrumentation, 14*(01), C01020. https://doi.org/10.1088/1748-0221/14/01/C01020
24. Iniewski, K. (2014). CZT detector technology for medical imaging. *Journal of Instrumentation, 9*(11), C11001. https://doi.org/10.1088/1748-0221/9/11/C11001
25. Chen, H., et al. (2008). Spectral response of THM grown CdZnTe crystals. *IEEE Transactions on Nuclear Science, 55*(3), 1567–1572. https://doi.org/10.1109/TNS.2008.924089
26. Zanettini, S., et al. (2018). Al2O3 coating as passivation layer for CZT-based detectors. In *2018 IEEE nuclear science symposium and medical imaging conference proceedings (NSS/MIC)*. https://doi.org/10.1109/NSSMIC.2018.8824623
27. Benassi, G., Nasi, L., Bettelli, M., Zambelli, N., Calestani, D., & Zappettini, A. (2017). Strong mechanical adhesion of gold electroless contacts on CdZnTe deposited by alcoholic solutions. *Journal of Instrumentation, 12*(02), P02018. https://doi.org/10.1088/1748-0221/12/02/P02018
28. Abbene, L., & Gerardi, G. (2015). High-rate dead-time corrections in a general purpose digital pulse processing system. *Journal of Synchrotron Radiation, 22*(5), 1190. https://doi.org/10.1107/S1600577515013776
29. Abbene, L., et al. (2018). Digital fast pulse shape and height analysis on cadmium–zinc–telluride arrays for high-flux energy-resolved X-ray imaging. *Journal of Synchrotron Radiation, 25*(1), 257. https://doi.org/10.1107/S1600577517015697
30. Gerardi, G., & Abbene, L. (2014). A digital approach for real time high-rate high-resolution radiation measurements. *Nuclear Instruments and Methods in Physics Research Section A: Accelerators, Spectrometers, Detectors and Associated Equipment, 768*, 46–54. https://doi.org/10.1016/j.nima.2014.09.047
31. Shockley, W. (2004). Currents to conductors induced by a moving point charge. *Journal of Applied Physics, 9*(10), 635–636. https://doi.org/10.1063/1.1710367
32. Ramo, S. (1939). Currents induced by electron motion. *Proceedings of the IRE, 27*(9), 584–585. https://doi.org/10.1109/JRPROC.1939.228757
33. He, Z. (2001). Review of the Shockley–Ramo theorem and its application in semiconductor gamma-ray detectors. *Nuclear Instruments and Methods in Physics Research Section A: Accelerators, Spectrometers, Detectors and Associated Equipment, 463*(1), 250–267. https://doi.org/10.1016/S0168-9002(01)00223-6
34. Guerra, P., Santos, A., & Darambara, D. G. (2008). Development of a simplified simulation model for performance characterization of a pixellated CdZnTe multimodality imaging system. *Physics in Medicine and Biology, 53*(4), 1099. https://doi.org/10.1088/0031-9155/53/4/019
35. Brambilla, A., Ouvrier-Buffet, P., Rinkel, J., Gonon, G., Boudou, C., & Verger, L. (2012). CdTe linear pixel X-ray detector with enhanced spectrometric performance for high flux X-ray Imaging. *IEEE Transactions on Nuclear Science, 59*(4), 1552–1558. https://doi.org/10.1109/TNS.2012.2206828
36. Bolotnikov, A. E., et al. (2016). Use of high-granularity CdZnTe pixelated detectors to correct response non-uniformities caused by defects in crystals. *Nuclear Instruments and Methods in Physics Research Section A: Accelerators, Spectrometers, Detectors and Associated Equipment, 805*, 41–54. https://doi.org/10.1016/j.nima.2015.08.051
37. Kim, J. C., et al. (2011). Charge sharing in common-grid pixelated CdZnTe detectors. *Nuclear Instruments and Methods in Physics Research Section A: Accelerators, Spectrometers, Detectors and Associated Equipment, 654*(1), 233–243. https://doi.org/10.1016/j.nima.2011.06.038
38. Bettelli, M., et al. (2020). A first principle method to simulate the spectral response of CdZnTe-based X- and gamma-ray detectors. *Nuclear Instruments and Methods in Physics Research Section A: Accelerators, Spectrometers, Detectors and Associated Equipment, 960*, 163663. https://doi.org/10.1201/9781003147633
39. Bolotnikov, A., Camarda, G., Carini, G., Cui, Y., Li, L., & James, R. (2007). Modeling the geometrical effects of Te precipitates on electron transport in CdZnTe. *Nuclear Instruments and Methods A, 579*, 125. Art. fasc. BNL-80682-2008-JA. https://doi.org/10.1016/j.nima.2007.04.022

40. Gehrels, N., et al. (2004). The swift gamma-ray burst mission. *Astrophysical Journal, 611*(2), 1005. https://doi.org/10.1086/422091
41. Winkler, C., et al. (2003). The INTEGRAL mission. *A&A, 411*(1), L1. https://doi.org/10.1051/0004-6361:20031288
42. Harrison, F. A., et al. (2013). The nuclear spectroscopic telescope array (nustar) high-energy x-ray mission. *Astrophysical Journal, 770*(2), 103. https://doi.org/10.1088/0004-637X/770/2/103
43. Frontera, F., et al. (2021). Understanding the origin of the positron annihilation line and the physics of supernova explosions. *Experimental Astronomy, 51*(3), 1175–1202. https://doi.org/10.1007/s10686-021-09727-7
44. Frontera, F., & Von Ballmoos, P. (2010). Laue gamma-ray lenses for space astrophysics: Status and prospects. *X-Ray Optics and Instrumentation, 2010*, 1–18. https://doi.org/10.1155/2010/215375
45. Frontera, F., et al. ASTENA, a new mission concept for an advanced surveyor of transient events and nuclear astrophysics. *Societa Astronomica Italiana. Memorie, 90*, 247–251.
46. Zoglauer, A., Andritschke, R., & Schopper, F. (2006). MEGAlib – The Medium Energy Gamma-ray Astronomy Library. *New Astronomy Reviews, 50*(7), 629–632. https://doi.org/10.1016/j.newar.2006.06.049
47. Caroli, E., et al. (2009). A polarimetric experiment with a laue lens and CZT pixel detector. *IEEE Transactions on Nuclear Science, 56*(4), 1848–1854. https://doi.org/10.1109/TNS.2009.2021475
48. Weisskopf, M., et al. (2022). Imaging X-ray Polarimetry Explorer: prelaunch. *Journal of Astronomical Telescopes Instruments and Systems, 8*(02). https://doi.org/10.1117/1.JATIS.8.2.026002
49. Soffitta, P., et al. (2023). *The Imaging x-ray polarimetry explorer (IXPE) at last! in* UV, X-ray, and gamma-ray space instrumentation for Astronomy XXIII. *Proceedings of SPIE, 12678*, 1267803. https://doi.org/10.1117/12.2677296
50. Del Sordo, S., et al. (2022). *The BADG3R prototype on board of a stratospheric balloon flight in the framework of the H2020/HEMERA program.* Proc. of the 25th ESA Symposium on European Rocket and Balloon Programmes and related Research, ESA SP-743 (ISBN 978-92-9221-308-4 44)
51. Fiore, F., Guzman, A., Campana, R., & Evangelista, Y. (2022). Springer. *HERMES-Pathfinder*. https://doi.org/10.1007/978-981-19-6960-7_35
52. Fatemi, S., et al. (2018). High performance 3D CZT spectro-imager for BNCT-SPECT: Preliminary characterization. In *2018 IEEE nuclear science symposium and medical imaging conference proceedings (NSS/MIC)*. https://doi.org/10.1109/NSSMIC.2018.8824381

Compton Imaging Systems Based on CdZnTe/CdTe Detectors

Mostafa Niknami, Seyed Abolfazl Hosseini, and Mahdi Valipour

1 Introduction

Nuclear imaging technologies have extensive applications across detection, diagnosis, therapy, and other domains. Within nuclear imaging, two primary methodologies emerge, categorized by the type of radioactive material employed. Single-photon emission computed tomography (SPECT) and the Compton imaging system are notable examples of single-photon imaging, focusing on radioactive nuclei's gamma-ray emissions [1, 2]. Another technique, recognized as Positron emission tomography (PET), utilizes radioisotopes that emit positrons as the radioactive material [3]. Due to their limitations, electronic collimators are replacing mechanical ones in modern nuclear imaging. SPECT uses mechanical collimators to detect gamma-ray direction, while PET detects the simultaneous emission of two 511 keV gamma rays from positron-electron annihilation [4]. In contrast, the Compton camera is an imaging system based on an electronic collimator [5]. It allows multinuclide imaging without a mechanical collimator, and several researchers have achieved in vivo multitracer imaging [6–8].

The Compton imaging method, developed in the mid-twentieth century, evolved into the Compton telescope and Compton spectrometer in the early 1960s [9–12]. Used in astronomy, COMPTEL was the first to orbit the Compton telescope [13–19]. Other Compton telescope models, including the nuclear Compton telescope (NCT), a balloon-borne Compton telescope (now Compton spectrometer and imager

M. Niknami (✉)
School of Cognitive Sciences, Institute for Research in Fundamental Sciences, Tehran, Iran

Department of Energy Engineering, Sharif University of Technology, Tehran, Iran
e-mail: mostafaniknami@ipm.ir; mostafa.niknami13@sharif.edu

S. A. Hosseini · M. Valipour
Department of Energy Engineering, Sharif University of Technology, Tehran, Iran
e-mail: sahosseini@sharif.edu; Mehdi.valipour@energy.sharif.ir

K. Iniewski (ed.), *CdTe and CdZnTe Materials*,
https://doi.org/10.1007/978-3-031-64521-1_10

(COSI)), were developed to overcome limitations [20, 21]. Compton cameras, valued for their detection capabilities, are also advancing astrophysical applications [22].

However, the application of Compton imaging expanded into other fields, which was initially suggested by Everett and colleagues for nuclear medical imaging [23]. In 1983, Singh et al. developed a prototype using semiconductor and thallium-doped sodium iodide (NaI(TL)) detectors [24]. Subsequent research enabled the imaging of three-dimensional (3D) cylindrical test phantoms [25]. Silicon detectors, notably stacks of silicon pad or silicon drift detectors, gained popularity in Compton cameras due to their enhanced efficiency in the following years [26–28]. Compton cameras simultaneously imaged multiple molecular targets within a living mouse [29].

Additionally, researchers proposed their utilization for treating radioisotopes with alpha particles [30–32]. A hybrid system, combining a pinhole and a Compton camera, actively detected X-rays and gamma rays within the same application region [33]. Researchers utilized both Compton and PET cameras in separate studies for imaging purposes [34–36]. Comparative analyses between Compton cameras, PET, and SPECT demonstrated the former's superior economic effectiveness and performance in medical applications across various investigations [37–39]. Additionally, several research groups explored the application of Compton cameras in ion beam therapy, employing scintillation and semiconductor detectors [40–42].

Organizations use Compton cameras to ensure safety and security against disasters or contaminations resulting from human or natural events. They commonly deploy portable Compton cameras to measure radiation levels in affected areas [43, 44]. Hmissi et al. evaluated a portable Compton camera with cerium bromide ($CeBr_3$) scintillation crystals and digital photon counter detectors, demonstrating acceptable energy and angular resolutions [45]. Gamma-ray detection aids in identifying and characterizing material assemblies [46]. In Fukushima, researchers developed and tested a Compton camera to reduce noise in gamma-ray images by incorporating a lead (Pb) layer into the rear panel [47]. Researchers tested a lightweight Compton camera mounted on a drone system in a Fukushima area divided into radiation zones [48]. They also explored an industrial-grade Compton camera with a ring-shaped NaI(TL) absorber detector, albeit with reduced efficiency [49].

The materials in scatterer and absorber detectors are crucial for achieving high-energy resolution in Compton imaging detectors as they directly impact spatial resolution. Semiconductor detector devices used in Compton cameras include several materials such as germanium (Ge), silicon (Si), cadmium telluride (CdTe), and cadmium zinc telluride (CdZnTe) (CZT) [50–53]. Scintillator materials like cerium-doped gadolinium gallium aluminum garnet (GAGG), cerium-doped lanthanum bromide ($LaBr_3$), cerium bromide ($CeBr_3$), and cerium-doped gadolinium orthosilicate (GSO) have also been utilized [54–57].

Considering the criticality of Compton cameras and their diverse applications, along with the significance of the detector material type discussed earlier in this section, the subsequent segment will delve into the foundational principles and influential parameters governing Compton cameras' functionality.

2 Compton Imaging System

Compton imaging, utilizing Compton scattering, produces gamma-ray images. These systems, known as Compton cameras, offer comprehensive energy coverage, construct 3D images without detector movement, and effectively reduce background radiation during scattering. Compton imaging systems feature two detectors (see Fig. 1) sensitive to scattered gamma rays' position and energy. These detectors operate in a coincidence mode, electronically detecting photons without path restrictions. Optimization aims to enhance Compton scattering probability in the prompt gamma energy range while minimizing Doppler broadening effects due to electron binding and motion [58]. The absorber detector identifies Compton-scattered photons using the photoelectric effect. In the Compton camera, the scattering angle defines the cone's opening angle. The cone axis extends perpendicularly from the circular base, which traces the initial gamma ray's path, with the first scatterer interaction determining the apex [24]. Correct events are those accurately reconstructed based on known interaction locations and scattering angles. Due to potential degeneracies, multiple cones may intersect at the emission point, requiring analysis to identify the correct events consistent with unique gamma-ray interactions. Equation 1 describes the Compton scattering angle θ [59]:

$$\theta = \cos^{-1}\left[1 - m_0c^2\left(\frac{1}{E_1} - \frac{1}{E_0}\right)\right] \tag{1}$$

where E_0 is the incoming gamma-ray energy, E_1 is the scattered gamma-ray energy immediately after contact, and m_0c^2 is the electron's rest mass energy.

Based on the scattered electron energy (E_{re}), Eq. 1 can be rewritten as Eq. 2:

$$\cos\theta = 1 - \left(\frac{E_{re}}{\frac{E_0}{m_0c^2}(E_0 - E_{re})}\right) \tag{2}$$

Critical parameters in Compton cameras include efficiency, energy, spatial, and angular resolution [60]. The following section will discuss these parameters further.

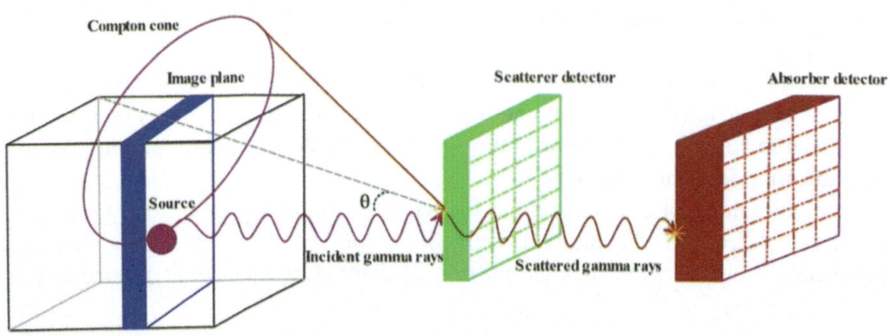

Fig. 1 Principles of the Compton camera

2.1 *Effective Parameters on the Performance of Compton Cameras*

The performance of Compton cameras hinges on spatial resolution and sensitivity to detect photons. Precision in identifying photon scatter angle determines the characteristics of the conic surface derived from tracing the emitted photon's original direction. The crucial factors defining Compton cameras' performance include angular, energy, and spatial resolution, and efficiency [61].

Angular resolution in Compton cameras, representing uncertainty in detecting photon incident direction, is affected by energy, spatial resolution, and detector geometry. Equation 3 quantifies these uncertainties [24, 53]:

$$\tan^2(\Delta\theta_{\text{overall}}) = \tan^2(\Delta\theta_1) + \tan^2(\Delta\theta_2) + \tan^2(\Delta\theta_3) + \tan^2(\Delta\theta_4) \quad (3)$$

where $\Delta\theta_1$ is related to energy resolution in the scatterer detector, $\Delta\theta_2$ and $\Delta\theta_3$ are related to geometry, and $\Delta\theta_4$ is related to spatial resolution.

A detector response function, known as the pulse height distribution, describes how the detector responds to a radiation source. The detector's energy resolution is determined by dividing the photopeak's full width at half maximum (FWHM) by the peak centroid [62]. A low energy resolution from the scatterer detector leads to higher energy uncertainty, impacting angular resolution and imaging reconstruction. Angular uncertainty decreases as gamma-ray energy increases, enhancing angular resolution in Compton cameras at medium to high energies [63]. Improved angular resolution correlates with the energy resolution of the scatterer detector. Energy resolutions for various materials were estimated using mathematical models from existing literature [28, 53, 64].

Doppler broadening affects Compton cameras' angular resolution. At the same time, the Klein-Nishina equation describes scattering angle distribution based on photon-electron collisions; interactions with moving electrons bound to atoms introduce complexities. The equation fails to accurately represent scattered electron energy distribution at low energy deposits or higher atomic numbers. Instead, a scattering angle distribution is centered around the central angle (see Fig. 2). This challenges unique angle determination based on stored energy, necessitating a consideration of a Lorentz distribution. Higher photon energies or lower atomic numbers reduce angular uncertainty, increasing Doppler broadening for larger angles [65].

Detector geometry in the Compton camera is crucial, affecting spatial and scattering angle uncertainty. Angular uncertainty arising from geometry contribution decreases inversely proportional to the square of the distance between the scatterer and absorber detectors [24, 63].

Radiation detector efficiency relies on geometric and intrinsic factors. Geometric efficiency represents the ratio of emitted photons to detected signals. Intrinsic efficiency, however, describes the portion of gamma rays reaching a detector that

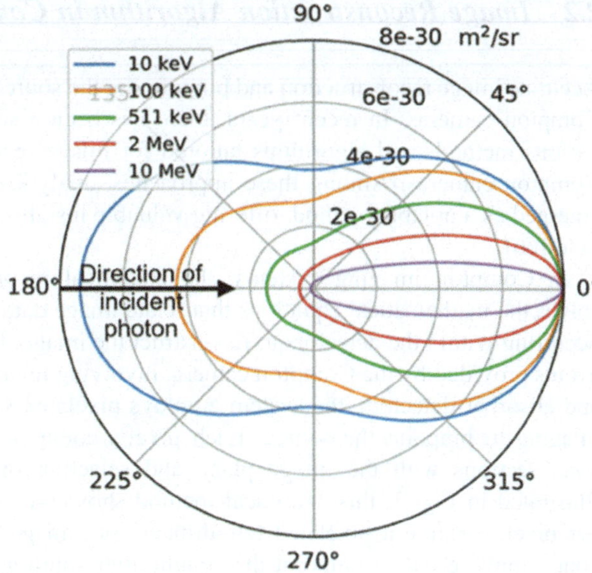

Fig. 2 Klein-Nishina differential cross-sections as a function of the scattering angle and incident photon energies. (The figure is reprinted from the work of Kierans, C., et al. [65])

produces valuable signals [62]. The intrinsic efficiency of a Compton camera refers explicitly to the fraction of photons entering the initial detector undergoing a single Compton scattering and subsequent photoelectric absorption in the second detector, as shown in Eq. 4 below [53]:

$$\varepsilon_{icc} = \frac{1}{\Omega_1} \int_{\text{det1}} dv_1 \int_{\text{det2}} dv_2 \Delta\Omega_1 \mu_c(E_0) e^{(-\mu_t(E_0)L_1)} \frac{d\sigma}{d\Omega} \times (\theta_{12}) \Delta\Omega_2 e^{(-\mu_t(E_0)L_{12})} \mu_p(E_1) \tag{4}$$

where $\Omega_1$is the solid angle subtended by the scatterer detector at the source; $\Delta\Omega_1$ is the solid angle for a differential volume element dv_1 in the scatterer detector, Ω_2 is the solid angle subtended by the absorber detector at the interaction position in the scatterer detector; L_1 is the attenuation length in the scatterer detector between the source and dv_1; L_{12} is the attenuation length between the two elements dv_1 and dv_2; μ_c, μ_p, and μ_t are the Compton, photoelectric, and total linear attenuation coefficients, respectively; and $\frac{d\sigma}{d\Omega}$ is the Klein-Nishina Compton scattering angular cross-section per electron.

As can be seen from Eq. 4, efficiency depends on the incoming photon energy, geometry, system, and detectors. The probability of interactions, the deposited energy in detectors, and the relation between irradiated photons are all stochastic processes. Therefore, calculating efficiency is more complicated than it seems.

2.2 *Image Reconstruction Algorithm in Compton Cameras*

Accurate image reconstruction and pinpointing the source location are paramount for Compton cameras. In recent years, there has been a surge in the development of various methods and algorithms tailored for image reconstruction in the realm of Compton cameras. Among these approaches, analytical image reconstruction has emerged as a notable method, offering valuable insights and applications within the field [66].

In Compton imaging systems, the analytical reconstruction method directly solves the mathematical equations that relate image data to the physical interactions occurring within the detectors to reconstruct the image. Utilizing a list of coincident events provided by the Compton camera, involving interactions in both the scatterer and absorber detectors, the system employs pixelated screens for two-dimensional imaging to pinpoint the source. Each pixel's count is determined by intersecting cone sections with the image plane and selecting only the correct events. As illustrated in Fig. 3, this analytical method showcases the number of cone passes per pixel, creating a pixelated two-dimensional image of four events. When four cones converge, they pinpoint the sought-after solution at their intersection point. However, this method's reconstructed images must be more apparent in practice [67].

Many other methods rely on iterative algorithms for image reconstruction [68]. In the early 1980s, researchers proposed a two-step iterative algorithm for Compton camera imaging. It first reconstructed a 2D projection image and then used it to reconstruct a 3D image [24]. Approximately in the year 2000, a method called Compton list-mode maximum likelihood expectation maximization (LM-MLEM) was introduced. This approach processes each measurement event individually without the need for histogramming, enabling the direct reconstruction of a 3D image (see Fig. 4) [69, 70]. Afterward, system development and simulation investigations have extensively utilized maximum likelihood (ML)-based image reconstruction techniques [71–73].

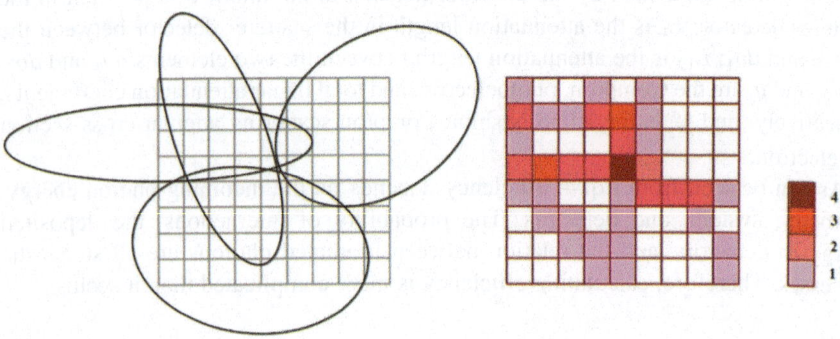

Fig. 3 Analytical image reconstruction for four events in a Compton camera

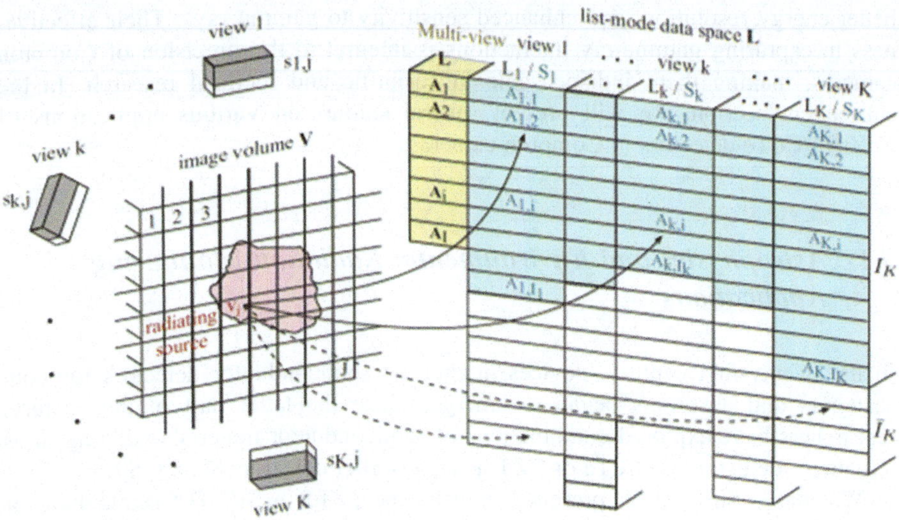

Fig. 4 LM-MLEM reconstruction 3D in a Compton camera. (The figure is reprinted from the work of Le, N., et al. [70])

An accelerated version, Compton list-mode ordered subset expectation maximization (LM-OSEM), divides events into subsets for faster convergence. Methods to boost Compton image quality involve integrating prior knowledge into reconstruction algorithms, enhancing the system matrix with considerations for blurring effects like energy resolution and Doppler broadening, and minimizing errors through correction techniques [74–81].

The stochastic origin ensemble (SOE) method is a popular iterative Compton image reconstruction technique. It utilizes the Metropolis-Hastings algorithm to estimate event origins by randomly shifting positions on the response-function distribution. This stochastic approach aims to reach an equilibrium state where the radioactivity distribution image is derived by normalizing the event set with detection probabilities at each location [82–86].

High computational complexity in ML-based and SOE methods, especially with high-precision system matrices, often leads to using graphics processing units for Compton image reconstruction [57, 84]. Research also leverages deep learning for the Compton camera image reconstruction [87].

3 Application of CdZnTe/CdTe Detectors in Compton Cameras

Due to their high energy resolution capacity, semiconductor detectors are anticipated to deliver excellent performance in Compton imaging. CdZnTe/CdTe detectors play a pivotal role in advanced Compton cameras, offering higher detection efficiency, better energy resolution, and enhanced sensitivity to gamma rays. Their effectiveness in capturing gamma-ray interactions is integral to the precision of Compton cameras, making them indispensable in scientific and medical research. In the subsequent section, we will present several studies on various applications of CdZnTe/CdTe detectors in Compton cameras.

3.1 Astrophysics and Environmental Radiation Monitoring Applications

In the twenty-first century, devices integrating semiconductor detectors for both scatterers and absorbers began to emerge. Since the latter half of the century, researchers have explored combinations of semiconductor materials utilizing Si as a scatterer detector and CdTe or CZT as an absorber detector [80, 88, 89].

Watanabe, Shin, et al. present a novel concept for a Si/CdTe semiconductor Compton camera, utilizing Si-strip detectors as scatterers and CdTe pixel detectors as absorber detectors. This camera is designed for gamma-ray photon detection ranging from tens of keV to several MeV. This research discusses the results of a prototype, evaluates our series of CdTe pixel detectors, and showcases their effective performance. The prototype comprises six deconvolutional single-shot detector (DSSD) layers and three CdTe detectors, demonstrating its potential for the next generation of Compton cameras (see Fig. 5) [90].

Subramanian, Mythili, et al. evaluated a Compton telescope with Si and CZT detectors, serving as a prototype for the Silicon Tower for an Advanced Compton Telescope (STACT) and the hard X-ray polarimeter (HX-POL). The Compton camera uses a thick double-sided Si-strip scatterer detector and a pixelated CZT

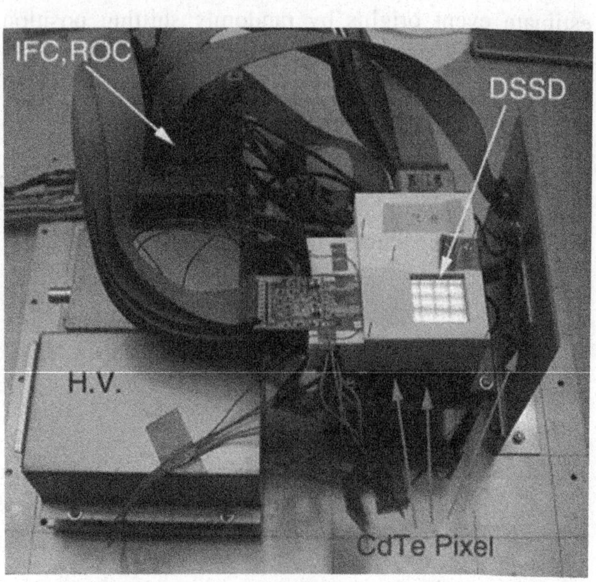

Fig. 5 The prototype of the Si/CdTe Compton camera. (The figure is reprinted from the work of Watanabe, S., et al. [90])

detector for calorimetry. Due to its low atomic number, Si offers room-temperature operation and superior scattering performance, while CZT detectors provide efficient photo absorption and effective calorimetry at room temperatures. The telescope demonstrated background suppression, isotope identification, and source localization. It achieved a 1σ angular resolution of 2.4°, consistent with simulations. Simulated energy resolution exceeded measured values, indicating additional noise. Increasing Si-CZT separation by 2.5 improved angular resolution by 50% but reduced detection efficiency [91].

A new Compton camera, including a silicon double-sided strip detector (Si-DSD) and a CdTe double-sided strip detector (CdTe-DSD), initially developed for the ASTRO-H satellite mission [92, 93], was tested for hotspot monitoring by Takeda, Shin'ichiro, et al. in 2012. The experiment demonstrated the camera's ability to image hotspots and identify radioisotopes due to its good energy resolution. Three radioisotopes were imaged individually with accurate gamma-ray overlays on optical pictures. Stacking five detector modules yielded a detection efficiency of 1.68×10^{-4} and an angular resolution of 3.8 degrees for a 662 keV gamma ray. The prototype Si/CdTe Compton camera, with one layer of Si-DSD and four layers of CdTe-DSDs, showed promising results, with a large field of view (FOV) and the ability to identify radioisotopes. It operated effectively at radiation levels of around 100 μSv/h, making it suitable for practical applications [51].

Harayama, Atsushi, et al. developed the first commercial Compton camera based on Si/CdTe technology. Their field tests in Fukushima demonstrated its effectiveness in hotspot detection and decontamination evaluation. Unlike traditional gamma cameras, their camera offers a wider field of view without collimators. The compact ASTROCAM 7000HS [94], derived from the ASTRO-H satellite, is scalable and customizable for various applications. It combines Si and CdTe detectors for excellent angular and energy resolution. Field tests in Fukushima validated its effectiveness, highlighting its potential for practical use (see Fig. 6) [43].

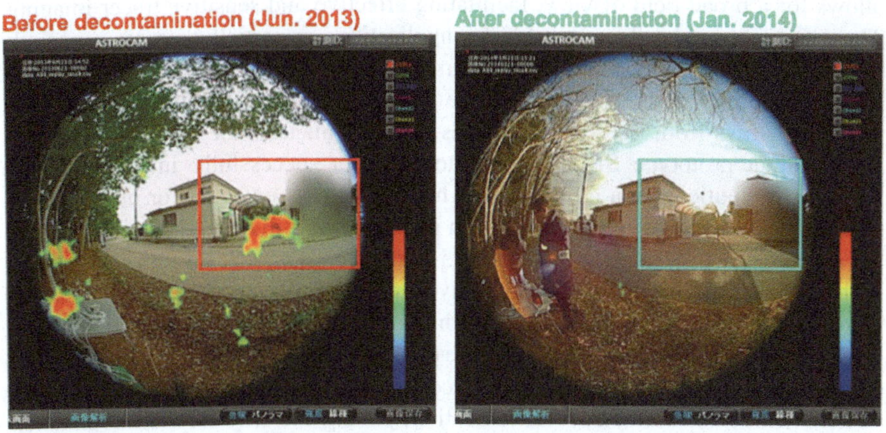

Fig. 6 One demonstration within the 20 km vicinity of the nuclear plant showcased ASTROCAM before (left) and after (right) radioactive decontamination in June 2013 and January 2014, respectively. (The figures are reprinted from the work of Harayama, A., et al. [43])

The other research, DTU Space, has developed a high-resolution 3D CZT drift-strip detector, providing sub-mm position resolution at energies exceeding 100 keV and high energy resolution. Researchers characterized prototype detectors using a fine-collimated high-energy (^{137}Cs) monochromatic beam and pulse shape analysis. The 3D CZT prototype detectors exhibited sub-mm resolutions in x, y, and z directions and an energy resolution of 1.6%. They devised methods for determining the positions of double events, enabling the successful reconstruction of weak point sources. The single 3D CZT detector prototype effectively demonstrated imaging capabilities [95].

Tsuzuki, Yutaka, et al. developed the electron beam ion trap Compton camera (EBIT-CC), integrating pixelated Si and CdTe counters for a precise 3D localization of Compton scattering and photoelectric absorption. Installed on the Tokyo electron beam ion trap (Tokyo-EBIT), it underwent evaluation during an experiment on radiative recombination x-rays emitted by highly charged krypton ions. The achieved polarization degree, with an absolute uncertainty of 0.02, enables probing differences between zero-frequency and full-frequency-dependent calculations for Breit interaction in dielectronic recombination x-rays of highly charged heavy ions [96].

In addition to ongoing research, CZT detectors continue to be utilized and advanced in Compton cameras for astrophysics and environmental radiation monitoring [97–99].

3.2 Medical Applications

Compton cameras have garnered significant interest in nuclear medical imaging due to their heightened sensitivity over collimated cameras and their capacity to simultaneously capture γ rays of various energies across a wide range. Their compact size allows for a broad field of view, facilitating effective and sensitive tracer imaging. These features render them suitable for applications in particle therapy, especially when employing CdTe and CZT detectors [100].

A pioneering study used a Si/CdTe Compton camera for in vivo imaging, detecting γ rays from injected substances with millimeter-scale resolution. Experiment with a prototype Si/CdTe Compton camera successfully imaged multiple radiopharmaceuticals in a living mouse, highlighting its potential for nuclear medicine. The DSSD scatterer had a 0.4 mm pitch, 0.5 mm thickness, and 25.6 × 25.6 area. The CdTe absorber, positioned 14.5 mm away, consisted of four layers, each with a 0.5 mm thickness, 13.2 × 13.2 active area, and 1.35 mm pitch. Measurements were taken over 6 hours from above an immobilized mouse (approx. 15 mm body thickness), with a DSSD-to-stage distance of 45 mm. Traditional sensor technologies have lacked the spatial resolution needed for medical imaging, but the Si/CdTe camera achieved millimeter-level resolution, paving the way for further enhancements in future iterations [101].

The new Si/CdTe Compton camera developed by Suzuki, Yoshiyuki, et al., with one layer of Si-strip detector and four layers of CdTe detectors, enabled highly resolved multiplanar and multinuclear gamma-ray imaging in ex vivo experiments. It successfully separated distributions of fluorodeoxyglucose (FDG), ^{131}I-methylnorcholestenol, and ^{67}Ga-citrate in test-tube-implanted models. In more physiologic models, it visualized distributions of ^{131}I-methylnorcholestenol and ^{64}Cu-chloride, capturing differences between isotopes, although imaging ^{111}In-chloride was challenging due to blurring [102].

Sakai et al. have developed a medical imaging technique using a Si/CdTe Compton camera to image ^{99m}Tc and ^{18}F simultaneously. Their camera showed precise imaging without significant cross-talk artifacts in simulated and in vivo rat experiments. Consistent with PET imaging and well-counter measurements, the results indicate its potential as a multitracer imager for various SPECT and PET probes. These findings support the possibility of future human trials [7, 103, 104].

A two-stage Compton camera prototype, with a CZT scatterer and three bismuth germanate (BGO) block absorber detectors, demonstrated a 10% energy resolution at 4.44 MeV for BGO detectors. Coincidence time spectra indicated a 4.9 ns (FWHM) resolution. Imaging tests successfully visualized the 4.44 MeV γ-ray emission scene and its lateral displacement. However, the Compton imaging efficiency was only 2.8×10^{-4}, suggesting a significant proton flux requirement for clinical application. Proposed enhancements include enlarging detector volumes and incorporating multiple prompt γ-ray lines, though these changes would increase the system's complexity and cost. Challenges persist near the Bragg peak due to dominant γ-ray lines, impacting the certainty of Compton imaging's efficacy. Golnik, C., et al. conducted the study [40].

Parajuli et al. employed a clinical carbon ion beam to investigate the real-time measurements of 511 keV annihilation γ-ray emissions from a polymethyl methacrylate (PMMA) target using a Si/CdTe Compton camera. This research is the first experimental use of a Compton camera in conjunction with carbon-ion radiotherapy (CIRT) [105]. The camera accurately traced the Bragg peak, confirming its performance. Shiba et al. used ASTROCAM to measure carbon-ion-induced annihilation γ rays in mice. They observed positron emitter transport post the C-ion irradiation of the mouse abdomen [106]. Parajuli et al. measured 718 keV prompt γ rays from a graphite phantom. Despite needing efficiency improvements, their study hints at the Si/CdTe Compton camera's potential in CIRT [107].

Other research combines Polaris-J CZT semiconductor detectors with LaBr3:Ce scintillation detectors to create a hybrid Compton camera for high-quality, prompt γ-ray imaging. The LaBr3:Ce detectors offer superior timing and energy resolution, addressing the limitations of the Polaris-J detectors [108]. CZT detectors continue to evolve within Compton cameras for medical and particle therapy, supported by ongoing research [109, 110].

4 Summary

The authors explored the evolution of Compton imaging within medical, astrophysics, and environmental radiation monitoring applications. They presented the latest research and development outcomes from active research groups worldwide, highlighting the pivotal role of CZT and CdTe detectors. Additionally, in this chapter, the authors provided a concise overview of the fundamental principles of the Compton imaging system, including its key parameters, effectiveness, and image reconstruction methods. Due to the characteristics of CZT/CdTe detectors, which play a fundamental role in advanced Compton cameras and provide higher detection efficiency, better energy resolution, and increased sensitivity to gamma rays, the new generation of portable or imaging Compton cameras should be given more attention.

References

1. Holly, T. A., et al. (2010). *Single photon-emission computed tomography*. Elsevier.
2. Tashima, H., & Yamaya, T. (2022). Compton imaging for medical applications. *Radiological Physics and Technology, 15*(3), 187–205.
3. Rong, J., et al. (2023). Radiochemistry for positron emission tomography. *Nature Communications, 14*(1), 3257.
4. Kim, D., et al. (2021). Evaluation of single scattering correction method in compton imaging system. *Nuclear Instruments and Methods in Physics Research Section A: Accelerators, Spectrometers, Detectors and Associated Equipment, 1010*, 165568.
5. Harkness, L., et al. (2012). A Compton camera application for the GAMOS GEANT4-based framework. *Nuclear Instruments and Methods in Physics Research Section A: Accelerators, Spectrometers, Detectors and Associated Equipment, 671*, 29–39.
6. Uenomachi, M., et al. (2021). Simultaneous in vivo imaging with PET and SPECT tracers using a Compton-PET hybrid camera. *Scientific Reports, 11*(1), 17933.
7. Sakai, M., et al. (2018). In vivo simultaneous imaging with 99mTc and 18F using a Compton camera. *Physics in Medicine & Biology, 63*(20), 205006.
8. Kim, S. M. (2015). Multi-tracer Imaging of a Compton Camera. *Progress in Medical Physics, 26*(1), 18–27.
9. Hofstadter, R., & McIntyre, J. A. (1950). Measurement of gamma-ray energies with two crystals in coincidence. *Physical Review, 78*(5), 619.
10. Kalish, Y., & Nardi, E. (1964). A three crystal NaI (Tl) gamma-ray scintillation spectrometer. *Nuclear Instruments and Methods, 26*, 329–332.
11. Peterson, L., & Howard, R. (1961). Gamma-ray astronomy in space in the 50-kev to 3-mev region. *IRE Transactions on Nuclear Science, 8*(4), 21–29.
12. Schönfelder, V., Hirner, A., & Schneider, K. (1973). A telescope for soft gamma ray astronomy. *Nuclear Instruments and Methods, 107*(2), 385–394.
13. McConnell, M., et al. (1993). COMPTEL gamma-ray observations of X-ray binaries. In *American Astronomical Society, 183rd AAS Meeting, id. 29.05; Bulletin of the American Astronomical Society*, 25, 1336.
14. Bloemen, H., et al., COMPTEL observations of the Orion complex: Evidence for cosmic-ray induced gamma-ray lines. Astronomy and Astrophysics 281, no. 1, p. L5-L8, 1994. (ISSN 0004-6361)
15. Schönfelder, V., et al. (1996). COMPTEL overview: Achievements and expectations. *Astronomy and Astrophysics Supplement, 120*, 13–21.

16. Kuiper, L., et al. (2001). The Crab pulsar in the 0.75–30 MeV range as seen by CGRO COMPTEL-A coherent high-energy picture from soft X-rays up to high-energy γ-rays. *Astronomy & Astrophysics, 378*(3), 918–935.
17. Iyudin, A., et al. (1994). COMPTEL observations of 44Ti gamma-ray line emission from Cas A. In *AIP Conference Proceedings*. American Institute of Physics.
18. Schönfelder, V. (2004). Lessons learnt from COMPTEL for future telescopes. *New Astronomy Reviews, 48*(1–4), 193–198.
19. Ryan, J. M., & Collmar, W. (2024). The COMPTEL experiment and its in-flight performance. In *Handbook of X-ray and gamma-ray astrophysics* (pp. 2281–2305). Springer.
20. Boggs, S., et al. (2001). The nuclear compton telescope: A balloon-borne soft γ-ray spectrometer, polarimeter, and imager. In *AIP conference proceedings*. American Institute of Physics.
21. Beechert, J., et al. (2022). Calibrations of the compton spectrometer and imager. *Nuclear Instruments and Methods in Physics Research Section A: Accelerators, Spectrometers, Detectors and Associated Equipment, 1031*, 166510.
22. Kierans, C., Takahashi, T., & Kanbach, G. (2024). Compton telescopes for gamma-ray astrophysics. In *Handbook of X-ray and gamma-ray astrophysics* (pp. 1711–1782). Springer.
23. Everett, D., et al. (1977). Gamma-radiation imaging system based on the Compton effect. In *Proceedings of the institution of electrical engineers*. IET.
24. Singh, M., & Doria, D. (1983). An electronically collimated gamma camera for single photon emission computed tomography. Part II: Image reconstruction and preliminary experimental measurements. *Medical Physics, 10*(4), 428–435.
25. Singh, M., & Brechner, R. R. (1990). Experimental test-object study of electronically collimated SPECT. *Journal of Nuclear Medicine, 31*(2), 178–186.
26. Studen, A., et al. (2004). First coincidences in pre-clinical Compton camera prototype for medical imaging. *Nuclear Instruments and Methods in Physics Research Section A: Accelerators, Spectrometers, Detectors and Associated Equipment, 531*(1–2), 258–264.
27. Llosa, G., et al. (2008). Last results of a first Compton probe demonstrator. *IEEE Transactions on Nuclear Science, 55*(3), 936–941.
28. Çonka-Nurdan, T., et al. (2005). First results on Compton camera coincidences with the silicon drift detector. *IEEE transactions on nuclear science, 52*(5), 1381–1385.
29. Motomura, S., et al. (2008). Multiple molecular simultaneous imaging in a live mouse using semiconductor Compton camera. *Journal of analytical atomic spectrometry, 23*(8), 1089–1092.
30. Seo, Y. (2019). Quantitative imaging of alpha-emitting therapeutic radiopharmaceuticals. *Nuclear medicine and molecular imaging, 53*(3), 182–188.
31. Roser, J., et al. (2021). Compton cameras and their applications. In *Radiation detection systems* (pp. 161–198). CRC Press.
32. Nurdan, T. C., et al. (2015). Design criteria for a high energy Compton camera and possible application to targeted cancer therapy. *Journal of Instrumentation, 10*(07), C07018.
33. Omata, A., et al. (2020). Performance demonstration of a hybrid Compton camera with an active pinhole for wide-band X-ray and gamma-ray imaging. *Scientific reports, 10*(1), 14064.
34. Llosá, G., & Rafecas, M. (2023). Hybrid PET/Compton-camera imaging: an imager for the next generation. *The European Physical Journal Plus, 138*(3), 214.
35. Yoshida, E., et al. (2020). Whole gamma imaging: a new concept of PET combined with Compton imaging. *Physics in Medicine & Biology, 65*(12), 125013.
36. Ogane, K., et al. (2021). Simultaneous measurements of single gamma ray of 131I and annihilation radiation of 18F with Compton PET hybrid camera. *Applied Radiation and Isotopes, 176*, 109864.
37. Han, L., et al. (2008). Statistical performance evaluation and comparison of a Compton medical imaging system and a collimated Anger camera for higher energy photon imaging. *Physics in Medicine & Biology, 53*(24), 7029.
38. Zhang, L., Rogers, W. L., & Clinthorne, N. H. (2004). Potential of a Compton camera for high performance scintimammography. *Physics in Medicine & Biology, 49*(4), 617.

39. Fontana, M., et al. (2017). Compton camera study for high efficiency SPECT and benchmark with Anger system. *Physics in Medicine & Biology, 62*(23), 8794.
40. Golnik, C., et al. (2016). Tests of a Compton imaging prototype in a monoenergetic 4.44 MeV photon field—A benchmark setup for prompt gamma-ray imaging devices. *Journal of Instrumentation, 11*(06), P06009.
41. Richard, M.-H. (2012). *Design study of a Compton camera for prompts-gamma imaging during ion beam therapy*. Université Claude Bernard-Lyon I.
42. Aldawood, S., et al. (2017). Development of a Compton camera for prompt-gamma medical imaging. *Radiation Physics and Chemistry, 140*, 190–197.
43. Harayama, A., et al. (2015). A portable Si/CdTe Compton camera and its applications to the visualization of radioactive substances. *Nuclear Instruments and Methods in Physics Research Section A: Accelerators, Spectrometers, Detectors and Associated Equipment, 787*, 207–211.
44. Wahl, C. G., et al. (2015). The Polaris-H imaging spectrometer. *Nuclear Instruments and Methods in Physics Research Section A: Accelerators, Spectrometers, Detectors and Associated Equipment, 784*, 377–381.
45. Hmissi, M. Z., et al. (2018). First images from a CeBr 3/LYSO: Ce Temporal Imaging portable Compton camera at 1.3 MeV. In *2018 IEEE nuclear science symposium and medical imaging conference proceedings (NSS/MIC)*. IEEE.
46. Ziock, K. (2018). Principles and applications of gamma-ray imaging for arms control. *Nuclear Instruments and Methods in Physics Research Section A: Accelerators, Spectrometers, Detectors and Associated Equipment, 878*, 191–199.
47. Nishiyama, T., et al. (2014). A novel Compton camera design featuring a rear-panel shield for substantial noise reduction in gamma-ray images. *Journal of Instrumentation, 9*(12), C12031.
48. Sato, Y., et al. (2020). Remote detection of radioactive hotspot using a Compton camera mounted on a moving multi-copter drone above a contaminated area in Fukushima. *Journal of nuclear science and technology, 57*(6), 734–744.
49. Martin, J., et al. (1993). A ring Compton scatter camera for imaging medium energy gamma rays. *IEEE transactions on nuclear science, 40*(4), 972–978.
50. Motomura, S., et al. (2007). Gamma-ray Compton imaging of multitracer in biological samples using strip germanium telescope. *IEEE Transactions on Nuclear Science, 54*(3), 710–717.
51. Takeda, S. i., et al. (2012). Applications and imaging techniques of a Si/CdTe Compton gamma-ray camera. *Physics Procedia, 37*, 859–866.
52. Calderón, Y., et al. (2014). Evaluation of Compton gamma camera prototype based on pixelated CdTe detectors. *Journal of Instrumentation, 9*(06), C06003.
53. Du, Y., et al. (2001). Evaluation of a Compton scattering camera using 3-D position sensitive CdZnTe detectors. *Nuclear Instruments and Methods in Physics Research Section A: Accelerators, Spectrometers, Detectors and Associated Equipment, 457*(1–2), 203–211.
54. Shimazoe, K., et al. (2020). Development of simultaneous PET and Compton imaging using GAGG-SiPM based pixel detectors. *Nuclear Instruments and Methods in Physics Research Section A: Accelerators, Spectrometers, Detectors and Associated Equipment, 954*, 161499.
55. Llosá, G., et al. (2012). Detector characterization and first coincidence tests of a Compton telescope based on LaBr3 crystals and SiPMs. *Nuclear Instruments and Methods in Physics Research Section A: Accelerators, Spectrometers, Detectors and Associated Equipment, 695*, 105–108.
56. Iltis, A., et al. (2018). Temporal imaging CeBr3 Compton camera: A new concept for nuclear decommissioning and nuclear waste management. In *EPJ web of conferences*. EDP Sciences.
57. Tashima, H., et al. (2020). 3D Compton image reconstruction method for whole gamma imaging. *Physics in Medicine & Biology, 65*(22), 225038.
58. Ordonez, C. E., Bolozdynya, A., & Chang, W. (1997). Doppler broadening of energy spectra in Compton cameras. In *1997 IEEE nuclear science symposium conference record*. IEEE.

59. Niknami, M., Hosseini, S. A., & Loushab, M. E. (2021). Sensitivity analysis of the efficiency of Compton camera to the detector parameters using the GEANT4 computer code. *Applied Radiation and Isotopes, 176*, 109883.
60. Alnaaimi, M. (2011). *Evaluation of the UCL Compton camera imaging performance*. UCL (University College London).
61. Phillips, G. W. (1995). Gamma-ray imaging with Compton cameras. *Nuclear Instruments and Methods in Physics Research Section B: Beam Interactions with Materials and Atoms, 99*(1–4), 674–677.
62. Knoll, G. F. (2010). *Radiation detection and measurement*. Wiley.
63. Niknami, M., & Valipour, M. (2023). Compton imaging systems based on semiconductor detectors. In *Gamma ray imaging: Technology and applications* (pp. 65–86). Springer.
64. Ordonez, C. E., Bolozdynya, A., & Chang, W. (1997). Dependence of angular uncertainties on the energy resolution of Compton cameras. In *1997 IEEE nuclear science symposium conference record*. IEEE.
65. Kierans, C., Takahashi, T., & Kanbach, G. (2022). Compton telescopes for gamma-ray astrophysics. In C. Bambi & A. Santangelo (Eds.), *Handbook of X-ray and gamma-ray astrophysics* (pp. 1–72). Springer.
66. Basko, R., Zeng, G. L., & Gullberg, G. T. (1997). Analytical reconstruction formula for one-dimensional Compton camera. *IEEE Transactions on Nuclear Science, 44*(3), 1342–1346.
67. Jorge, R., et al. (2020). Image reconstruction for a multi-layer Compton telescope: an analytical model for three interaction events. *Physics in Medicine & Biology 65*(14), 145005.
68. Ida, T., et al. (2018). Accurate modeling of event-by-event backprojection for a germanium semiconductor Compton camera for system response evaluation in the LM-ML-EM image reconstruction method. *Japanese Journal of Applied Physics, 58*(1), 016002.
69. Wilderman, S. J., et al. (2001). Improved modeling of system response in list mode EM reconstruction of Compton scatter camera images. *IEEE Transactions on Nuclear Science, 48*(1), 111–116.
70. Le, N., et al. (2023). An extended list-mode MLEM algorithm for 3D Compton image reconstruction from multi-view data. In *EPJ web of conferences*. EDP Sciences.
71. Mihailescu, L., et al. (2007). SPEIR: A Ge Compton camera. *Nuclear Instruments and Methods in Physics Research Section A: Accelerators, Spectrometers, Detectors and Associated Equipment, 570*(1), 89–100.
72. Kishimoto, A., et al. (2017). First demonstration of multi-color 3-D in vivo imaging using ultra-compact Compton camera. *Scientific Reports, 7*(1), 2110.
73. Seo, H., et al. (2009). Feasibility study on hybrid medical imaging device based on Compton imaging and magnetic resonance imaging. *Applied Radiation and Isotopes, 67*(7–8), 1412–1415.
74. Kolstein, M., De Lorenzo, G., & Chmeissani, M. (2014). Evaluation of list-mode ordered subset expectation maximization image reconstruction for pixelated solid-state compton gamma camera with large number of channels. *Journal of Instrumentation, 9*(04), C04034.
75. Nguyen, V.-G., Lee, S.-J., & Lee, M. N. (2011). GPU-accelerated 3D Bayesian image reconstruction from Compton scattered data. *Physics in Medicine & Biology, 56*(9), 2817.
76. Sakai, M., et al. (2020). Improved iterative reconstruction method for Compton imaging using median filter. *Plos one, 15*(3), e0229366.
77. Lojacono, X., et al. (2013). Low statistics reconstruction of the Compton camera point spread function in 3D prompt-$\gamma $ imaging of ion beam therapy. *IEEE Transactions on Nuclear Science, 60*(5), 3355–3363.
78. Feng, Y., et al. (2021). Influence of doppler broadening model accuracy in Compton camera list-mode MLEM reconstruction. *Inverse Problems in Science and Engineering, 29*(13), 3509–3529.
79. Mackin, D., et al. (2013). The effects of Doppler broadening and detector resolution on the performance of three-stage Compton cameras. *Medical physics, 40*(1), 012402.

80. Kim, S. M., et al. (2013). Resolution recovery reconstruction for a Compton camera. *Physics in Medicine & Biology, 58*(9), 2823.
81. Sakai, M., et al. (2020). Crosstalk reduction using a dual energy window scatter correction in Compton imaging. *Sensors, 20*(9), 2453.
82. Draeger, E., et al. (2018). 3D prompt gamma imaging for proton beam range verification. *Physics in Medicine & Biology, 63*(3), 035019.
83. Lee, Y. (2020). Preliminary evaluation of dual-head Compton camera with Si/CZT material for breast cancer detection: Monte Carlo simulation study. *Optik, 202*, 163519.
84. Andreyev, A., Sitek, A., & Celler, A. (2011). Fast image reconstruction for Compton camera using stochastic origin ensemble approach. *Medical Physics, 38*(1), 429–438.
85. Mackin, D., et al. (2012). Evaluation of a stochastic reconstruction algorithm for use in Compton camera imaging and beam range verification from secondary gamma emission during proton therapy. *Physics in Medicine & Biology, 57*(11), 3537.
86. Kohlhase, N., et al. (2019). Capability of MLEM and OE to detect range shifts with a Compton camera in particle therapy. *IEEE Transactions on Radiation and Plasma Medical Sciences, 4*(2), 233–242.
87. Kroiz, G. C., et al. (2021). Exploring deep learning to improve Compton camera based prompt gamma image reconstruction for proton radiotherapy. In *The 17th international conference on data science (ICDATA'21).*
88. Nakamura, Y., Shimazoe, K., & Takahashi, H. (2013). Design and fabrication of endoscope-type Compton camera. *Nuclear Instruments and Methods in Physics Research Section A: Accelerators, Spectrometers, Detectors and Associated Equipment, 731*, 283–287.
89. Turecek, D., et al. (2018). Compton camera based on Timepix3 technology. *Journal of Instrumentation, 13*(11), C11022.
90. Watanabe, S., et al. (2006). Development of CdTe pixel detectors for Compton cameras. *Nuclear Instruments and Methods in Physics Research Section A: Accelerators, Spectrometers, Detectors and Associated Equipment, 567*(1), 150–153.
91. Subramanian, M., et al. (2012). Compton imaging with thick Si and CZT detectors. *Nuclear Instruments and Methods in Physics Research Section A: Accelerators, Spectrometers, Detectors and Associated Equipment, 682*, 79–84.
92. Takahashi, T., Mitsuda, K., & Kelley, R. (2010). The ASTRO-H Mission. In *AIP Conference Proceedings*. American Institute of Physics.
93. Watanabe, S., et al. (2014). The Si/CdTe semiconductor Compton camera of the ASTRO-H soft gamma-ray detector (SGD). *Nuclear Instruments and Methods in Physics Research Section A: Accelerators, Spectrometers, Detectors and Associated Equipment, 765*, 192–201.
94. Ikeda, N., Sugano, T., & MHI-MS, Kobe. (2016). Residual radiation measurements at J-PARC MR using the ASTROCAM 7000HS newly developed radioactive substance visualization camera. In *Proceedings of IPAC2016 WEPOR022*.
95. Owe, S. H., et al. (2019). Evaluation of a Compton camera concept using the 3D CdZnTe drift strip detectors. *Journal of Instrumentation, 14*(01), C01020.
96. Tsuzuki, Y., et al. (2021). An application of a Si/CdTe Compton camera for the polarization measurement of hard x rays from highly charged heavy ions. *Review of Scientific Instruments, 92*(6), 063101.
97. Bolotnikov, A., et al. (2020). A 4× 4 array module of position-sensitive virtual Frisch-grid CdZnTe detectors for gamma-ray imaging spectrometers. *Nuclear Instruments and Methods in Physics Research Section A: Accelerators, Spectrometers, Detectors and Associated Equipment, 954*, 161036.
98. Takahashi, T. (2005). A Si/CdTe Compton Camera for gamma-ray lens experiment. *Experimental Astronomy, 20*(1), 317–331.
99. Du, Y. (2001). *Development of a prototype Compton scattering camera using three-dimensional position sensitive CZT detectors*. University of Michigan.

100. McCleskey, M., et al. (2015). Evaluation of a multistage CdZnTe Compton camera for prompt γ imaging for proton therapy. *Nuclear Instruments and Methods in Physics Research Section A: Accelerators, Spectrometers, Detectors and Associated Equipment, 785*, 163–169.
101. Takeda, S.i., et al. (2012). Demonstration of in-vivo multi-probe tracker based on a Si/CdTe semiconductor Compton camera. *IEEE Transactions on Nuclear Science, 59*(1), 70–76.
102. Suzuki, Y., et al. (2013). Three-dimensional and multienergy gamma-ray simultaneous imaging by using a Si/CdTe Compton camera. *Radiology, 267*(3), 941–947.
103. Sakai, M., et al. (2019). Compton imaging with 99mTc for human imaging. *Scientific Reports, 9*(1), 12906.
104. Sakai, M., et al. (2018). Effect of number of views on cross-sectional Compton imaging: A fundamental study with backprojection. *Physica Medica, 56*, 1–9.
105. Parajuli, R. K., et al. (2019). Annihilation gamma imaging for carbon ion beam range monitoring using Si/CdTe Compton camera. *Physics in Medicine & Biology, 64*(5), 055003.
106. Shiba, S., et al. (2020). Use of a Si/CdTe compton camera for in vivo real-time monitoring of annihilation gamma rays generated by carbon ion beam irradiation. *Frontiers in Oncology, 10*, 635.
107. Parajuli, R. K., et al. (2021). Carbon range verification with 718 keV Compton imaging. *Scientific Reports, 11*(1), 21696.
108. Hart, S., et al. (2023). A hybrid CZT-LaBr3: Ce Compton camera system for improving proton therapy imaging. In *Journal of Physics: Conference Series*. IOP Publishing.
109. Yao, Z., Chen, Z., & Xiao, Y. (2020). Origin ensemble algorithm based doppler broadening effect correction for CZT Compton camera in prompt gamma imaging: A Monte Carlo study. In *2020 IEEE nuclear science symposium and medical imaging conference (NSS/MIC)*. IEEE.
110. Yao, Z.-Y., Xiao, Y.-S., & Zhao, J.-Z. (2023). Dose reconstruction with Compton camera during proton therapy via subset-driven origin ensemble and double evolutionary algorithm. *Nuclear Science and Techniques, 34*(4), 59.

Enhanced γ-Ray Imaging Utilizing Coded Apertures with Pixelated Detectors

Ioannis Kaissas

1 Introduction

The pin-hole camera is the ancestor of both visual spectrum cameras and γ cameras. In the late '60 s, two astrophysicists [1, 5] had the idea of arranging in a unique way many holes, instead of just one, in order to get more photons on the film of the γ camera or telescope. The arrangement of the holes of coded apertures evolved through the years, starting with Dicke's Random Array and continuing with mathematically sophisticated arrays, like the nonredundant array [10], the uniformly redundant array [7], the modified uniformly redundant array (MURA) [11], and many more. Compared with collimators and pin holes, coded apertures exhibit a wider field of view (FOV) combined with great efficiency, great spatial resolution, and the accuracy of localizing radioactive sources. Imaging applications of ionizing radiation have been developed in observational astronomy, the mapping of the environmental radioactivity of an area, nuclear security [9], and nuclear medicine [2]. Also, coded apertures have been used in photography for deblurring out-of-focus regions [15].

Coded apertures comprise several pinholes arranged in a special pattern. The specialty lies in the fact that the arrangement of the holes is such that every different subarea of the coded aperture is unique at least by one hole. For instance, in Fig. 1 (top center), every single area of the coded aperture, encompassing 19 × 19 elements, differs from each of the other same-size areas at least by one element. This is the case for the MURA-coded aperture, which comprises 37 × 37 elements and with a basic (unique) pattern of 19 × 19 elements. The term elements replaces the term holes because instead of drilling holes on a lead plate, one can place lead elements on a

I. Kaissas (✉)
Department of Electrical & Computer Engineering, Aristotle University of Thessaloniki, Thessaloniki, Greece
e-mail: ikaissas@ece.auth.gr

K. Iniewski (ed.), *CdTe and CdZnTe Materials*,
https://doi.org/10.1007/978-3-031-64521-1_11

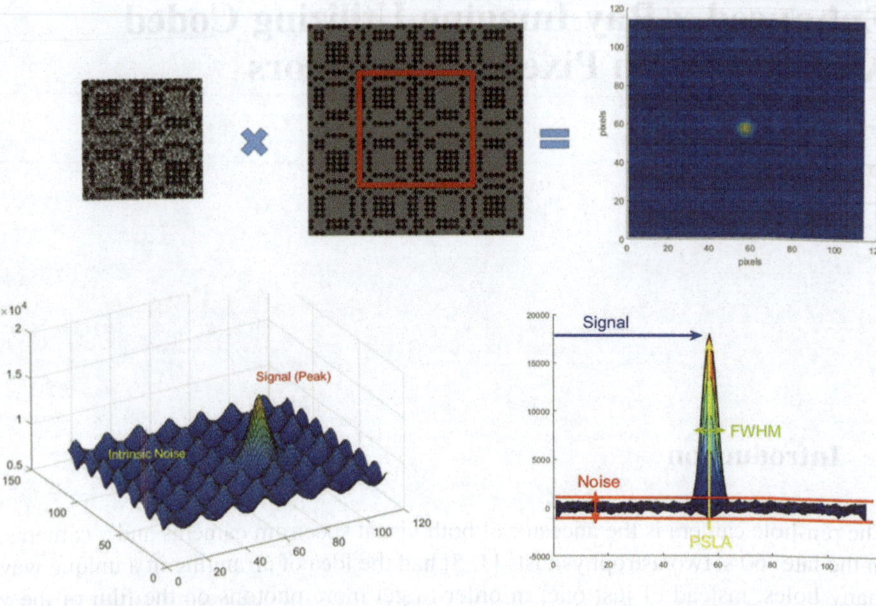

Fig. 1 *Top*: the shadowgram (left) of a point source projected through the MURA 37 × 37 coded aperture (center) on the pixelated detector, matches only with a unique surface (red square) for the unique location into the FOV. The correlation of the shadowgram with the G matrix yields the correlation matrix (right) with a peak known as a single-point spread function. The coordinates of the peak are correlated geometrically with the actual direction of the source. *Bottom left*: a correlation matrix with the characteristic peaks and lows of the intrinsic, element-wise noise. The tall peak is the PSF, and its diameter is proportional to the diameter of the elements. *Bottom right*: the basic figures of merit depicted on the face of a correlation matrix with a peak corresponding to a source. SNR is the signal-to-noise ratio

transparent plate. The benefit of such replacement is that the transparent surface of the coded aperture increases, and consequently, the photon flux reaching the pixelated detector increases, allowing for shorter collection time for photons.

The geometries of the coded aperture and the pixelated detector, particularly the distance between them and their sizes, define the size of the field of view (FOV). When a γ source is placed in the FOV, a fraction of the coded aperture is projected onto the detector's surface, forming a two-dimensional (2D) image known as a shadowgram. Subsequently, the shadowgram is cross-correlated (i.e., de-convoluted) with a predefined decoding function, known as the G matrix, which has the same configuration of elements as the actual-coded aperture [6, 8]. Then the direction of the source is revealed as a peak on the correlation matrix. The process is depicted in Fig. 1 for one source. Considering that only the basic pattern (i.e., 19 × 19 elements for the 37 × 37 MURA) has to be projected onto the pixelated detector, the physical dimensions of the detector can be about a quarter of the size of the coded aperture. This case is called in-focus imaging. When the sources are placed far away from the γ camera and hence no magnification takes place (far-field imaging), the number of elements of the basic pattern determines exactly how much of the quarter, mentioned in the previous sentence. For the 37×37 MURA-coded aperture, its surface must be $(37/19)^2 = 3.8$ times the surface of the detector.

When the near-field imaging comes to the fore (i.e., the ratio: coded aperture-detector distance/source-coded aperture distance is not negligible), the magnification of the shadowgram is affected by the distance between the coded aperture and the pixelated detector. As the source approaches the γ camera, the magnification of each element of the coded aperture on the detector increases, and thus the number of projections of elements constituting the shadowgram decreases, resulting in a shadowgram with fewer elements than what the basic pattern has. The correlation of such a shadowgram yields a correlation matrix with undesired side lobes beside the main peak on the correlation matrix. The first step in developing a γ camera is envisaging how far away the sources under investigation will be and appropriately regulating the geometry of the coded aperture-detector pair in order to obtain correlation matrices with no side lobes. This is the optimum distance of the source, and the imaging then is in focus. One can think of some tricks to extend the range of the optimum distance of the sources under investigation, for instance, choosing a quite larger detector for a short optimum distance and then cropping the shadowgram in the case of further sources. Another trick is to increase the element pitch (i.e., the interval between neighbor elements) as the source is placed further. A key limitation of these games is the matching of the digitized G matrix with the digitized shadowgram during the correlation process.

If each element of a coded aperture does not occupy the entire surface available for the element but only a part of it, the coded aperture is characterized as NTHT (not two holes touching) in the case of holes or NTOT (not two obscures touching) in the case of opaque elements; i.e., there is always a blank area between two neighbor elements. The MURA example in Fig. 1 (top) is an NTOT-coded aperture. In a close-up look, no element is in touch with the neighbor. This special configuration is used when the segmentation of the pixelated detector is more delicate than the segmentation of the coded aperture. The size of the elements can be reduced until their projection on the detector covers at least one pixel. The imaging capacities of these coded apertures remain ideal (as long as they are ideal for the original coded aperture), or even improved, as the G matrix is modified accordingly in a matrix of three levels [8]. An advantage is the improvement of the spatial resolution, inasmuch as the full width at half maximum of the point spread function shrinks while the element size is decreased [8]. Recent approaches utilize machine learning techniques, like decision trees and neural networks [14] applied directly on the shadowgram, without de-convolution. The results are satisfactory.

1.1 Metrology of Coded Aperture Imaging

The imaging capacities of a γ camera with a coded aperture are evaluated with specific figures of merit. They derived mostly from the correlation matrix or are, in fact, a special correlation matrix.

The autocorrelation function (ACF) is the correlation matrix produced by the correlation of the basic pattern with the coded aperture. For ideal-coded apertures, the result will be a δ-function on a correlation matrix without background noise. This figure of merit evaluates the coded aperture design and the correlation process. For no ideal apertures, there are some modifications that can be applied to the G matrix in order to diminish the background noise [7].

The point spread function (PSF) or single-point spread function (SPSF) is the correlation matrix produced when a single-point source irradiates the γ camera. In the case of an ideal experimental setup (i.e., coded aperture, pixelated detector, and point source), the result will also be a δ-function. This figure of merit evaluates the coded aperture in combination with the pixelated detector.

Point source localization accuracy (PSLA or SLA) is the difference between the actual and the detected coordinates of a point source in the FOV of the γ camera. The PSLA is related directly to the accuracy of calculating the exact coordinates of the peak of the PSF on the correlation matrix. The calculation is usually conducted with fitting methods on the PSF to find its mathematical (e.g., 3D Gaussian) function.

Spatial resolution (SR) or angular resolution (AR) is the minimum distance (or angle) between two sources by reason of which they appear as two distinctive peaks on the correlation matrix. As they approach each other, it is just before their two peaks join into one common peak. Each peak on the correlation matrix is characterized by its full width at half maximum (FWHM). If two peaks with the same height and FWHM approach each other closer than their FWHM, they become indistinguishable. According to Caroli et al. [4], AR (i.e., the FWHM) is the projection of an element of the coded aperture from the detector to infinite distance. In practice, SR is better when the elements of the coded aperture are smaller because the surface of their projection on the shadowgram is smaller. Also, SR is better when the coded aperture-detector distance is increased because the FOV is decreased and the details of the imaging depicted larger on the correlation matrix (i.e., the analogous of zoom-in with an optical camera).

The sensitivity of a γ-camera for a certain photon-energy window W can be determined experimentally accumulating for time T the counts C_W on the pixelated detector from a source with well-known activity A and the counts Bg_W accumulated on the detector for the same time T, with the absence of the source (i.e., only the background):

$$\text{Sensitivity}\,(W) = \frac{C_W - Bg_W}{T \cdot A} \tag{1.1.i}$$

The greater the transparent-to-opaque ratio of the coded aperture and the wider the FOV, the better the sensitivity of the γ camera.

The signal-to-noise ratio (SNR) is the counts under a source peak on the correlation matrix (e.g., the height of a PSF minus the height of the background noise) to the standard deviation of the background noise of the correlation matrix.

$$\text{SNR} = \frac{\text{Peak height} - \text{Mean value of noise}}{\text{Standard deviation of noise}} \tag{1.1.ii}$$

Also, other definitions of SNR can be found in bibliography as the case-specific definition might suit the characteristics of a special imaging. For instance, in near-field applications, Accorsi et al. [2] include the effect of the transparency of coded apertures. SNR is sensitivity-depended and determines the minimum activity A for which a source is detectable by the γ-camera. This value is usually SNR $\cong$ 10 [12], but not strictly, as it is related to the specific imaging application. The larger the accumulating time, the better the SNR. Increasing the source-γ-camera distance reduces the photons' flux on the detector, but the implied fluctuations in the peak's height and in the standard deviation of noise are the main factors that will determine whether the SNR will go up or down.

The noise or fluctuation in the background is due to:

- The radiation background or other sources outside the FOV.
- Other sources inside the FOV.
- Imperfections in the design and construction of the coded aperture, the pixelated detector, and their geometrical combination.
- The deconvolution procedure.
- The Poisson noise of the photons reaching the detector.
- The side lobes in the correlation matrix, which are due to:
  - The extension and position of the radioactive source.
  - Out-of-focus imaging.
  - The de-convolution procedure.

The correlation matrix (and the ACF) for a NTOT MURA 37 × 37 contains a systematic, element-wise noise, i.e., intrinsic noise in the surrounding background of the PSF. Figure 1 (bottom) shows the details of its shape, which reminds egg cartons. Its small peaks and valleys have a diameter proportional to the size of the elements of the coded aperture.

1.2 *From 2D to 3D γ Vision*

Let me assume that most living creatures are granted two eyes, for them to distinguish whether an object is close or far. The phenomenon is called parallax and is translated in mathematical terms as triangulation. Two γ-cameras combined with coded apertures, exploiting the triangulation method, localize with high accuracy radioactive sources or hot-spots in 3D space.

When two straight lines converge in a 3D space, there is a straight line segment, which is the minimum distance between two converging lines and is perpendicular to these two straight lines. Therefore, finding the straight line segment of the minimum distance between the lines can be reduced to finding the straight line perpendicular to these two lines. A convenient way is described by Bourke Paul [3].

In the case of more than one source, a PSF for each source arises on the correlation matrix. Consequently, a pair of convergent lines points to each source. In order not to pair two lines that do not belong to the same source, the minimum distances between all lines are calculated, and consequently, the long ones are rejected, with the hypothesis that they are not a pair pointing to a real source. The rejection is terminated when the number of remaining pairs is equal to the number of the PSFs on the correlation matrix.

In the case of extended sources, i.e., radioactive hotspots spatially distributed, the peaks on the correlation matrix will be also extended. Despite that, points P_2 and P_4 can be considered the central coordinates of the peaks; hence, lines P_1P_2 and P_3P_4 converge to the center of the spatial distribution of the hotspot.

Two of the figures-of-merit described in 1.1 can be defined also in 3D localization. In this case, the coordinates are the basis of the mathematical definitions.

The 3D source location accuracy (|R|) is the absolute value of the vector defined by the actual and estimated location of the source. It is directly related to the accuracy of calculating the exact coordinates of the peak of the PSF on the correlation matrices and to the deviations of the triangulation technique.

The 3D spatial or voxel resolution (VR) is the absolute value of the vector defined by the 3D coordinates of a source and its neighbor, inasmuch as they are so close to each other that the triangulation can distinguish them as two separate sources.

2 Evolution of an Actual γ Camera

In the following sections, the evolution of a system of two γ cameras is presented with its results. The system exploits the parallax phenomenon and combines the derived directions with triangulation to find the 3D coordinates of the radioactive sources [16]. Special interest emerged for the in-focus and out-of-focus imaging of the system [13] and for an "unexpected" reduction of the noise for extended sources that led to the maximization of the SNR. Both experiments and simulations evaluate the dependencies of the 3D SLA, the spatial resolution, and the SNR.

Each γ camera has a total active area of 44×44 mm^2. The hybrid pixelated detectors are CMOS ASICs with 1 mm thick CdTe crystal bumps bonded. The interval between two neighbor pixels, i.e., the pixel pitch, is 350 μm. For these CdTe detectors, the energy resolution of each pixel is 3–4 keV FWHM for a photo peak of 141 keV.

An NTOT MURA-coded aperture is placed parallel to the CdTe surface at a 2 cm distance and on top of each γ camera. Table 1 presents the specifications of two different MURA-coded apertures that were constructed to localize sources at short and medium distances. Their basic pattern consists of 19×19 elements, and they differ in the element pitch, which determines the optimum source distance (from the detector), for which a point source produces a PSF on the correlation matrix without side lobes.

Table 1 Specifications of the MURA coded apertures for optimum imaging. The optimum source distance from the detector is 16 cm and 31 cm

Coded aperture type	Number of elements	Element pitch (μm)	Optimum source distance (cm)
19R-1821	37 × 37	1821	16
19R-1958	37 × 37	1958	31

Table 2 The characteristics of the radioactive sources, which used in the actual experiments as extended sources, are presented

Source type	Type A	Type B	Type C
Isotope	Tc-99 m	Tc-99 m	Co-57
Activity (MBq)	~1.5	~1.5	1.5
Photopeak (keV)	141	141	122
Diameter of cylinder (mm)	24	11	30
Height of cylinder (mm)	9	9	32
Volume (cm^3)	4.072	0.855	22.619

For the experiments with the actual system, two cylindrical containers with a ^{99m}Tc uniform solution and a cylindrical uniform gel of ^{57}Co were arranged separately in several positions (X, Y, Z) within the FOV_S. Table 2 describes the specifications for these sources.

A simulation code was developed in MATLAB to investigate several geometric setups and factors affecting the performance of the actual system. The code does not consider scattered photons, in order to run fast. The comparison with a simulation that considers scattered photons confirms that this intentional omission does not affect the resulting shadowgrams, especially in cases where photoelectric phenomena prevail over Compton scattering.

Successive simulation experiments were performed, to produce figures-of-merit's statistics, like the standard deviation, the skewness and the kurtosis of the SLA or the degradation of the SNR as the source move away from the γ-cameras. The results are used to recognize "strange behaviors" like the maximization of the SNR, which is analyzed in Sect. 3.3.

Moreover, the simulation is used to produce shadowgrams of sources distributed at various positions as presented in Fig. 5, SNR-plot. The simulation experiment was repeated five times for every icremental step of the source positioning, to obtain good enough statistics.

3 Results and Discussion

This section presents the main results derived from virtual experiments with simulations and real experiments with the actual system, as shown in Fig. 2d. Three main studies are unfolded: the 3D SLA, the spatial resolution, and the SNR maximization phenomenon, along with a proposition for kernel filtering, which reduces the intrinsic noise on the correlation matrix.

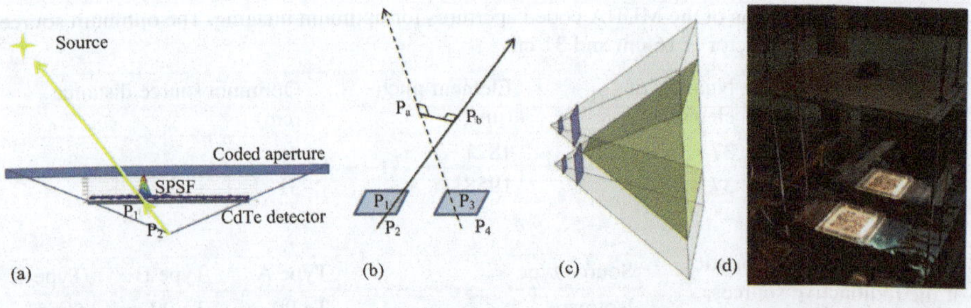

Fig. 2 (**a**) The correlation matrix with the PSF projected on the CdTe detector. P1 is the center of the peak, and P_2 is the vertex of the upside-down virtual pyramid, which confines the coded aperture and the detector. Line P_1P_2 points to the source that irradiates the γ camera. (**b**) Each γ camera finds independently the direction of the source. Lines P_1P_2 and P_3P_4 from the γ-camera pair converge to the location of the source. The middle of P_aP_b is the estimation of the location of the source. (**c**) The common field of view (FOV) of the two γ-camera systems is the common place of the FOV of each γ camera. (**d**) The actual system consists of two γ cameras

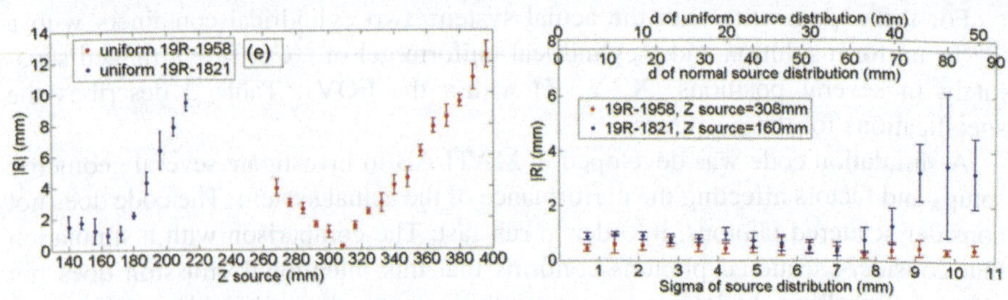

Fig. 3 *Left*: 3D SLA (|R|) dependence on the source distance from the detector (Z_{source}) for 19R-1821 and 19R-1958 coded aperture types. In-focus imaging stands for 160 mm and 308 mm, where the best SLA is achieved, i.e., the minimum of the curves, for the 19R-1821 and 19R-1958, respectively. For lower or higher Z_{source}, imaging turns into out-of-focus imaging, and |R| is degraded. *Right*: |R| dependence on the extension of a spatially distributed source

3.1 3D Source Localization Accuracy (3D SLA)

Via simulation, a source is placed at several distances from the detector plane in order to study the deterioration of the 3D SLA. Figure 3 (left) shows that the 3D SLA |R| becomes minimum for the optimum source to detector distance of Table 1 (i.e., in-focus imaging). For coded aperture type 19R-1821, |R| deteriorates with a source distance higher than 160 mm because, then, the shadowgram is comprised of more than 19 × 19 elements due to magnification.

|R| dependence on the standard deviation of the hotspot extension is shown in Fig. 3, right. For coded aperture type 19R-1821, |R| deteriorates when the source dimension becomes larger than 25 mm. This happens because the peak on the correlation matrix becomes too wide and the fitting procedure cannot accurately find its center.

3.2 Spatial Resolution for Extended Sources

With the purpose to study the spatial resolution of the two γ-cameras system, the shadowgrams produced by placing the same source at two neighbor positions, are summed into one which undergoes the correlation procedure. The experimental correlation matrices for the sources of types A and C are presented because they are more extended, and the effects due to this fact are more pronounced.

When the sources are not at the optimum distance from the detector plane, the resolution degrades, and side lobes appear beside the peaks. In addition, the presence of the side lobes of one source may coincide with the main peak of the other and in this way with its FWHM, resulting to the degradation of the spatial resolution. This can be seen by the comparison of the corellation matrices in Fig. 4b, c, as the volume and the lateral distance of the sources differ in such a way that their apparent angle remains constant although the height Z of the sources is different.

With the two sources placed at a 308 mm height and with the use of 19R-1958 coded apertures, the FWHM of the peaks on the correlation matrix decreases, so they become sharper due to the smaller magnification of the elements. The sharpening of the peaks is obvious in the comparison of Fig. 4a with Fig. 4b.

Implementing triangulation in the case of Fig. 4b results in the graphical representation of the 3D coordinates of the two hotspot centers (Fig. 4d). In this way, the 3D spatial resolution can be assumed.

3.3 Intrinsic Noise Reduction

In the current subsection, we focus on the systematic, element-wise noise appearing on the correlation matrix, which we call intrinsic noise of the coded aperture pattern (Fig. 1). The rest of the background noise is defined as random noise. The explanation of the origin of the intrinsic noise is attempted in the following text. Also, the optimum combination of element size and source-detector distance, which minimize the intrinsic noise for extended hotspots with certain dimensions, is presented. Finally, an image filtering method that reduces the intrinsic noise is introduced for the imaging of point sources.

The diameter of the intrinsic noise structure (peaks and valleys in Fig. 1) is proportional to the diameter of the projection of the elements on the pixelated CdTe detector, and consequently, it is also proportional to the FWHM of the PSF of the correlation matrix. As the element pitch is increased, the wavelength of the periodic structures of the shadowgram also increases. The SNR of the correlation matrix becomes maximum (Fig. 5, SNR plot) for a certain size of the extended source, for which the penumbra blurs the shade of each single-coded aperture element. As far as the extension of the hotspot eliminates, via the penumbra, the separate projection of each element on the shadowgram (it can be seen in the comparison between the left and middle shadowgrams in Fig. 5) but does not

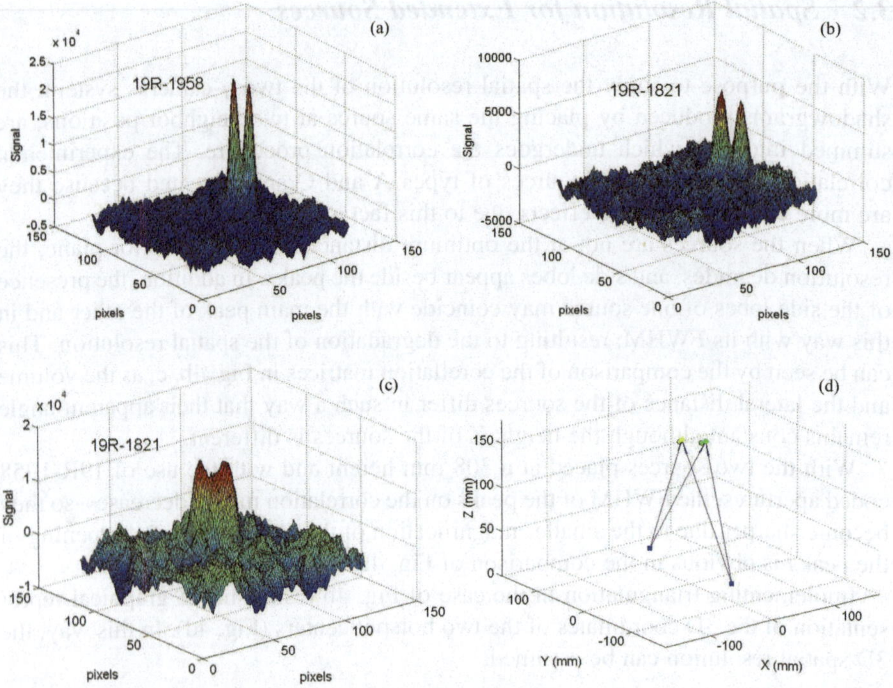

Fig. 4 The correlation matrix for two type A cylindrical sources positioned at the center of FCFOV$_S$: (**a**) at $Z = 308$ mm and with a distance of 42 mm between their centers and (**b**) at $Z = 160$ mm and with a distance of 28 mm between their centers. (**c**) The correlation matrix for two type C cylindrical sources positioned at $Z = 231$ mm and with a distance of 43 mm between their centers. (**d**) The experimental 3D localization of the centers of two type A sources for the case described in (**b**) with an accuracy of <5%, after triangulation. In each graph, the employed coded aperture is indicated

deteriorate the basic pattern of the coded aperture (this deterioration can be seen in the comparison between the middle and right shadowgrams in Fig. 5), the SNR of the correlation matrix is improved. A further extension of the hotspot's dimension deteriorates the basic pattern on the shadowgram by hiding the details in the middle range of the spatial frequencies, which are essential for a good correlation with the G matrix.

The 2D FFT of the shadowgram shows clearly the elimination of the high spatial frequencies of each element's projection. Further extension of the hotspot eliminates also the medium spatial frequencies of the basic pattern of the shadowgram and consequently deteriorates the SNR.

A detailed comparison of experimental data with simulation data is difficult due to the deviations in the activity and the positioning of the hotspots. The type A and type B sources were placed in the FOV for 184 sec each. The derived correlation matrices are presented in Fig. 6. The deviations in the activity and the positioning of the hotspots slip into the SNR, preventing the verification of the simulation results in

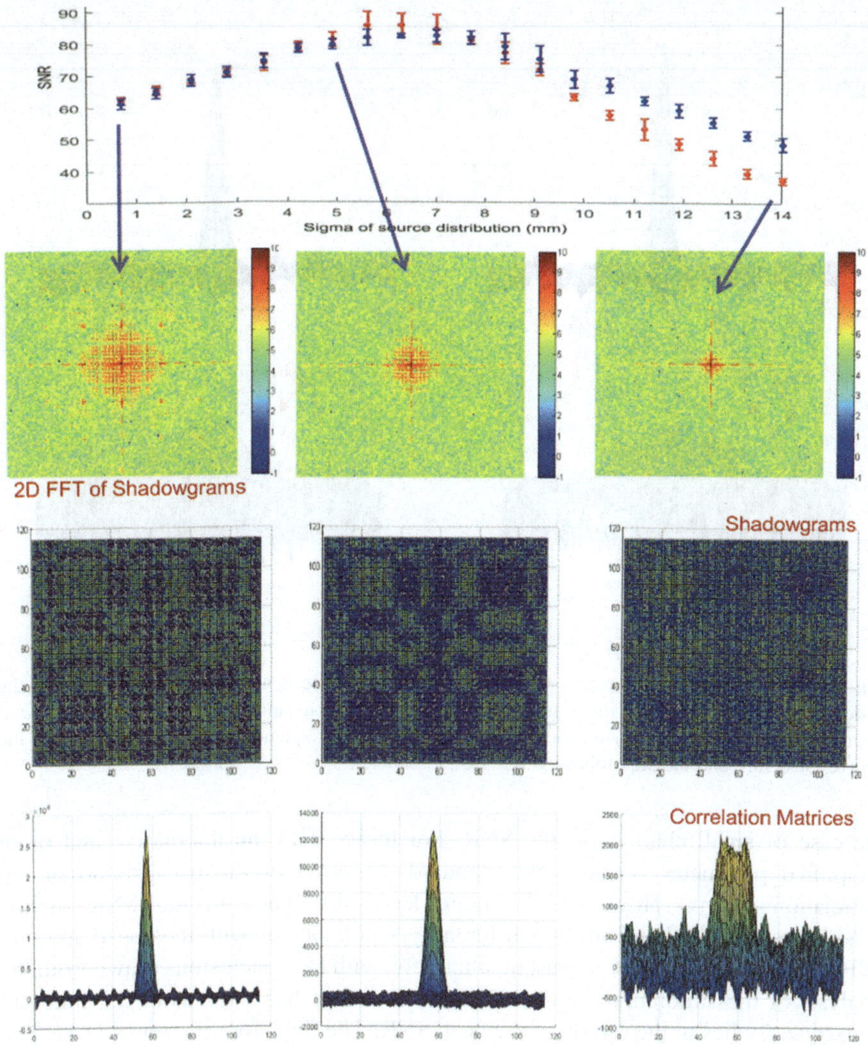

Fig. 5 *SNR plot*: the signal-to-noise ratio of the correlation matrix as a function of the extent of the hotspot, i.e., the standard deviation (sigma) of the source distribution. Blue: uniform distribution. Red: normal distribution. All the simulations were performed with coded aperture 19R-1821, and the hot spot was placed at a 160 mm distance from the γ camera. The error bars stand for the standard deviation of the results of five-time repeated simulation experiments. *2D FFT plots*: in every column (top to bottom): the XY projection of the fast Fourier transform (FFT) of the shadowgram, the XY projection of the actual shadowgram, and the YZ projection of the correlation matrix. In every row (left to right): extended Tc-99 m hotspot with 0.7 mm, 5 mm, and 14 mm standard deviation

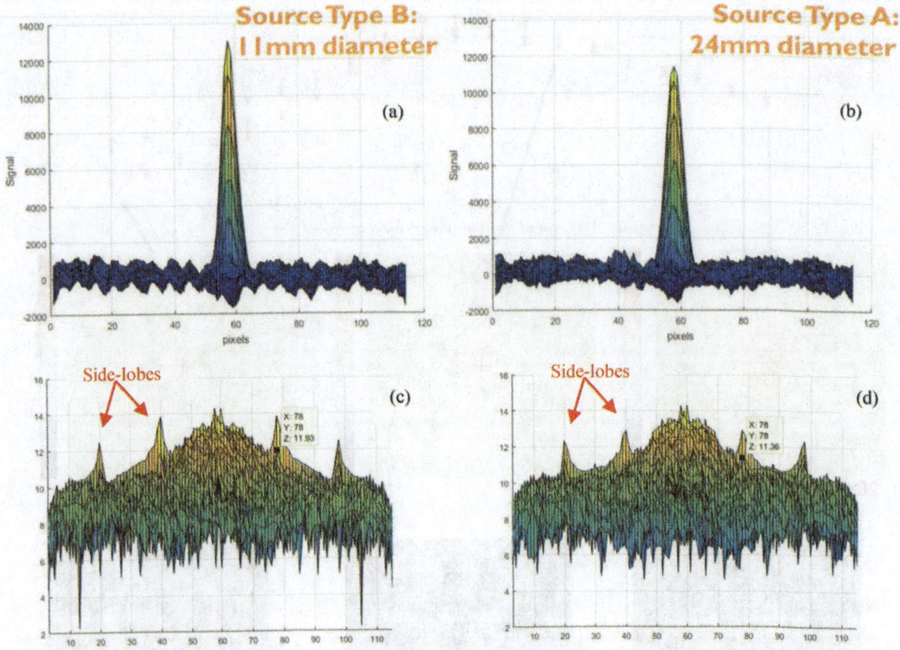

Fig. 6 *Top*: experimental correlation matrix for the ^{99m}Tc type B and type A hotspots, with the coded aperture 19R-1958 and the hotspots placed at a 308 mm distance from the γ camera. *Bottom*: their fast Fourier transforms with the indication of a spatial frequency, which corresponds to the "wave" structure and the side lobes

the case of small changes in the SNR. For these cases, an alternative and more simplified parameter is used—the standard deviation of the background of the correlation matrix. However, the quite large difference in the SNR of the shadowgrams recorded from two different types of sources with the coded aperture 19R-1821 is in agreement, at least qualitatively, with the conclusions drawn from the analysis of the simulation results. The type B source has a radius of 5.5 mm and consequently lies in the region of the peak SNR. The SNR for this source is 32. The type A source has a radius of 12 mm and lies in the region of the descending SNR, right to the peak. The SNR for this source is 20.

The standard deviation of the background of the correlation matrix, which includes the variation of the intrinsic noise, is slightly smaller for source type A than for source type B when the coded aperture 19R-1958 is used. For this combination of optimum source distance, spatial dimensions of hotspots, and element pitch, the type A source lies in the region of the peak of the SNR, and the type B source lies in the region of its left descending (Fig. 5, SNR plots). The standard deviation of the background of the correlation matrices presented in Fig. 6 is 333 for the type A source and 337 for the type B source, which is not a significant difference. However, the peaks and valleys of the intrinsic noise, like the ones appearing in Fig. 1 (bottom left), are evident in the background in Fig. 6a, while they are eliminated in the background in Fig. 6b.

Following the existence of the two components in the background, the fast Fourier transform of the correlation matrices also has two components: (i) the intrinsic and (ii) the random noise signal for each spatial frequency. The elimination of the peaks and valleys should appear as a reduction of the signal of their specific spatial frequencies, even though the random noise of these frequencies still exists. Indeed, the side lobes from Fig. 6c to Fig. 6d are reduced, and specifically, a reduction of 5% (from 11.93 to 11.36) is evident for the signal with a spatial frequency correlated with the "wave" structure of the background noise. Therefore, the intrinsic noise is reduced as the spatial dimensions of the source become larger, as shown in Fig. 6a to Fig. 6b.

3.4 The Optimum Geometry

The middle column of the 2D FFT plots in Fig. 5 indicates that there exists a geometrical setup that yields the maximum SNR of the correlation matrix. Specifically, for a source dimension of around 2 cm and with the coded aperture 19R-1958, the source has to be placed around a distance of 30 cm from the γ camera. For the coded aperture 19R-1821, the source must be placed around a distance of 16 cm from the γ camera, and its dimension has to be around 1 cm.

As an example of applying the optimum geometry, in nuclear medicine and intraoperative imaging, the objects under investigation have certain spatial dimensions. For example, the lymph nodes can be considered spheres with a diameter of around 1 to 2 cm. Therefore, different geometrical configurations of the coded aperture γ camera can be used for the imaging of sentinel lymph, or thyroid residue, in order to achieve the best SNR possible.

3.5 Kernel Filtering

The reduction in intrinsic noise resulting from the increase in the dimensions of the hotspot can be exploited to reduce this noise, also, in the correlation matrix of point-like sources, where this noise is the highest. Due to the periodicity of the peaks and valleys on the correlation matrix, scanning the whole surface of the correlation matrix with an appropriate kernel filter should reduce the intrinsic noise. Since the dimensions of the peaks and valleys are proportional to the FWHM of the PSF of the correlation matrix, this filter (Fig. 7) is chosen to be the cropped and normalized peak of the autocorrelation function (ACF) of each coded aperture. For the 19×19 basic pattern NTOT MURA, pixel size, and geometry configuration, it is a matrix of 6 by 6 pixels. The result is the slight expansion of the FWHM of the peak as the peak undergoes convolution with the normalized, cropped ACF and a significant reduction in intrinsic noise on the correlation matrix, as can be seen in Fig. 7.

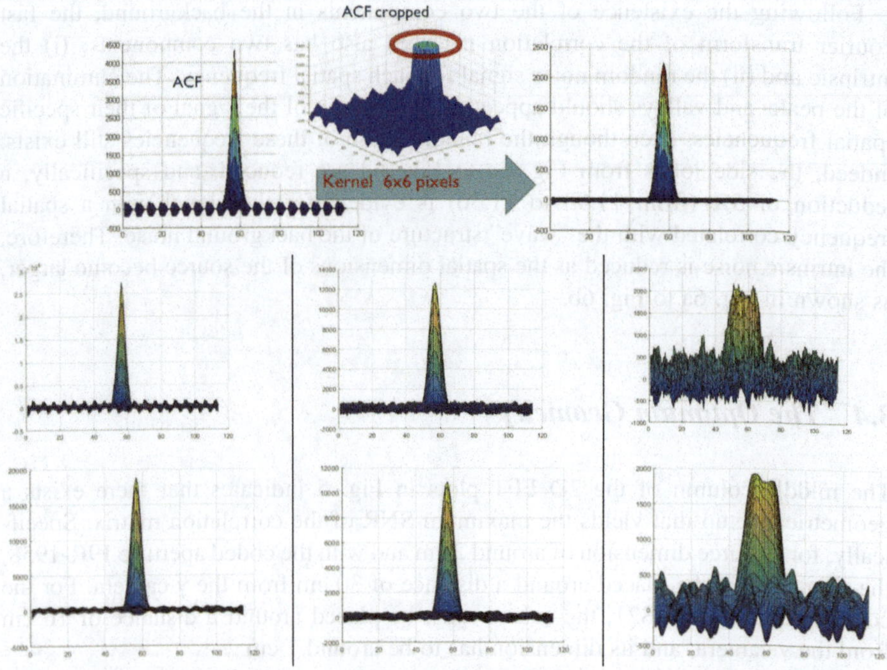

Fig. 7 The kernel filter consists of the cropped normalized peak of the ACF. Its application on the correlation matrix reduces intrinsic noise. Top: kernel filter is applied on the ACF. In each column from left to right: a kernel filter is applied on the correlation matrix of the extended Tc-99 m hotspot with 0.7 mm, 5 mm, and 14 mm standard deviation

In many astronomy or security applications of coded apertures, the sources under investigation are considered point like. The aforementioned filtering can be useful for smoothing the background of the correlation matrix, which includes one or more peaks related to point-like sources.

4 Summary

Coded apertures reveal characteristics unbeatable by collimators and pinholes, such as wide FOV, great sensitivity, and fine spatial resolution. Exploiting the parallax phenomenon with two coded aperture γ cameras can be considered the initial step for 3D γ-ray vision.

An important adjustment is the respective geometry of the aperture and the pixelated detector, for the imaging to be in focus, i.e., without side lobes in the correlation matrix. In near-field applications, the spatial resolution is strongly affected by the distance of the source from the γ camera.

NTOT-coded apertures are more luminous for γ rays than NTHTs. The proposed NTOT-coded apertures are simple in construction and less heavy than their NTHT counterparts.

The MURA pattern used in the present work reveals a periodic intrinsic noise on the correlation matrix. The investigation of the maximization of the SNR for spatially extended radioactive sources leads to the proposal of using specific setup geometry for similar-sized sources to reduce the intrinsic noise in the ultimate image. Exploiting this investigation, kernel filtering is proposed to achieve better SNR for the coded aperture imaging of point sources.

Acknowledgments My dedication to this topic would be insufficient without a team of brilliant researchers and friends. Namely thanks to: Charalambos P. Lambropoulos, Costas Karafasoulis, Costas Potiriadis, Christos Papadimitropoulos, Yiannis Kazas, Aris Kyriakis and Alexandros Clouvas.

References

1. Ables. (1968). Fourier transform photography: A new method for X-ray astronomy. *Proceedings Astronomical Society of Australia, 1*, 172–173.
2. Accorsi, R., Gasparini, F., & Lanza, R. C. (2001). Optimal coded aperture patterns for improved SNR in nuclear medicine imaging. *Nuclear Instruments and Methods in Physics Research, Section A: Accelerators, Spectrometers, Detectors and Associated Equipment, 474*(3), 273–284. https://doi.org/10.1016/S0168-9002(01)01326-2
3. Bourke, P. (1998). *The shortest line between two lines in 3D*. http://paulbourke.net/geometry/pointlineplane/
4. Caroli, E., Stephen, J. B., Di Cocco, G., Natalucci, L., & Spizzichino, A. (1987). Coded aperture imaging in X- and gamma-ray astronomy. *Space Science Reviews, 45*(3–4), 349–403. https://doi.org/10.1007/BF00171998
5. Dicke, R. H. (1968). Scatter-hole cameras for X-rays and gamma rays. *The Astrophysical Journal, 153*(2), 101–106. https://doi.org/10.1007/s13398-014-0173-7.2
6. Fenimore, E. E. (1978). Coded aperture imaging: Predicted performance of uniformly redundant arrays. *Applied Optics, 17*(22), 3562–3570.
7. Fenimore, E. E., & Cannon, T. M. (1978). Coded aperture imaging with uniformly redundant arrays. *Applied Optics, 17*(3), 337–347.
8. Fenimore, E. E., & Cannon, T. M. (1981). Uniformly redundant arrays: Digital reconstruction methods. *Applied Optics, 20*(10), 1858–1864. https://doi.org/10.1364/AO.20.001858
9. Gmar, M., Agelou, M., Carrel, F., & Schoepff, V. (2011). GAMPIX: A new generation of gamma camera. *Nuclear Instruments and Methods in Physics Research, Section A: Accelerators, Spectrometers, Detectors and Associated Equipment, 652*(1), 638–640. https://doi.org/10.1016/j.nima.2010.09.003
10. Golay, M. (1971). Point arrays having compact, nonredundant autocorrelations. *Journal of the Optical Society of America, 61*, 272–273.
11. Gottesman, S. R., & Fenimore, E. E. (1989). New family of binary arrays for coded aperture imaging. *Applied Optics, 28*(20), 4344–4352. https://doi.org/10.1364/AO.28.004344
12. Gros, A., Goldwurm, A., Soldi, S., Gotz, D., Caballero, I., Mattana, F., & Heras, J. A. Z. (2013). The IBIS/ISGRI source location accuracy. In *Proceedings of Science* (pp. 1–6).

13. Kaissas, I., Papadimitropoulos, C., Potiriadis, C., Karafasoulis, K., Loukas, D., & Lambropoulos, C. P. (2017). Imaging of spatially extended hot spots with coded apertures for intra-operative nuclear medicine applications. *Journal of Instrumentation, 12*(1). https://doi.org/10.1088/1748-0221/12/01/C01059
14. Karafasoulis, K., Kaissas, I., Papadimitropoulos, C., Potiriadis, K., & Lambropoulos, C. P. (2023). A machine learning approach in the estimation of a radioactive source position using a coded aperture device. *Journal of Instrumentation, 18*(1), C01062. https://doi.org/10.1088/1748-0221/18/01/C01062
15. Levin, A., & Freeman, W. T. (2007). Image and depth from a conventional camera with a coded. *Aperture, 26*(3), Article 70. https://doi.org/10.1145/1239451.1239521
16. Papadimitropoulos, C., Kaissas, I., Potiriadis, C., Karafasoulis, K., Loukas, D., & Lambropoulos, C. P. (2015). Radioactive source localization by a two detector system. *Journal of Instrumentation, 10*(12). https://doi.org/10.1088/1748-0221/10/12/C12022

Ring-Configured CZT SPECT-CT System: Clinical Experience and Implementation

Irma Ceric Andelius, Johan Gustafsson, Anna Stenvall, Bryan Haddock, Henrik Mosén, and Kristian Valind

1 Introduction

Since the first introduction of scintillation cameras in the 1950s, scintillation crystals with indirect conversion of gamma rays into an electrical signal have formed the basis for clinical nuclear medicine [1, 2]. The introduction of solid-state cadmium zinc telluride (CZT) detectors enables a direct conversion of gamma rays to signal within the crystal material [3–5]. CZT detectors are more compact, facilitating smaller and more diverse camera architectures such as C-shaped dedicated cardiac cameras and, more recently, ring-configured CZT single-photon emission computed tomography-computed tomography (SPECT-CT) systems [6–8]. Studies have shown that such systems demonstrate improved sensitivity and energy resolution compared with scintillation cameras, enabling reduced scan time or administered activity [9–11]. The CZT technology in itself offers several advantages in SPECT imaging, including better intrinsic spatial resolution, much-improved energy resolution and the capability to detect multiple-photon interactions simultaneously

I. Ceric Andelius (✉) · A. Stenvall
Radiation Physics, Department of Hematology, Oncology and Radiation Physics, Skåne University Hospital, Lund, Sweden
e-mail: irma.cericandelius@skane.se

J. Gustafsson
Medical Radiation Physics, Lund, Lund University, Lund, Sweden

B. Haddock
Department of Clinical Physiology, Nuclear Medicine & PET, Rigshospitalet, Copenhagen University Hospital, Copenhagen, Denmark

H. Mosén · K. Valind
Department of Clinical Physiology and Nuclear Medicine, Skåne University Hospital, Lund/Malmö, Sweden

K. Iniewski (ed.), *CdTe and CdZnTe Materials*,
https://doi.org/10.1007/978-3-031-64521-1_12

[5, 12, 13]. This chapter presents the current design of ring-configured CZT SPECT-CT systems, their technical specifications and resulting performance as well as pioneering clinical applications.

2 Camera Architecture

Anger systems rely on the scintillation process for their detection of ionising radiation and have a relatively poor energy resolution due to the inefficiency of the detection process, which causes a statistical variation in the estimation of the energy deposit. Semiconductor-based detectors offer an alternative to scintillation detectors with better energy resolution. A semiconductor crystal consists of valence and conduction bands separated by a band gap. The crystal is connected to an anode and a cathode. When a photon interacts in the crystal, electron-hole pairs will be created, causing a current. By integrating the current over time, a charge proportional to the energy deposit will result. This process generates a more accurate signal, better discrimination between different energy levels and less statistical variation, yielding an improved energy resolution compared to scintillation detectors.

Current commercially available ring-configured CZT SPECT systems consist of 12 detectors arranged in a ring, as illustrated in Fig. 1. Each detector consists of a column of CZT modules, placed along the system's axial field of view (FOV), where each module is composed of a CZT crystal with 16 × 16 anode elements, with a 2.46 mm quadratic shape. The Veriton™ system (Spectrum Dynamics, Caesarea, Israel) consists of eight modules giving an axial FOV of 32 cm, whereas the StarGuide™ system (GE Healthcare, Haifa, Israel) consists of seven modules giving an axial FOV of 27.5 cm. The detectors are equipped with permanently fixed tungsten parallel-hole collimators positioned in alignment with the anode-element array (Fig. 1). The systems detect photons in the energy range of 40–200 keV (Veriton™) or 40–279 keV (StarGuide™). The StarGuide™ can be run in a low-energy mode (40–200 keV) or in a medium-energy mode, with a coarser sampling of the energy spectrum if acquiring projections for photons with energies over 200 keV. The difference in energy range between the two vendors is due to the differences in CZT element thickness.

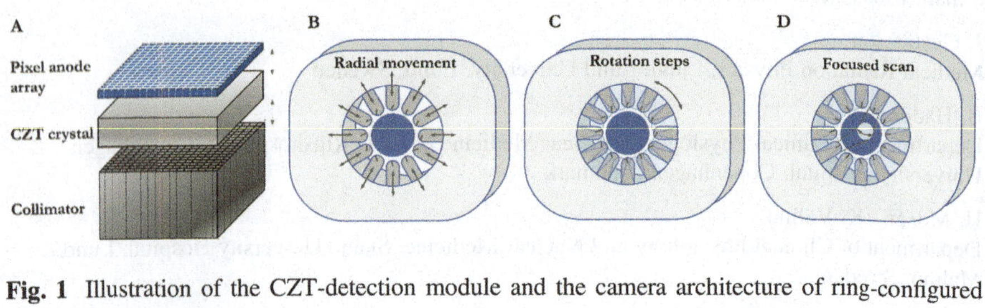

Fig. 1 Illustration of the CZT-detection module and the camera architecture of ring-configured CTZ gamma cameras

The coverage of the FOV while scanning is enabled by the detectors' ability to be rotated to different acquisition positions, their ability to move in and out radially and the ability for them to sweep the FOV. To enable a sufficient angular sampling of the FOV under consideration, each detector module performs a full sweeping motion at each acquisition position, combined with a gantry rotation of a few angular steps, typically four, to cover the entire patient volume. Each detector can move radially independent of the others. Thus, each detector module can be positioned close to the contour of the patient. The body contour is determined using body scanning with infrared transmitters and receiver diodes (StarGuide™) or sensors on each detector (Veriton™) to be positioned as close as possible to the imaged object. The acquisition can be performed in the focus mode, thus enabling data collection in a user-defined region of interest (ROI), as described in Fig. 1d. The fraction of the sweep time the detector spends in the focus region is set by the operator. The sweep time can be set to 1 s, enabling dynamic SPECT imaging with a frame time of as low as 1 s. During dynamic imaging, there is no gantry rotation.

Figure 1a illustrates a CZT module showing the CZT crystal between the tungsten collimator and the 16 × 16 anode array, which are aligned with each other. Figure 1b–d shows the degrees of freedom in detector motion. In B, the radial movement of each detector allows the individual positioning of the 12 detectors to maximise proximity to the scanned object. The gantry rotation (c) rotates all detectors simultaneously. The trans-axial sweep motion (d) of the detector heads enables focused scans, where the sweep is focused on the organ/area of interest, i.e. spends more acquisition time there.

The clinical application of CZT detectors was initially dedicated to cardiac perfusion imaging, utilising fixed gantries. Compared to conventional Anger systems, similar diagnostic accuracy could be seen but with the ability to reduce acquisition time or administered activity [14]. The introduction of dual-head CZT-based gamma cameras enabled a wider range of nuclear medicine applications and allowed for more effective workflows [15, 16]. However, the architecture of the dual-head camera systems does not take full advantage of the detector area, and the size of the detector heads prevents the detectors from being placed in close proximity to the patient, as illustrated in Fig. 2a, leading to a loss in spatial resolution. The architecture of the ring-configured CZT system facilitates the design of several smaller and more versatile detector heads, which can individually place themselves closer to the patient scanning at optimal geometries and even collectively focus on the organ of interest, as illustrated in Fig. 2b–d.

3 Performance

The commercialisation of CZT detectors provided a necessary building block for the development of ring-configured CZT SPECT-CT systems. The CZT technology itself offers several advantages in SPECT imaging compared to scintillation cameras. However, transitioning to CZT technology also entails certain challenges related to detector performance, such as charge sharing and defective pixels.

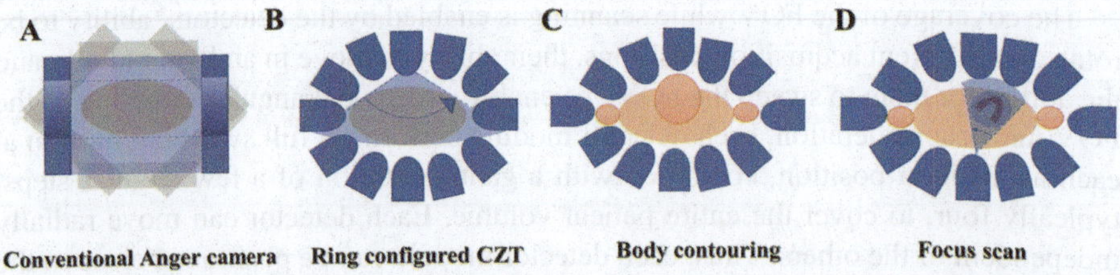

Fig. 2 Illustration of detector architecture. A dual-head Anger system (**a**) does not take full advantage of the detector area in each projection, whereas ring-configured CZT SPECT-CT systems' architecture (**b**) enables closer proximity to each detector due to its smaller area. Using body contouring (**c**) enables each detector to radially move independently farther, optimising the proximity. A focus scan (**d**) enables data acquisition for a user-defined ROI for organ-specific scans

3.1 Energy Spectrum

Semiconductor-based detector materials offer advantages due to their direct conversion of the ionising- radiation energy to electron-hole pairs. There are fewer steps in the generation of an electrical signal compared to a scintillation detector, and the energy required to create an information carrier is lower. Therefore, CZT detectors offer superior energy resolution compared to detectors made out of thallium-activated sodium iodide (NaI(Tl)). A NaI(Tl) detector relies on scintillation generation during the detection process, initiated when incident photons excite electrons within the crystal. Additionally, the efficiency of light collection and the characteristics of the photomultiplier tube play a crucial role in the detection process. The extended duration of this process introduces errors that contribute to a poor energy resolution. The energy resolution is approximately 5% at 140 keV energy for a CZT, whereas the corresponding resolution for a NaI(Tl) detector is approximately 10 % [17]. An illustration of the energy resolution for 140 keV photon energy for NaI (Tl) and CZT detectors is shown in Fig. 3.

The compound material cadmium-zinc-telluride can be operated at room temperature, which is an important practical advantage for its use in gamma cameras. However, unlike most single-material semiconductors such as high-purity germanium, the mobility is markedly different for electrons and holes. This leads to most of the induced signal being caused by the movement of electrons. Another issue related to semiconductor detectors is impurities in the manufacturing process, which cause poor charge transport, leading to a reduction in the induced current, energy resolution and the detection efficiency of the device [19, 20]. Also, in pixelated CZT detectors, an electron cloud from a single photon interaction can be created in the crystal between anode elements. Hence, current is induced in both elements as the electron cloud drifts towards the anode and a charge-shared event is created. The effect of charge sharing in a spectroscopic measurement is that full-energy depositions registered in the energy bins below the photopeak, thereby creating the tail of primary photons registered at lower energies [21, 22], as seen in Fig. 3.

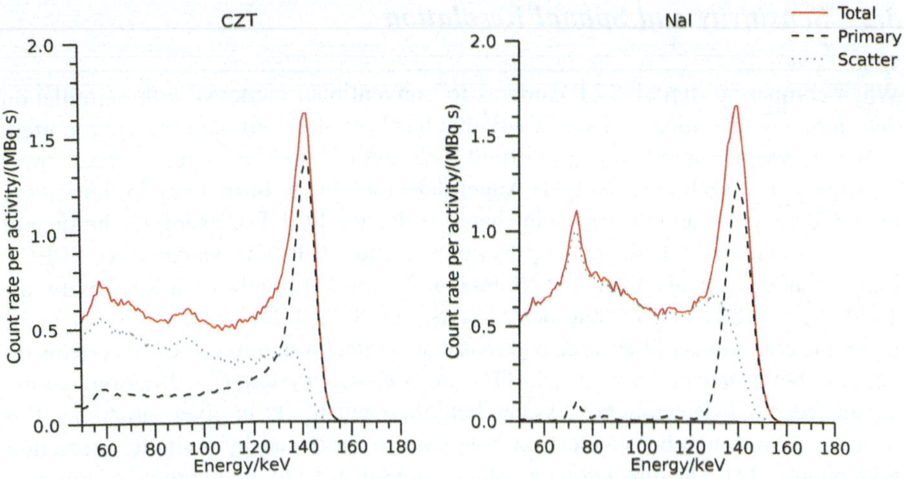

Fig. 3 Pulse height spectrum in a CZT detector equipped with a tungsten collimator and an NaI (Tl) detector equipped with a lead collimator for ^{99m}Tc. The source was a sphere with a diameter of 4 cm placed in a water cylinder with a diameter of 22 cm. Spectra were obtained using the SIMIND Monte Carlo programme [18], with the signal separated into primaries (counts from photons that are the direct result of decay) and scatter (counts resulting from a photon from a Compton-scatter event in the phantom)

The overall performance of gamma cameras with CZT detectors is intricately tied to the inherent characteristics of the detectors' material and the effectiveness of the collimator employed for projection acquisition. For a pixelated CZT system, it has been shown that collimator materials with high linear attenuation coefficients, such as tungsten, and pixel-geometry-matching architecture add to higher sensitivity [23, 24]. Ring-configured CZT systems are equipped with tungsten pixel-geometry-matching collimators, as illustrated in Fig. 1. However, using the same collimator across a wide range of energies can affect quantitative imaging since the performance of the collimator depends on photon energy. Another challenge associated with CZT detectors involves the presence of defective pixels, contributing to issues with detector performance. The term defective pixel aims to describe the effect that mainly occurs for pixels of small size when the interaction happens at the edge of a module. This can cause effects like the broadening of the spectral photopeak or unusually low or high response, which can have a negative impact on energy resolution [25]. While CZT detectors offer superior energy resolution, effects such as tailing, bad pixels and un-adjustable collimator combined with the sweeping motion of the ring-configured CZT SPECT-CT system need to be mapped out for the effective utilisation of the system in quantitative imaging.

3.2 Sensitivity and Spatial Resolution

When comparing digital CZT cameras to conventional cameras with scintillation detectors, it is the improved sensitivity that has been the dominant feature reported. This may seem contradictory as current CZT digital modules actually have lower sensitivity than traditional NaI(TI) Anger detectors due to limitations in fabricating thick CZT crystals and the losses in charge collection [26]. For example, the Imarad CdZnTe modules (Imarad Imaging Systems, Rehovot, Israel) designed for SPECT imaging have crystals with a thickness of 5 mm. These have a sensitivity for 140 keV, which is 71% of that measured using 3/8" NaI(TI) crystals [27]. In short, under the conditions of the same collimator and detector geometry, CZT systems do not have better sensitivity than a NaI(Tl) Anger camera system [27, 28]. Some of the advantages CZT detectors possess can help their sensitivity in given situations. For instance, a high number of electron-hole pairs is produced by a single interaction with photons [4]. The direct conversion is more sensitive to the photons' energy and is more localised than the process of detecting scintillation photons. This gives CZT detectors the advantage of both higher energy and intrinsic spatial resolution than usual scintillation detectors based on NaI(Tl). This advantage can translate to modest relative improvements in net counts when using a narrow energy window or when correcting for scatter. Improved intrinsic resolution can also allow for slightly less restrictive collimation when imaging a source close to the collimator face. These advantages are modest, however, and cannot fully explain the reported several-fold increases in sensitivity that CZT SPECT cameras can achieve with a similar or even smaller total detector area than conventional Anger SPECT systems [8, 10, 12, 29]. The principle advantage of the CZT detector modules used in digital SPECT systems is that they are small in size and free of useless detector areas at the edges; see Fig. 2. Ensuring that the total detector surface area is engaged in detecting photons at an optimised proximity and angle during the entire scan time enables a better compromise between sensitivity and resolution. Furthermore, the ability to perform imaging with several detectors simultaneously focused on a predefined region increases the number of counts acquired and, thus, the relative sensitivity to events in that region. Dedicated cardiac CZT-based cameras were among the first digital SPECT systems available and have been on the market for more than a decade. These cameras offered higher sensitivity and better energy resolution than Anger systems [30, 31], enabling reduced scan times and less activity administered to patients while providing images with better contrast and resolution [30]. Thorough studies using both phantom and clinical SPECT images have demonstrated the performance of CZT cameras for myocardial imaging to be superior to Anger cameras, even when equipped with dedicated cardiac collimators [10, 29, 30, 32]. An in-depth explanation of the improvements in myocardial SPECT with the arrival of dedicated digital cardiac cameras is presented by Ben-Haim et al. [32].

The multipurpose ring-configured CZT SPECT-CT systems present new opportunities with a larger diversity in potential clinical applications, but the degree of improved sensitivity is equally diverse. For instance, when evaluating the physical performance of the Veriton CZT camera, Desmonts et al. [10] reported superior sensitivity, higher energy resolution and better image contrast than the conventional Anger camera. The superiority of the sensitivity was, however, dependent on the scanning field of view. Using a point source in the air, the ring-configured system achieved an increase in reconstructed counts (counts·s^{-1}·MBq^{-1}) that was a factor of 1.6 times higher than the conventional Anger system. When scanning with focus over the point source, with no trans-axial motion of the detectors needed, a factor of 8 was achieved. Placing the point source in a head phantom, a factor of 1.2 (without focus) and 6 (with focus) was obtained. Scanning the point source in a body phantom, a factor of 6 (with focus) was also obtained, whereas without focus, the reconstructed counts were unchanged when compared to an Anger system. These measurements demonstrate the dependence of the scanning geometry for the ring-configured system [10]. For example, when performing a SPECT scan on a NEMA system volume sensitivity phantom uniformly filled with activity the StarGuide had a sensitivity 2.1 times higher than the two headed Anger SPECT camera and only 1.1 times higher when scanning the same phantom within a field of view 45 cm in diameter [33]. These findings demonstrate that as the scanning field of view increases, the sensitivity advantage provided by the ability of the smaller versitile CZT detector heads to have 100% of its area angled towards the scanning volume decreases. This is simply because the unused detector area of the conventional two-headed camera detector is less when scanning objects with a larger cross-sectional area.

3.3 *Focus Scan*

The body contouring and the detectors' sweeping motion, combined with gantry movement, aim to optimise the scan mode on a ring-configured CZT system, as illustrated in Fig. 1. The successful implementation of CZT detectors in nuclear cardiology is believed to have been caused by the combination of CZT detectors, no gantry movement and focused collimation, which are features that should preferably be preserved when going from a dedicated system to a multipurpose system. To maintain high sensitivity on the multipurpose CZT system, a focused scan mode can be enabled. A focused scan mode is not possible when scanning with parallel-hole collimators on a conventional Anger system or a two-headed CZT system, which makes this feature unique for this type of architecture. The small format of the detector and the sweep motion make it possible to control the speed of the movement of the detector. The user also has the flexibility to specify one or more regions of interest (ROIs), allowing for customised and targeted imaging. The user also has control over the temporal allocation within the ROI by setting the fraction of the total acquisition time dedicated to the focus region. Having, for instance, 2° between each angular step in the sweeping motion means the time spent acquiring data in each step will differ depending on the distance from each detector to the focus region. When scanning with the focus set to 100%, truncation artefacts may be introduced at the edge of the defined ROI because of the incomplete tomographic acquisition. To prevent truncation artefacts, scanning with a maximum of around 95% is preferable. When using a focused scan (<100%), the data outside the focus region is normalised to the counts within the focus region.

The increase in reconstructed counts, described by Desmont et al. when studying the physical performance of a focus mode, would not be achievable in a clinical setting [10]. An organ-specific scan is generally larger than the detectors' width, and therefore some movement is required. For this reason, the sweeping movement of the detectors may prevent these systems from achieving the performance obtained with a dedicated cardiac system.

3.4 *Scatter Compensation Methods*

The improved energy resolution of the CZT system translates to a better ability to reject scattered photons compared with an Anger system. Nevertheless, scatter is still present in the main energy window of CZT gamma cameras [34], and consideration of this effect is mandatory for quantitative studies [35, 36]. Also, diagnostic tasks for CZT gamma cameras have been described to benefit from explicit scatter compensation, such as cardiac imaging [37, 38] and cerebral imaging [35].

The most commonly used methods for scatter estimation in SPECT in clinical routine utilise the spectroscopic capabilities of the gamma camera by applying the dual-energy window (DEW) or triple-energy window (TEW) methods [39–41]. The resulting scatter estimates are either subtracted from the projections prior to reconstruction or, preferably, added to the forward projection in the iterative reconstruction. The low-energy tail of primary photons present in the spectra of pixelised CZT detectors (see Fig. 3) is a potential issue when implementing window-based scatter compensation methods since any scatter window below the full-energy peak will contain both primary and scattered photons. The conventional DEW method assumes proportionality between the scatter in the main window and the number of counts recorded in a lower scatter window [40], i.e.:

$$N^i_{\mathrm{sc}} = kN^i_{\mathrm{scw}}, \tag{1}$$

where N^i_{scw} is the recorded number of counts in the scatter window for pixel i, which are assumed to result from scattering; N^i_{sc} is the estimated scatter in the main window for pixel i; and k is an empirical constant. For CZT detectors, the scatter window will also contain primary photons due to the low-energy tail [37, 42]:

$$\begin{cases} N^i_{\mathrm{scw}} = k'N^i_{\mathrm{prim}} + N^i_{\mathrm{sc,scw}} \\ N^i = N^i_{\mathrm{prim}} + N^i_{\mathrm{sc}} \\ N^i_{\mathrm{sc}} = k''N^i_{\mathrm{sc,scw}} \end{cases} \Rightarrow N^i_{\mathrm{scw}} = k''\frac{N^i_{\mathrm{scw}} - k'N^i}{1 - k'k''} \tag{2}$$

where N_i is the number of counts recorded in the main energy window, N^i_{prim} is the number of primary photons in the main energy window, and $N^i_{\text{sc,scw}}$ is the number of scattered photons in the scatter window.

Much of the efforts of scatter compensation for CZT-based gamma cameras have stemmed from the requirements of dual-isotope imaging, e.g., the simultaneous imaging of ^{99m}Tc and ^{201}Tl, or ^{99m}Tc and ^{123}I. In addition to scatter, the low-energy tail will also lead to the crosstalk of primary photons between energy windows, which needs to be accounted for in quantitative evaluation. Kacperski et al. [43] developed a de-convolution method for the estimation of primary photons in projections from the simultaneous imaging of ^{99m}Tc and ^{201}Tl using a dedicated cardiac CZT gamma camera, further refined by Holstensson et al. [44] and Fan et al. [45] and applied to the simultaneous imaging of ^{99m}Tc and ^{123}I. Another option is the direct incorporation of scatter and the tailing phenomenon directly into the forward-projection model, as explored by Suzuki et al. [46] using a fast Monte Carlo simulation code for estimating the scatter component.

4 Clinical Applications

In a clinical setting, the improved energy resolution enables acquisition with a narrower energy window, which enhances scatter rejection and could optimise image contrast and the simultaneous acquisition of two radionuclides with adjacent photo peaks, such as ^{99m}Tc and ^{123}I. High photon efficiency translates into reduced scan time, which can be beneficial for dynamic SPECT. At the moment this chapter is being written, the majority of the studies reported assess the performance of phantom data, whereas patient data are scarce. The advantage of the ring-configured architecture in the clinical environment is, to a large extent, yet to be evaluated.

With the introduction of the ring-configured CZT SPECT system into clinical settings, it is imperative to verify the system against the ones used in clinical routine. This is necessary to safely implement the system, assuring similar diagnoses for patients on the new system versus the method currently used in clinical routine. The next step hereafter is to optimise the use of the new system. Duan et al. [47] had the opportunity to perform a head-to-head comparison for a variety of radioisotopes (^{67}Ga, ^{99m}Tc, ^{111}In, ^{123}I and ^{177}Lu) for different clinical indications. They compared an Anger system, a dual-head CZT system and a ring-configured CZT system and reported significantly improved performance when it comes to visual image quality for the ring-configured CZT SPECT system, vis-à-vis the Anger system and the dual-head CZT system.

4.1 *Single-Bed-Position SPECT Imaging*

The majority of examinations performed with ^{99m}Tc-labelled radiopharmaceuticals can be transitioned to a ring-configured CZT system, provided that clinicians accept a transition to SPECT-CT acquisition, since these systems are not able to perform planar imaging. However, multiple nuclear medicine sites still acquire planar images combined with SPECT-CT, and many guidelines still recommend the use of planar imaging [48–50]. To ease this transition, both commercial-ring-configured CZT SPECT-CT systems enable the re-projection of the 3D data set to planar images with the estimated attenuation, going from reconstructed 3D images to 2D anterior and posterior images that imitate the images that planar imaging would have produced. Ceric Andelius et al. [51] verified the diagnosis of cardiac transthyretin amyloidosis when using re-projected planar images from a ring-configured system compared to planar scintigraphy from a conventional Anger system, and Hagerman et al. [52] verified the use of heart-to-mediastinum ratio when diagnosing myocardial sympathetic innervation using ^{123}I-mIBG scintigraphy.

As previously mentioned, the multi-detector architecture equipped with body contouring allows the detectors to maintain a position close to the patient, especially when scanning in an extended mode, i.e. when scanning beyond the bed with the head extended. Since the bed does not obstruct the detectors, the detectors are able to be optimally positioned close to the brain. Bordonne et al. [53] studied camera performance for ^{99m}Tc-hexamethylpropyleneamine oxime (HMPAO) brain perfusion SPECT using a Hoffman 3D brain phantom and scan-rescan patient data. They reported a twofold increase in sensitivity compared to an Anger system as well as improved visual image quality, thus enabling reduced scan time with maintained image quality. ^{99m}Tc-HMPAO brain perfusion SPECT can be used to diagnose patients with neurological diseases, such as localising the foci of epilepsy in the ictal (seizure) state, where areas of reduced perfusion indicate chronic cerebral ischemia with acetazolamide challenge and neurodegenerative disorders. Imaging ictal areas is impossible with other nuclear imaging modalities, such as positron emission computed tomography-computed tomography (PET-CT), and therefore, an improved SPECT-CT system could be of great benefit for this patient population.

The improved resolution also enables the acquisition of dual isotopes with adjacent energy peaks in close proximity to each other. One of the most prevalent is ^{99m}Tc (140 keV) and ^{123}I (160 keV), where the energy resolution with an Anger system (FWHM ~10%) creates a large overlap between the peaks, and the clinical energy window used for the two isotopes (10%–20%) cannot properly separate them. The combined acquisition of ^{99m}Tc and ^{123}I is advantageous for simultaneous co-registered imaging from ictal and postictal (non-seizure) scans to localise the foci of epilepsy. ^{99m}Tc/^{123}I simultaneous acquisitions are also used to localise parathyroid adenoma. Additional dual isotopes that could benefit from the improved energy resolution are ^{81m}Kr (190 keV) and ^{99m}Tc-MAA (140 keV) for ventilation-perfusion scintigraphy and ^{201}Tl (70 keV and 167 keV) and ^{99m}Tc for rest-stress cardiac scintigraphy.

For their initial clinical experience when transitioning from a conventional Anger system to a ring-configured CZT SPECT-CT system for lung scintigraphy, Ceric Andelius et al. [11] reported no difference in visual image quality but stated that there is a possibility to reduce scan time further while maintaining image quality. They also emphasized the disadvantage of the relatively short axial FOV, which is

27.5 cm for the StarGuide™ system where patients with lungs larger than 25 cm, need to be scanned with more than one bed position extending the scan time. This would not be the case for a FOV of 32 cm, which is the case for the other ring-configured system on the market, Veriton™. The improved sensitivity can be used to reduce acquisition time when scanning more than one bed position. Callaud et al. [54] presented the advantages of dual-isotope lung imaging and the possibility of ring-configured CZT-SPECT systems to reduce scan time by 50% while maintaining visual image quality without a quantitative impact. Ventilation-perfusion SPECT/SPECT-CT has multiple clinical indications, including the diagnosis of acute and chronic pulmonary embolism and the quantification of lung function before pulmonary surgery, and it shows promise as a prognostic and diagnostic tool in heart failure, as demonstrated by Jögi et al. [55].

^{99m}Tc-dimercaptosuccinic acid (^{99m}Tc-DMSA) for imaging renal cortex is a nuclear medicine application that would benefit from reducing scan time since the majority of patients are children. Bahloul et al. [56] reported higher diagnostic quality and improved sensitivity using ring-configured CZT-SPECT compared to planar conventional scintigraphy from an Anger system [56]. The fact that SPECT outperforms planar scintigraphy is not new, nor is it specific to a ring-configured CZT system. However, the ability to reduce the acquisition time of SPECT to that of traditional planar imaging or less due to improved sensitivity would be beneficial. Dietz et al. [57] reported the feasibility of short dimercaptosuccinic acid (DMSA) examinations for a dual-headed CZT system that maintained image quality and reproducibility while the total acquisition time was reduced from 15 min to 5 min [57]. Bahloul et al. showed a 70% increase in sensitivity when assessing images generated for the ring-configured CZT system compared to planar scintigraphy, enabling further reduction in scan time, and also reported a high agreement when calculating relative kidney function when comparing the systems [56].

With no gantry movement and a temporal resolution of 1 s, 4D SPECT-CT could be one of the main advantages of ring-configured CZT systems. Two appointed images of the month by the European Journal of Nuclear Medicine and Molecular Imaging (2022) illustrate the possibilities and advantages of 4D SPECT-CT with ring-configured CZT systems. One highlights dynamic cardiac SPECT and flow measurements in myocardial perfusion imaging. The authors state that the quality of 4D SPECT-CT from the ring-configured system was comparable to that of myocardial perfusion PET [58]. Mairall et al. [59] reported a three-phase bone quantitative SPECT for analysing Muller Weiss syndrome, a rare condition characterised by a misshaped navicular bone in the foot, which could lead to degenerative arthritis in the neighbouring joints [59]. The authors state that the three-phase 4D SPECT can be used to indicate inappropriate blood flow. Thus, 4D SPECT can be used to identify the pathogenesis of Muller-Weiss syndrome and can also be useful when detecting necrosis and ischemia-related bone disease.

4.2 Whole-Body SPECT Imaging

The acquisition of SPECT-CT on a dual-headed camera is time-consuming, especially when using more than one bed position. Improved sensitivity enables reduced scan times and makes it possible to perform whole-body scans within a reasonable acquisition time. Skeletal imaging using ^{99m}Tc-labelled diphosphonates (hydroxydiphosphonate (HDP)) lends itself well to this type of camera geometry, where it can be performed as whole-body SPECT-CT studies. Performing bone scans this way instead of planar whole-body acquisitions, often combined with single-bed position SPECT-CT, simplifies workflows as no decision on if and where to perform tomographic imaging needs to be made during the examination. Always having a whole-body SPECT-CT provides more data for interpretation and generally reduces ambiguity.

Ceric Andelius et al. [11] showed that the acquisition time can be reduced by 70% for one-bed position with a ring-configured CZT system compared to an Anger system, showing that a whole-body SPECT can be performed within a reasonable acquisition time [11]. Figure 4 shows an example of [^{99m}Tc]Tc-HDP scan for a patient on both a ring-configured CZT SPECT system and an Anger system where the time per bed position ranges from 1 min and 6 min. Bahloul et al. [60] studied the monitoring of vertebral fractures and the reproducibility of standardised uptake values using the system (Veriton™) in a PET-like way, showing promising results. The low- and medium-energy range and the improved sensitivity also enable rapid whole-body ^{177}Lu scans.

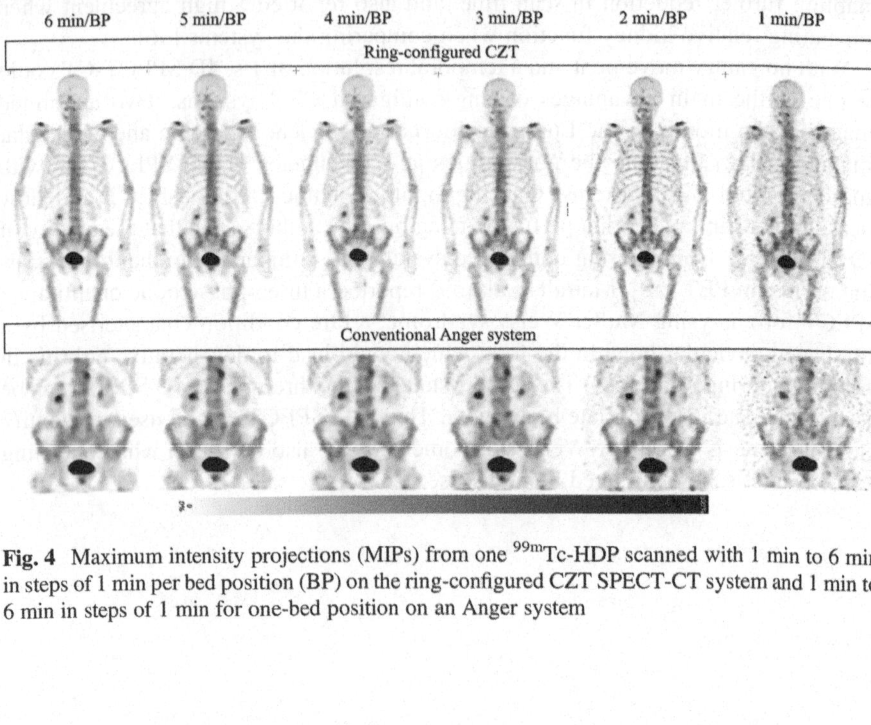

Fig. 4 Maximum intensity projections (MIPs) from one ^{99m}Tc-HDP scanned with 1 min to 6 min in steps of 1 min per bed position (BP) on the ring-configured CZT SPECT-CT system and 1 min to 6 min in steps of 1 min for one-bed position on an Anger system

It is important to keep in mind that when assessing image quality, the reconstruction algorithm has a critical role. Recent years have been marked by the introduction of maximum á posteriori (MAP) reconstruction methods, which incorporate prior information about the reconstruction problem [61–65]. Regularised reconstruction algorithms have played a crucial role in enhancing image quality, and recently, artificial intelligence (AI)-driven approaches have been introduced in post-reconstruction improvements to images but are also now being incorporated into reconstruction algorithms [66]. The combination of a ring-shaped CZT system and deep-learning noise reduction algorithms can further reduce acquisition time by 40% for whole-body bone scans [67]. The benefit of integrating AI into the steps of the reconstruction process remains to be seen.

4.3 *Focused Scan*

As previously mentioned, applying the focus mode enables the user to define an ROI that focuses tomographic acquisition on the region of a specific organ. The acquisition is then performed with the detectors spending more time acquiring data for the defined region. The fraction of the acquisition time to be spent within the focus region is set by the user, typically within the range of 50–90%.

The most prevalent nuclear medicine examination performed worldwide is myocardial perfusion imaging (MPI), which has a key role in diagnosing cardiovascular disease. When CZT detectors were introduced in the nuclear medicine field, the target application was initially MPI. The dedicated cardiac CZT systems showed improved performance with improved sensitivity and spatial resolution, enabling the reduction of administered activity and/or reduced scan time. When imaging with the ring-shaped architecture, myocardial perfusion SPECT is generally acquired with the focus mode. Figure 5 shows a bullseye plot from a patient stress scan performed on a conventional Anger system (Discovery 670, GE Healthcare), a ring-configured CZT SPECT-CT system (StarGuide™, GE Healthcare) with focus mode and a dedicated CZT cardiac system (Discovery 530c, GE Healthcare). Some small variations can be seen, which is to be expected due to variations in the camera architecture, collimators and detector material used. However, the clinical assessment does not differ between examinations. In the example, the scan time was 15 min with the conventional system and 8 min with the ring-configured system, and for the dedicated CZT system, the patients were scanned 8 min in the supine position (patient lying on their back) and 8 min in the prone position (patient lying on their stomach).

Ceric Andelius et al. [11] presented their first clinical experience using a focus scan on a ring-configured CZT system in myocardial perfusion SPECT compared to a dedicated C-shaped CZT system with pinhole collimation. They reported similar diagnostic accuracy between the systems with the possibility of matching the scan time used on a dedicated CZT cardiac system. The examinations also benefit from attenuation correction with the addition of CT on the ring-configured CZT system, as opposed to the dedicated CZT cardiac system, where additional imaging would be

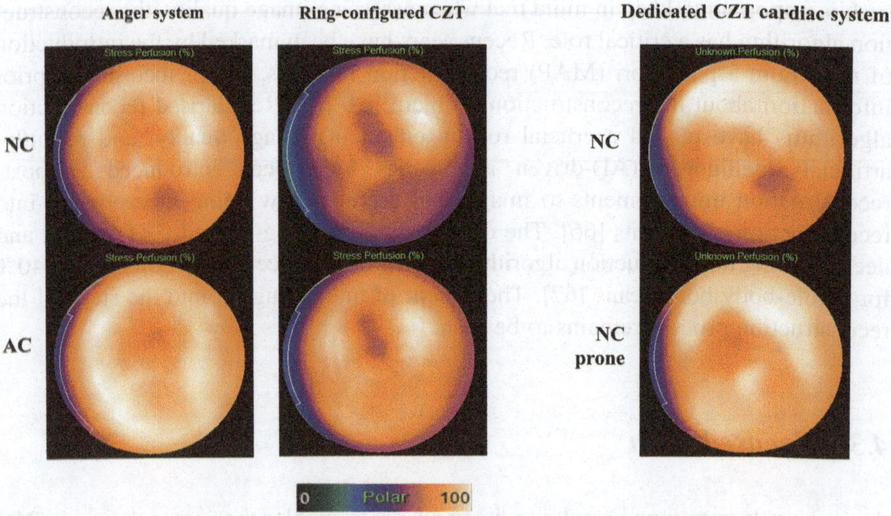

Fig. 5 Bullseye plots from a patient stress scan performed on a conventional Anger system, ring-configured CZT system and dedicated CZT system. For the Anger and ring-configured CZT SPECT-CT systems non-corrected (NC) and attenuation corrected (AC), images are presented, whereas NC and NC prone images are shown for the dedicated CZT system

required. In total, this makes acquisition on the ring-configured CZT system 50% shorter. Figure 6 shows one patient's stress and rest scan acquired with full scan time (100%) and reductions of 80%, 60%, 50% and 40% of the total scan times.

Bahloul et al. [56] performed DMSA renal scans with and without a focus mode, reporting an increase in sensitivity from 70% to 180% for the focused scan compared to planar scintigraphy on a conventional Anger system. The authors also reported two cases indicating a difference in the resulting diagnosis between the systems, where the assessment made on the ring-configured focused scan was ruled to be the correct diagnosis. Figure 7 shows DMSA SPECT images from one patient scanned on a ring-configured CZT system and an Anger system. For the ring-configured CZT system, the scan is performed in focus mode where 100% corresponds to a total acquisition time of 10 min. A coronal slice through the kidney for both camera systems for an acquisition time 4 min (40%) for the image generated from focus scan on the ring-configured system.

Goshen et al. [8] reported on the assessment of focused tomographic acquisition of the lumber spine for one patient, stating both superior spatial resolution and contrast when compared to non-focused scans. Bailly et al. [68] investigated the image quality for SPECT imaging of ^{123}I for dopamine transporters also reporting an improved visual image quality and improved resolution. These results are promising but need to be confirmed in a larger patient cohort. These initial comparisons do, however, show a potential advantage of these systems for cardiac, head and DMSA scan applications.

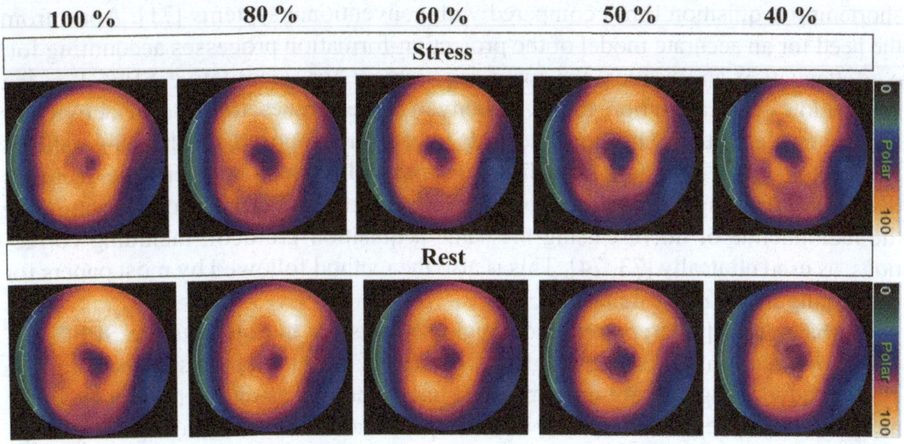

Fig. 6 Cardiac stress and rest images with scan times ranging from 100% to 40% (100% corresponds to 16 min stress scan and 6 min and 40 s rest scan). Reduced perfusion is seen in the basal and midventricular inferolateral and lateral walls of the left ventricle

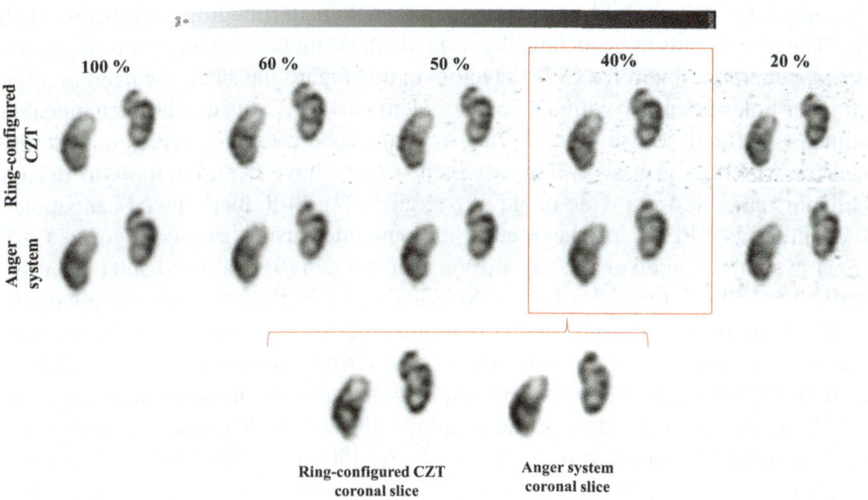

Fig. 7 A DMSA scan from one patient scanned on an Anger system and a ring-configured CZT system. Maximum intensity projection (MIP) images from a scan time of 10 min (100%) and reduced scan times are presented. For the 4 min (40%) scan time, a coronal slice is shown for both the Anger system and ring-configured CZT system

4.4 *Quantitative Imaging*

A range of publications have proven the quantitative ability of SPECT using Anger cameras [69, 70], and there has been an interest in studying the quantitative accuracy also of ring-configured CZT systems. A point of interest is whether quantitative SPECT can also benefit from the presumed better system sensitivity, thereby shortening acquisition times compared with conventional systems [71]. Apart from the need for an accurate model of the projection-formation processes accounting for, e.g. attenuation and scatter, the main requirement for quantitative SPECT is the application of a calibration factor to convert voxel values to units of activity or activity concentration. Different methods have been proposed for deriving calibration factors [72], but most guidelines recommend a method where the factor is derived by imaging a water phantom with a well-known activity concentration for the radionuclide of interest using the same acquisition protocol, including corrections, as used clinically [73, 74]. This is also the method followed by most papers for ring-configured CZT systems [60, 75–77].

Dorbala et al. [75] reported encouraging results for the quantitative SPECT of ^{99m}Tc-pyrophosphate for the diagnosis of cardiac amyloidosis using the Veriton system, advocating activity-concentration-based metrics as an alternative to visual analysis [75]. Likewise, the possibility for quantitative SPECT imaging using ring-configured CZT-based gamma cameras with results presented in terms of SUVs was reported by Bahloul et al. [60] for skeletal imaging.

One example of quantitative SPECT tasks is monitoring and performing dosimetry of radionuclide therapy. Since the choice of radionuclide for these applications is governed by its suitability for therapy rather than its imaging properties, such SPECT imaging tends to be technically more demanding than imaging for diagnostic tasks. A general problem for CZT detectors in this regard has been the need to align collimator holes precisely with the detector elements [78]. This makes exchangeable collimators difficult to use, thus forcing systems to be used at energies outside the range for which they were optimised, even if there have been attempts to design collimators suitable for a wide range of energies [24]. Still, there have been studies for quantitative SPECT of therapeutic radionuclides using ring-configured CZT SPECT systems. Danieli et al. [76] and Vergnaud et al. [79] evaluated and optimised quantitative SPECT for ^{177}Lu for the StarGuide and Veriton systems, respectively [76, 77]. Song et al. [71] compared patient imaging by ^{177}Lu-labelled radiopharmaceuticals using conventional gamma cameras to a ring-configured system, claiming the possibility of a major shortening of acquisition time using the ring-shaped system [71]. Examples of other therapeutic radionuclides for which imaging with ring-shaped systems has been demonstrated are ^{153}Sm [80] and ^{223}Ra [10]. For the latter of these radionuclides, SPECT is rarely performed using conventional systems due to the low count rates and the resulting poor repeatability of activity-concentration estimates [81]. Another issue of the SPECT imaging of therapeutic radionuclides using CZT-based systems has been problems associated with scatter compensation for this detector type. For quantitative SPECT in particular, scatter compensation is important to suppress spatial variation in the estimated activity concentrations. Huh et al. [82] compared the use of the original DEW scatter compensation method and the modified version thereof, detailed in Eq. 2, for ^{177}Lu and ^{99m}Tc. The modified version resulted in a better contrast recovery, though at the expense of slightly worse noise properties [82].

References

1. Anger, H. O. (1964). Scintillation camera with multichannel collimators. *Journal of Nuclear Medicine, 5*, 515–531.
2. Anger, H. O. (1958). Scintillation camera. *Review of Scientific Instruments, 29*, 27–33.
3. Madsen, M. T. (2007). Recent advances in SPECT imaging. *Journal of Nuclear Medicine, 48*(4), 661–673.
4. Peterson, T. E., & Furenlid, L. R. (2011). SPECT detectors: The Anger camera and beyond. *Physics in Medicine and Biology, 56*(17), R145–R182.
5. Ljungberg, M., & Pretorius, P. H. (2018). SPECT/CT: An update on technological developments and clinical applications. *The British Journal of Radiology, 91*(1081), 20160402.
6. Slomka, P. J., Patton, J. A., Berman, D. S., & Germano, G. (2009). Advances in technical aspects of myocardial perfusion SPECT imaging. *Journal of Nuclear Cardiology, 16*(2), 255–276.
7. Slomka, P. J., Miller, R. J. H., Hu, L. H., Germano, G., & Berman, D. S. (2019). Solid-state detector SPECT myocardial perfusion imaging. *Journal of Nuclear Medicine, 60*(9), 1194–1204.
8. Goshen, E., Beilin, L., Stern, E., Kenig, T., Goldkorn, R., & Ben-Haim, S. (2018). Feasibility study of a novel general purpose CZT-based digital SPECT camera: initial clinical results. *EJNMMI Physics, 5*(1), 6.
9. Sharir, T., Ben-Haim, S., Merzon, K., Prochorov, V., Dickman, D., Ben-Haim, S., et al. (2008). High-speed myocardial perfusion imaging initial clinical comparison with conventional dual detector anger camera imaging. *JACC: Cardiovascular Imaging, 1*(2), 156–163.
10. Desmonts, C., Bouthiba, M. A., Enilorac, B., Nganoa, C., Agostini, D., & Aide, N. (2020). Evaluation of a new multipurpose whole-body CzT-based camera: Comparison with a dual-head Anger camera and first clinical images. *EJNMMI Physics, 7*(1), 18.
11. Andelius, I. C., Minarik, D., Persson, E., Mosén, H., Valind, K., Trägårdh, E., et al. (2024). First clinical experience of a ring-configured cadmium zinc telluride camera: A comparative study versus conventional gamma camera systems. *Clinical Physiology and Functional Imaging, 44*(1), 79–88.
12. Morelle, M., Bellevre, D., Hossein-Foucher, C., Manrique, A., & Bailliez, A. (2020). First comparison of performances between the new whole-body cadmium-zinc-telluride SPECT-CT camera and a dedicated cardiac CZT camera for myocardial perfusion imaging: Analysis of phantom and patients. *Journal of Nuclear Cardiology, 27*(4), 1261–1269.
13. Arvola, S., Seppanen, M., Timonen, K. L., Rautio, P., Ettala, O., Anttinen, M., et al. (2022). Detection of prostate cancer bone metastases with fast whole-body (99 m)Tc-HMDP SPECT/CT using a general-purpose CZT system. *EJNMMI Physics, 9*(1), 85.
14. Oddstig, J., Hedeer, F., Jögi, J., Carlsson, M., Hindorf, C., & Engblom, H. (2013). Reduced administered activity, reduced acquisition time, and preserved image quality for the new CZT camera. *Journal of Nuclear Cardiology, 20*(1), 38–44.
15. Gregoire, B., Pina-Jomir, G., Bani-Sadr, A., Moreau-Triby, C., Janier, M., & Scheiber, C. (2018). Four-minute bone SPECT using large-field cadmium-zinc-telluride camera. *Clinical Nuclear Medicine, 43*(6), 389–395.
16. Bani Sadr, A., Testart, N., Tylski, P., & Scheiber, C. (2019). Reduced scan time in 123I-FP-CIT SPECT imaging using a large-field cadmium-zinc-telluride camera. *Clinical Nuclear Medicine, 44*(7), 568–569.
17. Ritt, P., Sanders, J., & Kuwert, T. (2014). SPECT/CT technology. *Clinical and Translational Imaging, 2*, 445–457.
18. Ljungberg, M., & Strand, S. E. (1989). A Monte Carlo program for the simulation of scintillation camera characteristics. *Computer Methods and Programs in Biomedicine, 29*(4), 257–272.
19. McGregor, D. S., & Hermon, H. (1997). Room-temperature compound semiconductor radiation detectors. *Nuclear Instruments & Methods A, 395*(1), 101–124.

20. Bolotnikov, A. E., Camarda, G. S., Cui, Y., Yang, G., Hossain, A., Kim, K., et al. (2013). Characterization and evaluation of extended defects in CZT crystals for gamma-ray detectors. *Journal of Crystal Growth, 379*, 46–56.
21. Pretorius, P. H., Liu, C., Fan, P., Peterson, M., & Ljungberg, M. (2015). Monte Carlo simulations of the GE discovery alcyone CZT SPECT systems. *IEEE Transactions on Nuclear Science, 62*(3), 832–839.
22. Kim, J. C., Anderson, S. E., Kaye, W., Zhang, F., Zhu, Y. F., Kaye, S. J., et al. (2011). Charge sharing in common-grid pixelated CdZnTe detectors. *Nuclear Instruments & Methods A, 654*(1), 233–243.
23. Weng, F., Bagchi, S., Huang, Q., & Seo, Y. (1997). Design studies of a CZT-based detector combined with a pixel-geometry-matching collimator for SPECT imaging. *IEEE Nuclear Science Symposium Conference Record, 2013*(2013), 1–4.
24. Weng, F. H., Bagchi, S., Zan, Y. L., Huang, Q., & Seo, Y. H. (2016). An energy-optimized collimator design for a CZT-based SPECT camera. *Nuclear Instruments and Methods A, 806*, 330–339.
25. Xie, Z. H., Li, S. Y., Yang, K., Xu, B. X., & Ren, Q. S. (2016). Evaluation of a wobbling method applied to correcting defective pixels of CZT detectors in SPECT imaging. *Sensors-Basel, 16*(6), 772.
26. Bolotnikov, A. E., Cook, W. R., Harrison, F. A., Wong, A. S., Schindler, S. M., & Eichelberger, A. C. (1999). Charge loss between contacts of CdZnTe pixel detectors. *Nuclear Instruments and Methods A, 432*(2–3), 326–331.
27. Wagenaar, D. J., Chowdhury, S., Engdahl, J. C., & Burckhardt, D. D. (2003). Planar image quality comparison between a CdZnTe prototype and a standard NaI(T1) gamma camera. *Nuclear Instruments and Methods A, 505*(1–2), 586–589.
28. Amrami, R. Shani, G., Hefetz, Y., Blevis, I., & Pansky, A. (2000). A comparison between the performance of a pixellated CdZnTe based gamma camera and Anger NaI(TI) scintillator gamma camera. In *Proceeding of the 22nd annual international conference of the IEEE; 2000 23–28 July 2000*; Chicago, IL, USA: Engineering in Medicine and Biology Society.
29. Nkoulou, R., Pazhenkottil, A. P., Kuest, S. M., Ghadri, J. R., Wolfrum, M., Husmann, L., et al. (2011). Semiconductor detectors allow low-dose-low-dose 1-day SPECT myocardial perfusion imaging. *Journal of Nuclear Medicine, 52*(8), 1204–1209.
30. Imbert, L., Poussier, S., Franken, P. R., Songy, B., Verger, A., Morel, O., et al. (2012). Compared performance of high-sensitivity cameras dedicated to myocardial perfusion SPECT: A comprehensive analysis of phantom and human images. *Journal of Nuclear Medicine, 53*(12), 1897–1903.
31. Takahashi, Y., Miyagawa, M., Nishiyama, Y., Ishimura, H., & Mochizuki, T. (2013). Performance of a semiconductor SPECT system: Comparison with a conventional Anger-type SPECT instrument. *Annals of Nuclear Medicine, 27*(1), 11–16.
32. Ben-Haim, S., Kennedy, J., & Keidar, Z. (2016). Novel cadmium zinc telluride devices for myocardial perfusion imaging-technological aspects and clinical applications. *Seminars in Nuclear Medicine, 46*(4), 273–285.
33. Haddock, B., & Ceric Andelius, I. (2022). Performance of ring configured CZT SPECT at small and large fields of view. Manuscript unpublished.
34. Erlandsson, K., Kacperski, K., van Gramberg, D., & Hutton, B. F. (2009). Performance evaluation of D-SPECT: A novel SPECT system for nuclear cardiology. *Physics in Medicine and Biology, 54*(9), 2635–2649.
35. Izawa, D., Ito, T., Sanada, H., Oku, T., Nakamoto, A., & Hiraki, H. (2023). Verification of optimal conditions for the scattering correction of ^{123}I-FP-CIT SPECT on a single-photon emission tomography system with a two-detector whole-body cadmium-zinc-telluride semiconductor detector. *Radiological Physics and Technology, 4*, 569–577.
36. Mann, S., Perez, K., & Tornai, M. (2011). Comparison of dual-window scatter correction and effective attenuation coefficients for quantification in dedicated breast SPECT. In *IEEE Nuclear Science Symposium Conference Record* (pp. 3760–3763).

37. Pourmoghaddas, A., Vanderwerf, K., Ruddy, T. D., & Wells, R. G. (2015). Scatter correction improves concordance in SPECT MPI with a dedicated cardiac SPECT solid-state camera. *Journal of Nuclear Cardiology, 22*(2), 334–343.
38. D'estanque, E., Hedon, C., Lattuca, B., Bourdon, A., Benkiran, M., Verd, A., et al. (2017). Optimization of a simultaneous dual-isotope ^{201}Tl/^{123}I-MIBG myocardial SPECT imaging protocol with a CZT camera for trigger zone assessment after myocardial infarction for routine clinical settings: Are delayed acquisition and scatter correction necessary? *Journal of Nuclear Cardiology, 24*(4), 1361–1369.
39. Ichihara, T., Ogawa, K., Motomura, N., Kubo, A., & Hashimoto, S. (1993). Compton scatter compensation using the triple-energy window method for single-isotope and dual-isotope SPECT. *Journal of Nuclear Medicine, 34*(12), 2216–2221.
40. Jaszczak, R. J., Greer, K. L., Floyd, C. E., Harris, C. C., & Coleman, R. E. (1984). Improved SPECT quantification using compensation for scattered photons. *Journal of Nuclear Medicine, 25*(8), 893–900.
41. Ogawa, K., Harata, Y., Ichihara, T., Kubo, A., & Hashimoto, S. (1991). A practical method for position-dependent compton-scatter correction in single photon-emission Ct. *IEEE Transactions on Medical Imaging, 10*(3), 408–412.
42. Mann SD, Tornai MP. Initial evaluation of a modified dual-energy window scatter correction method for CZT-based gamma cameras for breast SPECT. Proceedings of SPIE 2015;9413.
43. Kacperski, K., Erlandsson, K., Ben-Haim, S., & Hutton, B. F. (2011). Iterative deconvolution of simultaneous ^{99m}Tc and ^{201}Tl projection data measured on a CdZnTe-based cardiac SPECT scanner. *Physics in Medicine and Biology, 56*(5), 1397–1414.
44. Holstensson, M., Erlandsson, K., Poludniowski, G., Ben-Haim, S., & Hutton, B. F. (2015). Model-based correction for scatter and tailing effects in simultaneous ^{99m}Tc and ^{123}I imaging for a CdZnTe cardiac SPECT camera. *Physics in Medicine and Biology, 60*(8), 3045–3063.
45. Fan, P., Hutton, B. F., Holstensson, M., Ljungberg, M., Pretorius, P. H., Prasad, R., et al. (2015). Scatter and crosstalk corrections for ^{99m}Tc/^{123}I dual-radionuclide imaging using a CZT SPECT system with pinhole collimators. *Medical Physics, 42*(12), 6895–6911.
46. Atsuro Suzuki, W. T., Ueno, Y., Kobashi, K., Toyonaga, T., Shiga, T., & Tamaki, N. (2016). Monte Carlo-based scatter correction considering the tailing effect of a CdTe detector for dual-isotope brain SPECT imaging. *Biomedical Physics & Engineering Express, 2*(4), 045010.
47. Duan, H., Ferri, V., Castaneda, P., Visser, T., Luong, K., Davidzon, G., et al. (2022). Head-to-head comparison of a conventional Or CZT-based SPECT/CT with a next generation multidetector CZT based SPECT/CT system. *Journal of Nuclear Medicine, 63*, 3324.
48. Vali, R., Armstrong, I. S., Bar-Sever, Z., et al. (2022). SNMMI procedure standard/EANM practice guideline on pediatric [99mTc]Tc-DMSA renal cortical scintigraphy: An update. *Clinical and Translational Imaging, 10*, 173–184.
49. Van den Wyngaert, T., Strobel, K., Kampen, W. U., Kuwert, T., van der Bruggen, W., Mohan, H. K., et al. (2016). The EANM practice guidelines for bone scintigraphy. *European Journal of Nuclear Medicine and Molecular Imaging, 43*(9), 1723–1738.
50. Dorbala, S., Ando, Y., Bokhari, S., Dispenzieri, A., Falk, R. H., Ferrari, V. A., et al. (2021). ASNC/AHA/ASE/EANM/HFSA/ISA/SCMR/SNMMI expert consensus recommendations for multimodality imaging in cardiac amyloidosis: Part 1 of 2-evidence base and standardized methods of imaging. *Circulation: Cardiovascular Imaging, 14*(7), e000029.
51. Andelius, I. C., Fridriksdóttir, R., Minarik, D., Hedeer, F., Stenvall, A., Trägårdh, E., & Oddstig, J. (2024). Verification of reprojected planar images generated from a ring-configured cadmium zinc telluride gamma camera in scintigraphy for diagnosing transthyretin cardiac amyloidosis. *European Heart Journal – Imaging Methods and Practice, 2*(1), qyae051.
52. Hagerman, J., Minarik, D., Smith, R., Stormrud, E., Hall, S., Nahimi, A., Hansson, O., & Hedeer, H. (2023). Comparing the heart-to-mediastinum ratio of 123I-mIBG myocardial scintigraphy between a 2D image and a pseudo planar image converted from a 3D image acquired with a conventional and a ring CZT gamma camera. *Annual Congress of the European Association of Nuclear Medicine*, September 9–13, 2023, Vienna, Austria.

53. Bordonne, M., Chawki, M. B., Marie, P. Y., Zaragori, T., Roch, V., Grignon, R., et al. (2020). High-quality brain perfusion SPECT images may be achieved with a high-speed recording using 360 degrees CZT camera. *EJNMMI Physics, 7*(1), 65.
54. Callaud, A., Metrard, G., & Bailly, M. (2023). 3D-ring CZT system with new low- and medium-energy range ultrafast dual-isotope lung SPECT/CT improvement. *Clinical Nuclear Medicine, 48*, 716–718.
55. Jögi, J., Al-Mashat, M., Rådegran, G., Bajc, M., & Arheden, H. (2018). Diagnosing and grading heart failure with tomographic perfusion lung scintigraphy: Validation with right heart catheterization. *ESC Heart Failure, 5*(5), 902–910.
56. Bahloul, A., Imbert, L., Karcher, G., Marie, P.-Y., & Verger, A. (2020). DMSA renal investigations performed with 360 whole-body CZT SPECT camera as compared with the convetional method based on planar images from an Anger camera. *Journal of Nuclear Medicine., 61*, 244.
57. Dietz, M., Jacquet-Francillon, N., Bani Sadr, A., Collette, B., Mure, P. Y., Demede, D., et al. (2023). Ultrafast cadmium-zinc-telluride-based renal single-photon emission computed tomography: Clinical validation. *Pediatric Radiology, 53*(9), 1911–1918.
58. Bailly, M., Callaud, A., & Metrard, G. (2023). Dynamic cardiac SPECT with flow measurement using 3D-ring CZT: when SPECT is inspired by PET. *European Journal of Nuclear Medicine and Molecular Imaging, 50*(6), 1837–1839.
59. Mairal, E., Imbert, L., Marie, P. Y., & Bahloul, A. (2022). Three-phase bone quantitative-SPECT of navicular bones with a high-sensitivity whole-body CZT-camera in a Mueller-Weiss syndrome. *European Journal of Nuclear Medicine and Molecular Imaging, 49*(9), 3295–3296.
60. Bahloul, A., Verger, A., Blum, A., Bilal Chawki, M., Perrin, M., Melki, S., Karcher, G., Marie, P. Y., & Imbert, L. (2021). Bone scintigraphy of vertebral fractures with a whole-body CZT camera in a PET-like utilization. *Frontiers in Nuclear Medicine, 1*, 1–8.
61. Green, P. J. (1990). Bayesian reconstructions from emission tomography data using a modified EM algorithm. *IEEE Transactions on Medical Imaging, 9*(1), 84–93.
62. Shepp, L. A. V. Y. (1982). Maximum likelihood reconstruction for emission tomography. *IEEE Transactions on Medical Imaging, 1*,113–122.
63. Hudson, H. M., & Larkin, R. S. (1994). Accelerated image-reconstruction using ordered subsets of projection data. *IEEE Transactions on Medical Imaging, 13*(4), 601–609.
64. Lange, K., & Carson, R. (1984). Em reconstruction algorithms for emission and transmission tomography. *Journal of Computer Assisted Tomography, 8*(2), 306–316.
65. Ahn, S., & Fessler, J. A. (2003). Globally convergent image reconstruction for emission tomography using relaxed ordered subsets algorithms. *IEEE Transactions on Medical Imaging, 22*(5), 613–626.
66. Reader, A. J., & Pan, B. L. (2023). AI for PET image reconstruction. *British Journal de Radiologie, 96*(1150).
67. Bahloul, A., Verger, A., Lamash, Y., Roth, N., Dari, D., Marie, P. Y., & Imbert, L. (2023). Ultra-fast whole-body bone tomoscintigraphies achieved with a high-sensitivity 360° CZT camera and a dedicated deep-learning noise reduction algorithm. *European Journal of Nuclear Medicine and Molecular Imaging, 5*, 1215–1220.
68. Bailly, M., Merman, M., Le Rouzic, G., & Metrad, G. (2021). Dopamine transporter imaging using 3D-ring CZT StarGuide SPECT/CT: First phantom and clinical results. *Journal of Nuclear Medicine., 62*, 1126.
69. Shcherbinin, S., Celler, A., Belhocine, T., Vanderwerf, R., & Driedger, A. (2008). Accuracy of quantitative reconstructions in SPECT/CT imaging. *Physics in Medicine and Biology, 53*(17), 4595–4604.
70. Zeintl, J., Vija, A., Yahil, A., Hornegger, J., & Kuwert, T. (2010). Quantitative accuracy of clinical Tc-99 m SPECT/CT using OSEM-3D with attenuation and scatter correction. *Journal of Nuclear Medicine, 51*, 583.
71. Song, H., Ferri, V., Duan, H., Aparici, C. M., Davidzon, G., Franc, B. L., Moradi, F., Nguyen, J., Shah, J., & Iagaru, A. (2023). SPECT at the speed of PET: a feasibility study of CZT-based whole-body SPECT/CT in the post 177Lu-DOTATATE and 177Lu-PSMA617 setting. *European Journal of Nuclear Medicine and Molecular Imaging, 50*, 2250–2257.

72. Zhou, Y., Lu, J., Ran, C. Z., & Zhang, J. H. (2018). Recent advances in quantitative nuclear medicine and molecular imaging. *Contrast Media & Molecular Imaging, 2018*, 1–2.
73. Dewaraja, Y. K., Frey, E. C., Sgouros, G., Brill, A. B., Roberson, P., Zanzonico, P. B., et al. (2012). MIRD pamphlet No. 23: quantitative SPECT for patient-specific 3-dimensional dosimetry in internal radionuclide therapy. *Journal of Nuclear Medicine, 53*(8), 1310–1325.
74. Ljungberg, M., Celler, A., Konijnenberg, M. W., Eckerman, K. F., Dewaraja, Y. K., Sjögreen-Gleisner, K., et al. (2016). MIRD Pamphlet No. 26: Joint EANM/MIRD Guidelines for Quantitative 177Lu SPECT Applied for Dosimetry of Radiopharmaceutical Therapy. *Journal of Nuclear Medicine, 57*(1), 151–162.
75. Dorbala, S., Park, M. A., Cuddy, S., Singh, V., Sullivan, K., Kim, S., et al. (2021). Absolute Quantitation of Cardiac (99 m)Tc-Pyrophosphate Using Cadmium-Zinc-Telluride-Based SPECT/CT. *Journal of Nuclear Medicine, 62*(5), 716–722.
76. Danieli, R., Stella, M., Leube, J., Tran-Gia, J., Marin, C., Uribe, C. F., et al. (2023). Quantitative (177)Lu SPECT/CT imaging for personalized dosimetry using a ring-shaped CZT-based camera. *EJNMMI Physics, 10*(1), 64.
77. Vergnaud, L., Badel, J. N., Giraudet, A. L., Kryza, D., Mognetti, T., Baudier, T., et al. (2023). Performance study of a 360 degrees CZT camera for monitoring (177)Lu-PSMA treatment. *EJNMMI Physics, 10*(1), 58.
78. Weinmann, A. L., Hruska, C. B., & O'Connor, M. K. (2009). Design of optimal collimation for dedicated molecular breast imaging systems. *Medical Physics, 36*(3), 845–856.
79. Vergnaud, L., Badel, J. N., Giraudet, A. L., Kryza, D., Mognetti, T., Baudier, T., Rida, H., Dieudonné, A., & Sarrut, D. (2023). Performance study of a 360° CZT camera for monitoring 177Lu-PSMA treatment. *EJNMMI Physics, 10*, 1–18.
80. Bailly, M., Le Rouzic, G., & Metrard, G. (2023). Ultra-fast high resolution whole-body SPECT after treatment with (153)Sm-EDTMP using 3D-ring CZT: Applying new technology to an old tracer. *European Journal of Nuclear Medicine and Molecular Imaging, 50*(3), 968–969.
81. Gustafsson, J., Rodeno, E., & Minguez, P. (2020). Feasibility and limitations of quantitative SPECT for (223)Ra. *Physics in Medicine and Biology, 65*(8), 085012.
82. Huh, Y., Caravaca, J., Kim, J., Cui, Y., Huang, Q., Gullberg, G., et al. (2023). Simulation studies of a full-ring, CZT SPECT system for whole-body imaging of (99 m) Tc and (177) Lu. *Medical Physics, 50*(6), 3726–3737.

CdZnTe-Based Photon-Counting Detector Modeling and Calibration for Nondestructive Testing (NDT)

Pierre-Antoine Rodesch, Devon Richtsmeier, Elmaddin Guliyev, Kris Iniewski, and Magdalena Bazalova-Carter

1 Introduction

1.1 Motivation

There are several algorithms and methods used for material discrimination, depending on the specific application and the type of data being analyzed. X-ray diffraction (XRD), Raman spectroscopy, or magnetic resonance imaging (MRI) can be utilized in laboratory settings. XRD techniques analyze the diffraction patterns produced when X-rays interact with crystalline materials. Mathematical algorithms and pattern-matching techniques can be applied to XRD data to identify and discriminate between different crystalline structures. Raman spectroscopy measures the scattering of laser light to determine the vibrational modes of molecules in a material. Algorithms can be used to analyze Raman spectra and identify the unique molecular fingerprints of different materials. Magnetic resonance imaging (MRI) utilizes magnetic fields and can provide detailed information about the composition and structure of materials, such as in medical imaging. Image processing and pattern recognition algorithms are often employed to analyze MRI data and discriminate between different tissues or materials. Despite their powerful potential, XRD, Raman spectroscopy, and MRI are not suitable for mass production needs due to their slow speed, high cost, and special requirements. Only X-ray detection that relies on analyzing object attenuation can be effectively used in applications such as food scanning.

P.-A. Rodesch · D. Richtsmeier · M. Bazalova-Carter
Department of Physics and Astronomy, University of Victoria, Victoria, BC, Canada

E. Guliyev · K. Iniewski (✉)
Redlen Technologies, Saanichton, BC, Canada

K. Iniewski (ed.), *CdTe and CdZnTe Materials*,
https://doi.org/10.1007/978-3-031-64521-1_13

Single-energy (SE) X-ray testing has been widely used in a variety of applications since X-rays were discovered by Roentgen [1]. X-ray computed tomography (CT) was first described in the early 1970s by Hounsfield, which opened the field to three-dimensional (3D) reconstruction. Dual-energy (DE) CT for material decomposition was first hypothesized [2] and demonstrated [3] shortly after. Unfortunately, its use was limited clinically due to technological constraints. However, as CT technology has advanced, DECT has become more widely available and is used in a range of clinical applications [4–6]. With the two sets of linear attenuation information given by the two sets of energy data, two subsequent sets of data can be decomposed or extracted, usually referred to as the basis functions in the decomposition formula. For example, as shown in the seminal DECT paper by Alvarez et al. [2], the photoelectric and Compton scattering components of the linear attenuation coefficient can be determined with DECT material decomposition. Though there are many decomposition algorithms [2, 3], one of the most well-studied sets of basic functions is the effective atomic number (Z_{eff}) and electron density relative to that of water (ρ_{e}), first demonstrated by Rutherford et al. [3] in 1976. Since then, many algorithms have been developed to extract these values, including using the effective energy of two X-ray beams [7] and monochromatic X-rays [8, 9]. The decomposition of data into Z_{eff} and ρ_{e} has several uses, including increased diagnostic capability by distinguishing between materials with similar attenuation curves. We will review the application of these algorithms in this chapter.

Photon-counting detectors (PCDs) are the future of X-ray detection [7]. Commercial devices equipped with PCDs are now used for medical diagnosis [10], airport security [11], and nondestructive testing (NDT) [12]. PCDs are based on the direct conversion of X-rays to electrical charges when interacting with a semiconductor material [13, 14]. The electric charges are then transformed into a voltage pulse height signal by a pulse-shaping application-specific integrated circuit (ASIC). In the processing step of the ASIC, the voltage pulses are processed by analog comparators to be sorted according to their amplitude [15]. The different comparator levels, or the thresholds, can be changed between acquisitions according to the desired imaging protocol.

Now, with the advent of high-flux photon-counting detector (PCD) technology [16–18] that offers energy windowing, multienergy (ME) is becoming a clinical possibility as well. PCDs can offer up to eight energy bins [19] that are, in theory, nonoverlapping, which significantly increases the options for material decomposition. Additionally, each energy range is captured simultaneously, mitigating any issues with coregistration between the different energy data sets.

1.2 Photon-Counting Detectors (PCDs)

A typical photon-counting detector system, shown in Fig. 1, is used to measure the X-ray tube spectra after attenuation, caused by the tube filter, phantom, and/or the actual object in the scanner. It consists of (a) a semiconductor sensor, which uses

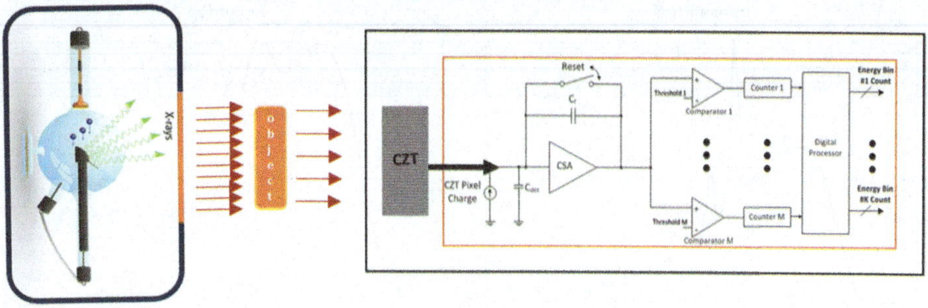

Fig. 1 Schematic representation of the photon-counting subsystem

cadmium zinc telluride (CZT) or another high atomic number (high Z) material, that converts photon energy to electron charge and (b) an ASIC (application-specific integrated circuit) electronic circuit, which amplifies the CZT-generated signal and assigns it to the appropriate energy bin on a photon-by-photon basis. We will assume that the detector material is CZT, which is the most used sensor material for high-flux X-ray applications. However, most of the material in this chapter is equally applicable to cadmium telluride (CdTe) as well or to other semiconductor direct conversion materials. We will not discuss ASIC operation in detail here, but ASIC information can be found in many references [20]. Some of the distortions caused by ASIC operation that affect detector characteristics will be discussed in this paper.

Let us annotate F_0 as a flux from an X-ray generator (XRG) before it is attenuated while passing through an object and F_1 as the flux that is being received by the detector after passing through the object. They are, in theory, both related through the standard attenuation formula:

$$F_1 = F_0^* \ \exp.(-\mu \times T) \tag{1}$$

where μ is the linear attenuation coefficient and T is the thickness of the object [20]. Unfortunately, the transmitted X-ray spectrum F_1 experiences significant distortion due to the detector system's nonidealities, as depicted in Fig. 2. The objective of the modeling effort here is to analyze these distortions and identify their impact on material discrimination.

The distortions observed in the system indicated in Fig. 2 can be attributed to the following effects:

- Incomplete photon absorption due to the finite thickness of the CZT sensor.
- Beam hardening effects, where low-energy photons are absorbed by the filter or phantom or device under test.
- Influence of finite energy resolution (ER).
- Charge-sharing events and X-ray fluorescence.
- Pile-up effect.
- Ballistic deficit resulting from a too-slow ASIC response.

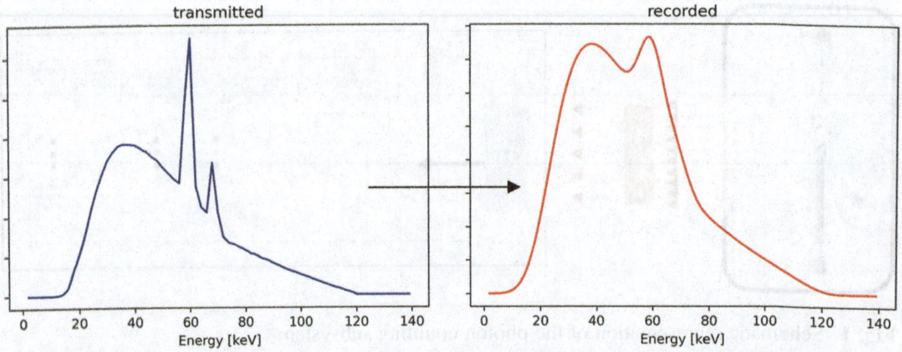

Fig. 2 Schematic illustration of transmitted spectra (F0, blue, left) and recorded spectra. The recorded spectra are based on a 330 PCCT detector that uses a 330 um CZT pixel pitch

In the following sections, we will study these effects to arrive at a clear understanding of what the received spectrum should look like, depending on the energy of the X-ray tube, its current, and the filter placed between the X-ray tube and the detector.

2 Models of Transmitted/Received X-Ray Spectra

2.1 X-Ray Transmitted Spectra

The shape of the transmitted spectra from the X-ray tube is well known. Figure 3 illustrates typical theoretical spectra generated at tube voltages of 80 kVp, 100 kVp, 120 kVp, and 140 kVp, where kVp is the kilovoltage peak. The X-ray spectral line was obtained from the TASMIP simulator, a diagnostic tool used in medical imaging and material analysis [21]. Tungsten anodes produce two characteristic X-ray peaks at 59.32 and 67.24 keV. These characteristic peaks (K-alpha and K-beta) occur at the same energy positions regardless of the X-ray tube voltage and could be used for detector energy calibration if they could be detected. However, as discussed below, achieving an energy resolution of 4 keV or better, which is required for this peak detection, might be difficult in real PCD systems. On the other hand, the endpoint of the spectrum coincides with the maximum voltage of the tube and could potentially be used for energy calibration as well (with some limitations as the transition to zero counts is not smooth due to the finite ER of the system).

2.2 Finite Absorption of Sensor Material (CdTe/CZT)

If the CdTe/CZT sensor were thick enough, it would be capable of stopping all incoming radiation. However, in practice, the thickness of CZT is limited to a value that avoids polarization effects. A typical CZT thickness for high-flux applications is

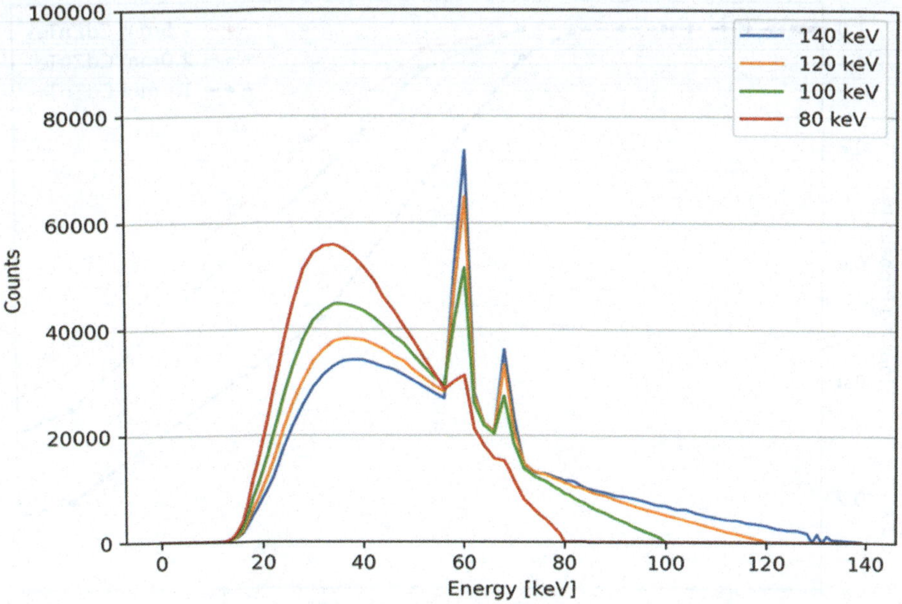

Fig. 3 Transmitted X-ray spectra for different X-ray tube voltages, from 80 kVp (keV) to 140 kVp (keV)

between 1.5 mm and 2.0 mm, while sensor thickness for CdTe is between 0.75 mm and 1.5 mm as it is difficult to fabricate thicker CdTe sensors due to polarization issues. In Fig. 4, the CZT attenuation coefficient is shown for sensor thicknesses ranging from 0.75 mm to 2 mm. At 100 keV, the stopping probability is only 50% for a thickness of 0.75 mm, but it increases to 76% at 1.5 mm and reaches 80% at 2 mm. Clearly, for X-ray testing at energies of 100 keV and above, a CZT thickness of at least 2 mm is strongly desirable.

The plot presented in Fig. 4 was calculated using the total absorption coefficient. If only the photoelectric effect is considered, the plots would exhibit a similar shape but with slight changes due to the negation of the Compton effect. Compton effects become a dominant source of photon interactions with CZT above 250 keV.

Incomplete absorption leads to distortion in the measured spectra, as shown in Fig. 5. Higher-energy photons are more likely to be transmitted through the finite thickness of CZT sensors and not stopped for detection. The degree of this spectral distortion is strongly dependent on in-line filter/object characteristics and is discussed in more detail in the next section. At 120 keV, a 2 mm thick CZT loses 32% of the photon flux, while a 1 mm sensor loses 67%. Therefore, sensor thickness is a critical parameter in detector selection.

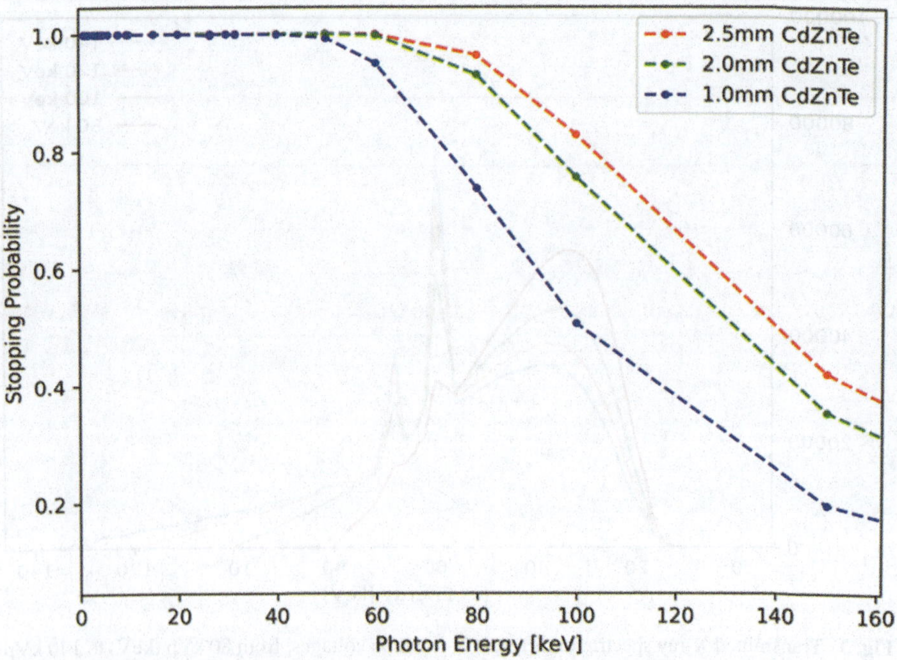

Fig. 4 Stopping probability for sensors with different CZT thicknesses from 1.0 mm to 2.5 mm within the energy range of 0 to 160 keV. This calculation involves utilizing the total absorption coefficient obtained from the NIST XCOM database [22]

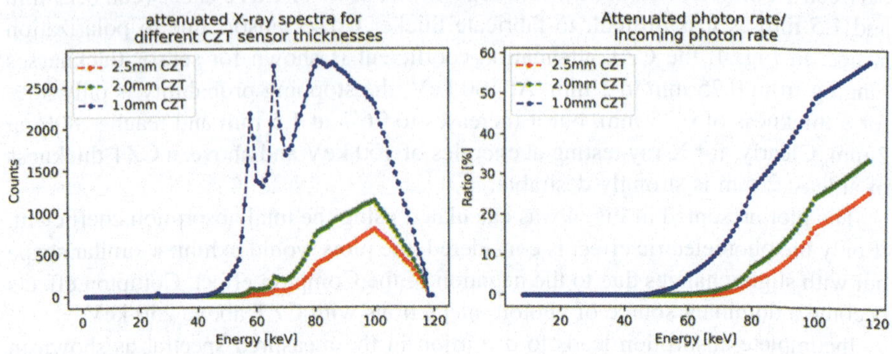

Fig. 5 The portion of the tungsten-generated spectra that is not being absorbed by CZT. Calculated for different CZT thicknesses expressed in absolute numbers of counts (left) and as a percentage of the incoming photon count (right)

2.3 Attenuation by Filters and Objects

Using various filters to attenuate X-ray signals is common in CT scanning to eliminate low-energy photons that are not going to pass through the object. Attenuation filters are used to regulate the dose, bow-tie filters to limit exposure at the edges of the field-of-view (FOV), or K-edge filters to separate low- and high-energy photons. In X-ray laboratory practice, simple filters such as copper foil, aluminum, or other materials, such as lead, are also frequently employed. All filters alter X-ray spectra to some extent. In this section, we will illustrate how X-ray spectra change when being attenuated by different objects through food scanning examples. An example of food scanning in NDT applications is studied here, and we will examine attenuation through water, proteins, bones, and lipids.

Figure 6 shows the portion of the X-ray spectra that is absorbed by 5 cm of various materials and, thus, not absorbed by the detector. The spectra for different CZT thicknesses are expressed both in absolute numbers of counts and as a percentage of the incoming photon count. Five centimeters of water attenuate 20% of the X-ray signal at 30 keV and up to 60% at 120 keV. Attenuation through 5 cm of protein, 5 cm of lipid (fat), and 5 cm of bone looks qualitatively the same, but attenuation is the highest for the bone, as expected. This is a direct result of the differences between the linear attenuation coefficients plotted in Fig. 7, in which bone has the highest attenuation. While National Institute of Standards and Technology (NIST) tables may not explicitly define lipid, chicken protein, and bone, the ICRU-44 serves as a valuable reference providing guidance on the properties and composition of various tissues, including bones, muscles, and lipids. To perform the calculations, we used adipose tissue (ICRU-44) data for lipids, muscle-skeletal (ICRU-44) for proteins, and cortical bone (ICRU-44) for bones [22].

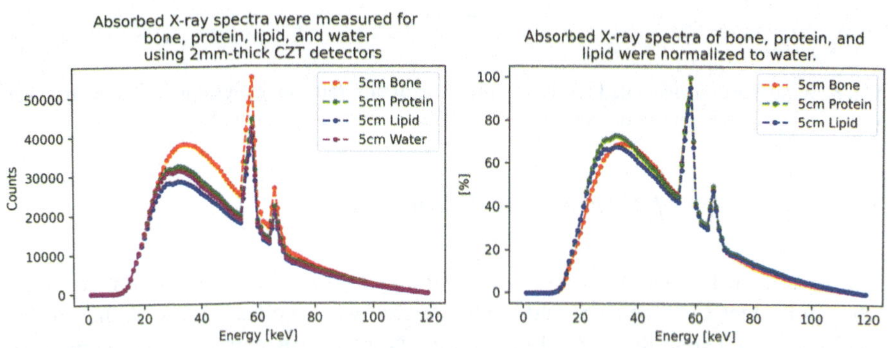

Fig. 6 The absorbed X-ray spectra with 2 mm CZT thicknesses were simulated with a 5 cm thickness of bone, protein, lipid, and water placed between the X-ray source and the detector (on the left), while the absorbed photons were simulated for bone, protein, and lip

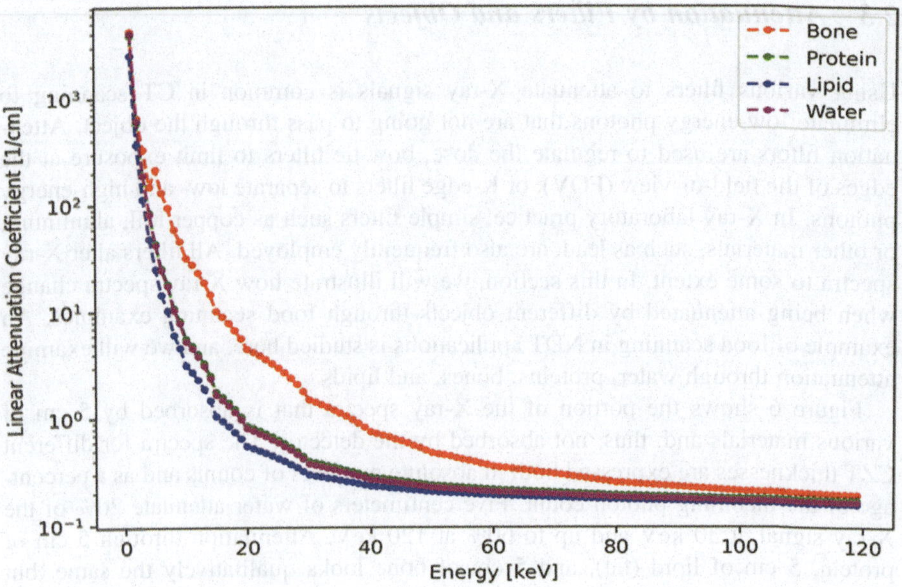

Fig. 7 Linear attenuation coefficients for bone, protein, lipid, and water were calculated using values from the NIST database [22]

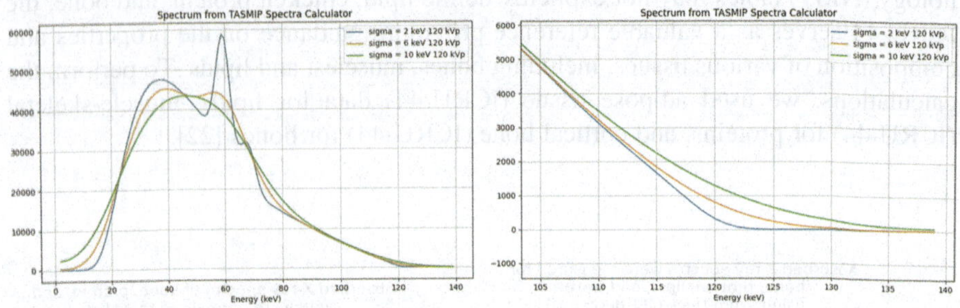

Fig. 8 X-ray spectra calculated for the no filter case of different energy resolutions. Note that even without pile-up, the spectra do not finish at the kVp location due to finite ER

2.4 Influence of Energy Resolution (ER)

The spectral CT detector subsystem is limited by the energy resolution resulting from imperfect CZT sensors and ASIC electronics. Typical values of the energy resolution (ER) span the 6 to 12 keV range. Nonzero ER causes the broadening of energy bins, as shown in this section. Figure 8 shows X-ray spectra calculated with Gaussian kernel representing finite ER value for the no filter case. As expected, the discrete characteristic peaks disappear at ER larger than 6 keV. The impact of finite ER on energy bin counts will be discussed in the next section.

3 Charge-Sharing Effects

3.1 Introduction

Charge sharing is a well-recognized effect in pixelated sensors that leads to spectral degradation in small-pixel PCDs. While the physics of charge sharing is well understood, the techniques for charge-sharing compensation are still being developed. At low count rates, typical for precise spectroscopy like single-photon emission computed tomography (SPECT), the compensation techniques are more developed as there is ample time between consecutive photons to allow for the readout electronics to react and monitor charge-sharing events precisely. At high count rates, typical for CT or NDT, the small amount of time between consecutive incoming photons leads to design trade-offs in the readout electronics hardware.

When ignored, the charge-sharing effect leads to significant spectral distortions. To correct for charge sharing, the detected charges in neighboring pixels can be summed to avoid spectral distortion. However, this summing results in a significant penalty to the maximum count rate, which can be processed, leading to increased X-ray scanning time, which may not be acceptable in practice. The alternative approach is to correct for charge sharing in the digital domain. We will discuss the impact of charge sharing on NDT testing results and suggest some ways of correcting for charge sharing.

3.2 Charge-Sharing Physics

If the charge created by the interaction of the X-ray photon with the CZT was infinitesimal and remained that way, it would induce a charge on one pixel only. That pixel will be where the trajectory of the charge terminates after traversing the thickness of the detector (assuming perfect interpixel gap passivation, meaning all field lines will eventually end on a pixel pad). However, the charge created by the absorption is a sphere of finite dimensions, which is created almost instantaneously after the X-ray interaction takes place. Moreover, this sphere will expand from the moment it is generated. If the charge is produced near the border between two, or more, pixels, the expanded charge sphere will separate into two charge clouds and terminate on these two or more pixels (Fig. 9). Such a separation will end up in multiple charge signals in the detector, one on each pixel, each of which will be pushed toward the low-energy tail of the energy histogram. Since this phenomenon is inherent to pixel detectors, it is important to estimate this effect in order to know what to evaluate their performance.

In the following discourse, we assume that electrons are the only charge carriers contributing to the charge buildup on the pixel anodes. This is true even though electron-hole pairs are created at the instance of an X-ray interaction, and those pairs are separated into an e-sphere and an h-sphere moving in opposite directions—the

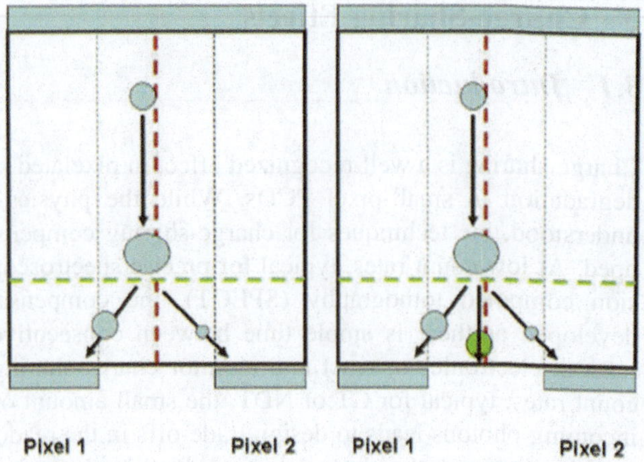

Fig. 9 Schematic illustration of the ideal charge-sharing effect (left) and realistic (right) when surface passivation is not ideal. The middle cloud indicates a charge that might be lost due to surface recombination

former toward the anode and the latter toward the cathode. In principle, both are contributing to the charge buildup. However, due to the small pixel effect that overstresses the contribution of charge carriers moving near the anode, it is the e-buildup that prevails. If we assume that the surface gap between the pads of the pixels is perfectly passivated, the electric field lines (which delineate the trajectories of the electrons) will terminate on the pads and not on the streets, where they would recombine with surface traps and thus will not fully contribute to the charge buildup.

This complex chain of events can be tackled only by a Monte-Carlo simulation, which indicates that the primarily created electron sphere is quite small for the photon energies relevant in CT. Other processes, namely, diffusion and electric repulsion, contribute to a stronger spread of the sphere, so that we can almost neglect the size of the sphere created by electron-scattering only.

Once the electron sphere is created, the electrons diffuse outward due to thermal agitation, where the driving force is the gradient in free electron density across the generated charge cloud. The analytical model by Iniewski [12] indicates that for a typical CT detector, with a 2 mm crystal thickness and a sensor high voltage (HV) of 1000 V, we obtain an electron sphere of about 30 μm in diameter. The contribution due to Coulomb repulsion can usually be neglected.

The physical space between the two anode pads is usually referred to as "street," and the distance between the two pixel pads is often referred to as the street width. To ensure the two contacts are electrically separated, a passivation layer between the two pads is used to ensure that the surface of the CZT in the street has a higher resistivity than the bulk. Thus, the electric field lines will converge on the pads and be diverted from the streets. For perfect passivation, all the field lines will end up on the pixel pads, and consequently, all X-rays absorbed in this region will be collected. As such, all charges created on the left of the theoretical center line in Fig. 9 would be collected in pixel 1, while all charges created on the right-hand side of the line would be collected in pixel 2. In modern state-of-the-art sensors, the charge loss at the street is made negligible.

3.3 Charge-Sharing Corrections

To illustrate the techniques used for charge-sharing correction, it is helpful to review charge-sharing effects using the diagrams in Fig. 10. Suppose that a pixelated detector is divided into 9 pixels, as shown in Fig. 10a. If an X-ray photon strikes pixel 5 near the center, as in Fig. 10b, then most of the charge will be localized to that pixel, and very little charge sharing will occur. However, if the photon lands near the pixel boundaries, as in Fig. 10c, then the charge will be shared among pixels 1, 2, 4, and 5. Assuming the thresholds of the pixels are set low enough to detect the charge, low-energy events will be detected in each of these pixels, creating the low-energy tail in the spectrum and shifting events from higher-energy bins to lower ones.

If all individual charges are denoted as Q1, Q2, Q4, and Q5 and the corresponding voltages after the analog front-end processing are denoted as V1, V2, V4, and V5, various readout schemes will report different numerical results.

In practical implementations, there are multiple challenges with the charge-sharing correction (CSC) schemes. The first is the selection of the threshold voltage (VTH) used to detect the events in the readout chain. To be able to detect small fractions of the split charge, the VTH value needs to be as small as possible. However, a too-small VTH value will increase the amount of noise counts detected.

For a given photon energy (EPH) corresponding to the comparator input voltage, only photons corresponding to energies E in [VTH, EPH] can be detected. For example, at $EPH = 60$ keV and $VTH = 20$ keV, only 1/3 of the events can be corrected for charge sharing regardless of which scheme is being used as the smaller components of the original charge with energies below 20 keV will be ignored completely.

Another challenge is the time window used to monitor coincidence detection. In a charge-sharing event, the time window at which the signal in both pixels crosses the VTH varies both with energy and with a charge-sharing ratio. For example, for a high-energy photon and a small sharing ratio, the window within which both signals cross the VTH may be large, but for a 50:50 charge-sharing ratio, the time window is almost zero as both electron charge clouds arrive at their corresponding pixels at the same time. Therefore, it is difficult to define a single window that meets all conditions and does not confuse true parallel events occurring in the neighboring

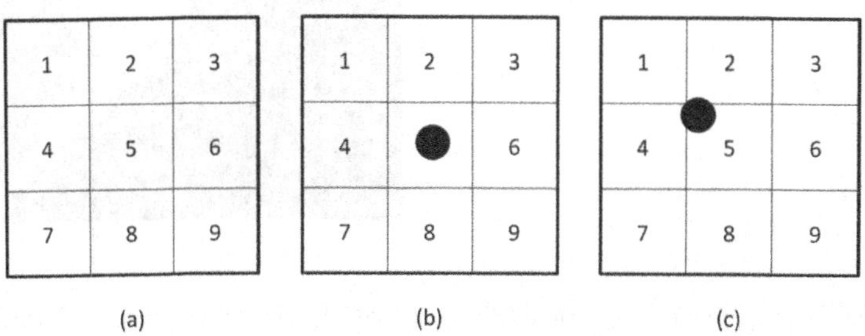

Fig. 10 Schematic illustration of the charge-sharing effect

pixels. If the event falls outside the user-specified time window, the event may be incorrectly counted as two separate events. This problem is not easily addressable and requires careful trade-off evaluations.

4 Energy Calibration

4.1 Definition

Energy calibration is a fundamental necessity for ASICs in photon-counting detectors because, at its core, an ASIC lacks intrinsic knowledge of physical energy units like kiloelectronvolts (keV). This limitation is significant in applications where precise energy measurements are crucial, such as in medical diagnostics, nuclear physics, and astrophysics.

ASICs process electrical signals generated when photons hit the detector material. These signals are indicative of the energy of the incoming photons, but the ASIC itself only understands these signals in terms of their electrical characteristics. More specifically, the signal intensity corresponds to voltage levels. Without calibration, the ASIC would be unable to accurately translate these voltage values into meaningful energy values.

During energy calibration, X-rays at different energies can generate the same voltage amplitude. This means that the relationship between output voltage amplitude and incoming photon energy is not unique. However, the response of PCD is characterized by a main peak, as shown in Fig. 12. This peak represents the detected energy that links an output voltage to an incident energy. The relation between the detected energy and the voltage level can be assumed linear.

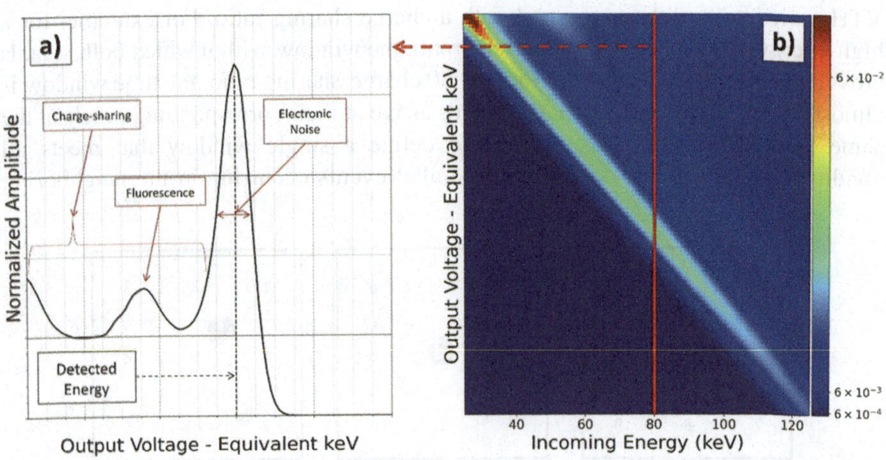

Fig. 12 The average PCD response to an 80 keV photon. b) The detector response matrix (DRM). The data were calculated based on the model by Tanguay et al. [9]. The model has been normalized: the integral of profile *a* is equal to 1, with a being perfectly calibrated

Energy calibration involves exposing the detector to known energy sources and adjusting the ASIC's interpretation of the electrical signals accordingly. This process establishes a correlation between the signal output by the ASIC and the actual energy of the photons. For instance, if a detector is exposed to a source-emitting photon of a known energy and the ASIC outputs mainly a certain voltage level in response, the calibration process links this signal level to the corresponding energy value.

CT relies on accurately energy-calibrated PCDs to produce detailed and reliable images. The quality of the photon-counting computed tomography (PCCT) images is directly influenced by how well these detectors interpret photon energies, which is why the energy calibration of ASICs for PCDs is so crucial. There are certain factors for PCCT influenced by energy calibration:

1. Linearity of response.
2. Energy discrimination for multienergy CT.
3. Energy-dependent artifact correction.
4. Consistency of the scanner across scans.
5. Enhanced image resolution and contrast.

This calibration is not a one-time procedure but rather an ongoing requirement. Factors like temperature changes, aging components, and environmental influences can affect the detector's response over time. Regular re-calibration ensures that the ASIC continues to accurately interpret the electrical signals as the correct energy values. In this chapter, we discuss different methods to perform energy calibration for PCDs.

4.2 Two-Isotope Calibration

The traditional method for energy calibration involves using radioactive sources that emit radiation with well-established energy peaks. In this process, the PCDs are exposed to these sources, and the ASICs within the detectors process the resultant electrical signals. The calibration involves correlating the signals generated by the ASICs with the known energy values emitted by the radiation sources. This method is particularly useful for establishing a baseline calibration and often serves as the initial step in a comprehensive calibration process. As illustrated in Fig. 13, an Am241 source is used, which has an energy peak at 59.5 keV and a Co57 source, which exhibits a peak at 122 keV. The figure demonstrates the spectrum acquired for a 330 μm pixel size CZT PCD module.

4.3 K-Edge Filter Calibration

The use of radiation sources for energy calibration can be challenging within the manufacturing process for PCDs due to the associated costs, the time requirements, and the need for specific licenses, not to mention safety concerns. To address these

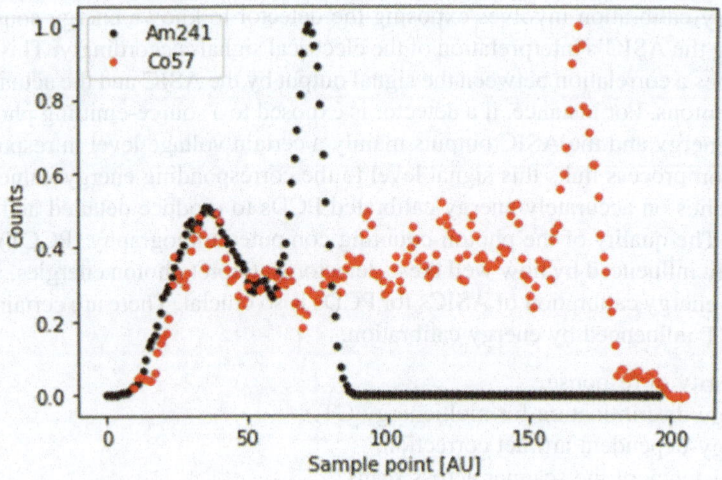

Fig. 13 Am241 and Co57 spectral lines for a 330 um PCD module

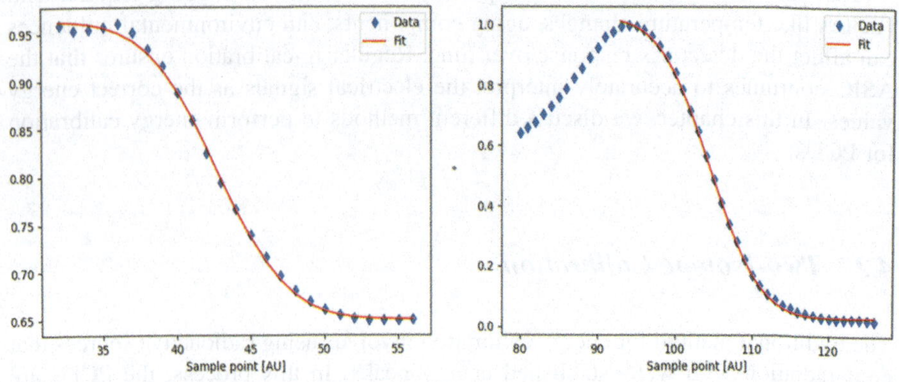

Fig. 14 K-edge spectral attenuation for Ce (left) and Pb (right)

issues, Redlen Technologies has developed and patented a novel method [23] that utilizes two different K-edge filters, specifically those made of cerium (Ce) and lead (Pb). This innovative approach simply requires the acquisition of two distinct K-edge spectra and the extraction of the K-edge locations using an error function fit. Figure 14 illustrates the K-edge spectrum for both filters. This method represents a significant advancement in PCD energy calibration, offering a more practical and safer alternative to traditional calibration methods.

4.4 kVp Calibration

K-edge filter-based energy calibration is effective for production lines and laboratory testing but is not ideally suited for clinical applications. The primary reason is the

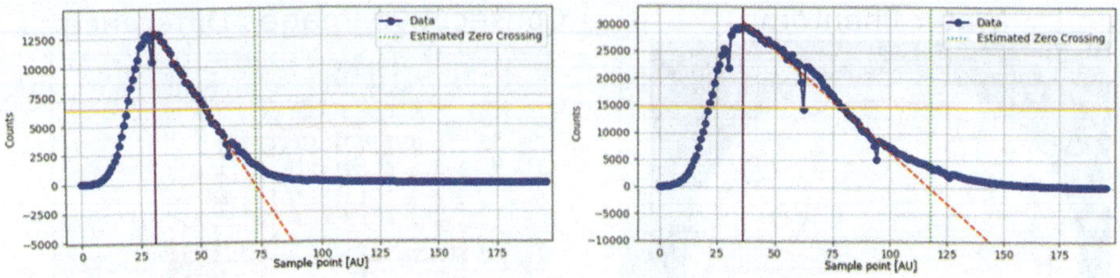

Fig. 15 Spectral X-ray tube lines for 70 kVp (left) and 120 kVp (right)

requirement of an X-ray tube integrated with K-edge filters, such as cerium (Ce) and lead (Pb), which can significantly increase the cost of the X-ray tube. To address this limitation, we have developed a third method that is compatible with any type of X-ray tube. This approach involves fitting the endpoint of the X-ray tube spectrum for two different tube voltages and determining the zero-crossing point. Using these two points, energy calibration can be effectively performed. This method is notably simple and does not require any hardware modifications. For illustration, refer to the spectra obtained at 70 and 120 kVp, as shown in Fig. 15. This approach offers a practical and versatile solution for energy calibration in diverse settings, including clinical applications.

In conclusion, the diverse energy calibration methods developed and implemented for photon-counting detectors showcase a comprehensive approach, each tailored to suit different application environments. Beginning with the baseline calibration provided by the traditional radiation source method, which effectively correlates ASIC outputs with known radiation energies, we move to the innovative K-edge filter method. This patented approach, designed for production and laboratory settings, simplifies the calibration process and addresses safety and practicality concerns. For clinical applications, the adaptable X-ray tube endpoint fitting method offers a straightforward and universally compatible solution. A comparison of the different calibration methods can be found in the Ref. [24].

4.5 *The Impact of Energy Calibration Methods on CT Image Reconstruction*

To quantify the effect of energy calibration on image quality, a comparison of image quality was performed with three different energy calibrations. The different methods led to different energy calibration accuracy levels: methods 1, 2, and 3 had an accuracy of, respectively, 2.0, 1.4, and 0.9 keV. The image quality evaluation was based on two consecutive CT acquisitions of a 90 mm diameter water-filled cylinder with 720 views. Both acquisitions were reconstructed, and the difference between them was computed. This process resulted in a white noise image, which was free of beam hardening and had reduced ring artifacts. The 2D noise power spectrum (NPS) was evaluated on these images from 24-square regions

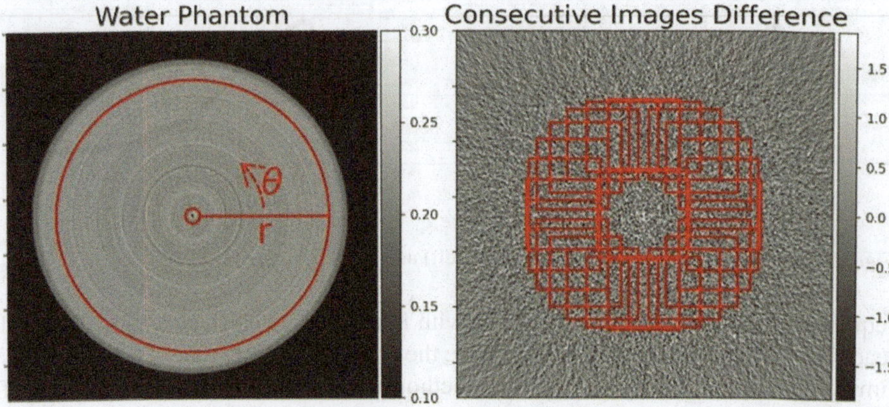

Fig. 16 (**a**) The water phantom reconstruction (unit cm^{-1}). The polar coordinates (r, θ) are represented. The center of the water cylinder and the rotation axis are not perfectly aligned; this is the rotation axis that defines the origin of the coordinate system. (**b**) The white noise image computed from the subtraction of two consecutive acquisitions (unit cm^{-1})

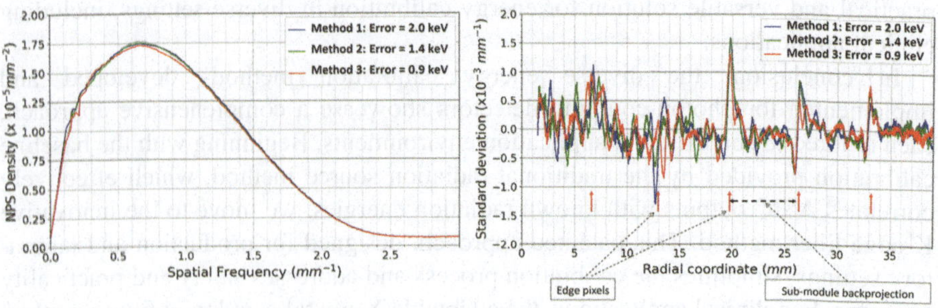

Fig. 17 (**a**) The noise power spectrum evaluated for each energy calibration. (**b**) The radial profile inside the water phantom average on 15 slices

of interest (ROIs), each measuring 128 × 128 voxels, as illustrated in Fig. 16b. These ROIs were averaged from 15 slices. Additionally, the intensity of ring artifacts was evaluated in the water phantom reconstructions. To eliminate the cupping artifact, a second-order polynomial was fitted after subtracting the inner part of the phantom. This value was then averaged along the angular coordinate, defined with the rotation axis at the origin (Fig. 16a). The standard deviation was assessed along the polar coordinate and across the slices, excluding voxels close to the iso-center.

The results of the NPS and ring artifact deviation are presented in Fig. 17. We can see that despite different energy calibration accuracy levels, the images' properties are very close. Even with a good energy calibration, the radial profile displays the remaining ring artifact caused by the heterogeneity between the pixels of a PCD. The pixels at the edge of each detector submodule are specifically visible (Fig. 17b). Further corrections and calibrations need to be applied to reconstruct artifact-free images.

5 Spectral Properties

5.1 Spectral Distortion

The physical effects occurring in the detector are a source of spectral distortion. The main consequence of the spectral distortion is that the spectral distribution of counts is different from the shape of the incident spectrum (Fig. 2). Figure 18 displays the number of counts detected in ten energy bin configurations for a polyenergetic spectrum. As the spectrum was generated from a tungsten anode, the bin expected with the largest number of counts should be the one between 55 and 65 keV.

5.2 Energy Bin Properties

In an X-ray imaging PCD system, the number of counts will be different in each energy bin. This will impact the noise in each bin image, which is directly linked to the number of detected photons. However, the noise is also impacted by pixel heterogeneity, which is more important at lower energies (bin 16–25 keV in Fig. 19b).

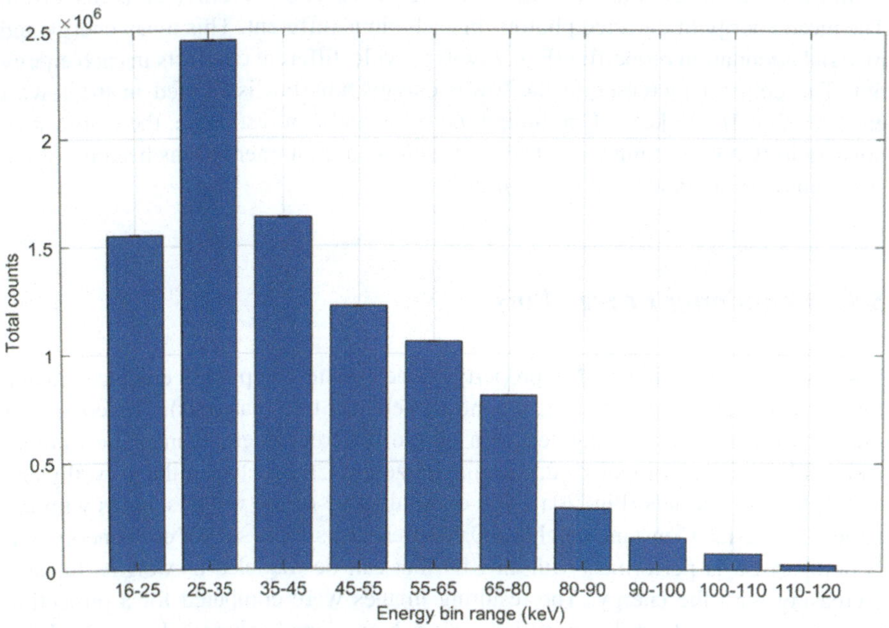

Fig. 18 Counts detected from a 120 kVp polyenergetic spectrum with a tungsten anode and 3 mm aluminum filtration

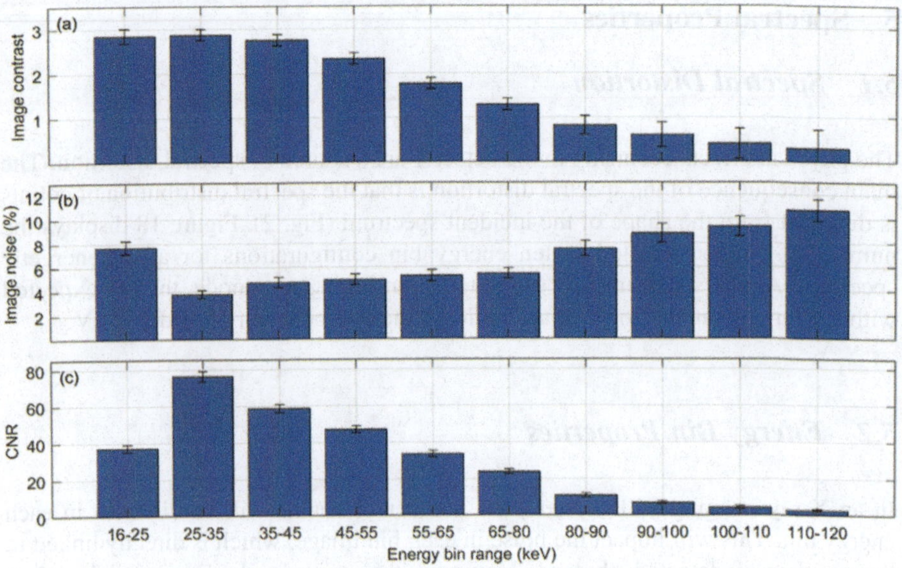

Fig. 19 Contrast, noise, and CNR measured for glass in a polyethylene background. The bin 16–25 keV suffers from a high noise level caused by important pixel heterogeneities at these energies

In Fig. 19a, the contrast measured between glass and polyethylene is displayed. The mean energy of detected photons in each bin is different. This mean energy and material attenuation properties (Fig. 7) will provide different contrasts in each energy bin. The contrast increases in the lower energy bins but is limited in the lowest energies (bin 16–25 keV). Combining the noise and contrast value, the contrast-to-noise ratio (CNR) is computed. The CNR is low in high-energy bins because of the small number of counts and low contrast.

5.3 *Combining Energy Bins*

It is possible to exploit the CNR property of each bin to compute an enhanced image for a selected imaging task. (i.e., a contrast between two materials). The computed image is obtained as a weighted sum of the different energy bins. The optimal weights have been computed [25] to maximize the CNR between the polyethylene and different materials (Fig. 20a). The optimal value of the weights varies with the targeted material. Constant weights at 0.1 represent a simple sum of each energy bin. A comparison is performed with an EID that can be modeled as weights linearly increasing with the energy. The resulting images were computed for a projection image of glass beads of different sizes in a polyethylene background. A normalized profile has been drawn with the maximal value set to 1. It is visible that the contrast is

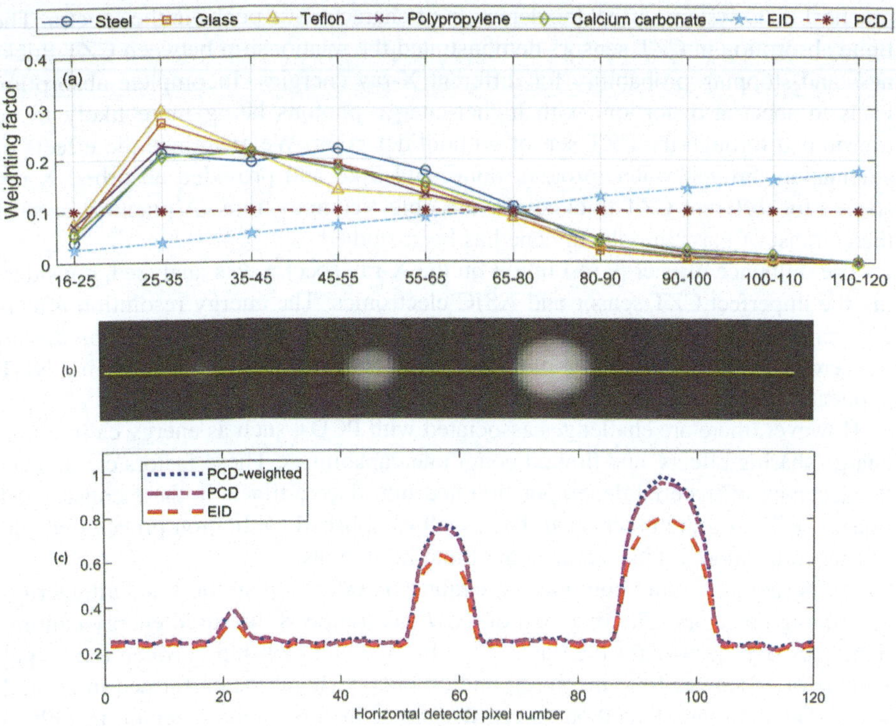

Fig. 20 (**a**) Weighting coefficients optimizing the CNR of the glass/polyethylene CNR. (**b**) and (**c**) Image and profile for 1, 3, and 5 mm diameter glass beads

enhanced with the PCD-weighted image compared to the EID and a simple PCD image (constant weight). Therefore, the energy bins can be exploited to increase the CNR. In parallel, they can be used to perform material quantification.

6 Summary and Conclusions

NDT is used routinely in industrial testing to check the quality, content, and contamination of various objects, materials, and items. NDT heavily exploits various material discrimination algorithms to analyze the object in question. Material discrimination algorithms refer to computational methods and techniques used to classify or identify different materials based on their properties or characteristics. These algorithms were developed to analyze data obtained from different methods, such as spectroscopy, imaging, or other physical measurements.

In this chapter, we discussed the shape of transmitted X-ray spectra from X-ray tubes and highlighted the characteristics of typical theoretical spectra generated at different tube voltages. The limitations of achieving high-energy resolution in practical PCD systems affect the detection of characteristic peaks in the spectra. The finite absorption in CZT sensors demonstrated the relationship between CZT thickness and stopping probability for different X-ray energies. Incomplete absorption leads to spectral distortion, with higher-energy photons being more likely to be transmitted through the CZT sensor without detection. We examined the effects of attenuation through water, protein, bone, and lipid and provided absorbed X-ray spectra for different CZT thicknesses and object compositions. A practical case of the contrast of glass in polyethylene has been studied.

The influence of energy resolution on the X-ray spectra was analyzed, considering the imperfect CZT sensor and ASIC electronics. The energy resolution affects the shape of the spectra, and we discussed the broadening of energy bins due to nonzero energy resolution. The optimization of energy bins for food scanning NDT applications was also discussed.

However, there are challenges associated with PCDs, such as energy calibration, charge-sharing effects, and limited count rate capabilities. These factors can impact the accuracy of material decomposition and introduce artifacts in the reconstructed images. EIDs, on the other hand, have well-established calibration procedures and are generally more robust against count rate limitations.

PCDs require a more complicated calibration process than the standard energy-integrating detectors. The first step of PCD calibration is threshold energy calibration. This step determines the linear pixel-by-pixel relationship between the output voltage of a PCD and the incoming photon energy. It can be performed in several ways. Three different methods were discussed in this chapter, requiring only a standard X-ray source or an X-ray source combined with K-edge materials or a PCD response model. It has also been demonstrated that increasing threshold calibration errors below a certain level does not improve image quality. PCDs require further calibration steps to correct for pixel heterogeneity. Finally, the long-term behavior of PCDs remains unknown, as for example the need to update the threshold calibration in the long term.

In subsequent calibration steps, optimal energy bin weights can be computed for a targeted application. The result is an increased CNR value. With PCD-weighted images, the PCD outperforms the EID. It also can provide material-specific images through the computation of material decomposition.

As a conclusion, we have shown that PCDs achieve superior performance compared to EIDs. We investigated the modeling of X-ray spectra and their distortions in PCD systems and the various factors that contribute to the distortions, including incomplete photon absorption, beam-hardening effects, finite energy resolution, charge-sharing events, X-ray fluorescence, pile-up effects, and ballistic deficit. The study aimed to analyze these effects and propose methods for their correction within several examples of NDT applications.

References

1. Glasser, O. (1995). W. C. Roentgen and the discovery of the roentgen rays. *American Journal of Roentgenology, 1033*, 164.
2. Alvarez, R. E., & Macovski, A. (1976). Energy-selective reconstructions in X-ray computerized tomography. *Physics in Medicine and Biology, 21*, 733–744.
3. Rutherford, R. A., Pullan, B. R., & Isherwood, I. (1976). Measurement of effective atomic number and electron density using an EMI scanner. *Neuroradiology, 11*, 15–21.
4. Albrecht, M. H., et al. (2019). Review of clinical applications for virtual monoenergetic dual-energy CT. *Radiology, 293*, 260–271.
5. Van Elmpt, W., Landry, G., Das, M., & Verhaegen, F. (2016). Dual energy CT in radiotherapy: Current applications and outlook. *Radiotherapy and Oncology, 119*, 137–144.
6. Goo, H. W., & Goo, J. M. (2017). Dual-energy CT: New horizon in medical imaging. *Korean Journal of Radiology, 18*, 555–569.
7. Taguchi, K., Blevis, I., & Iniewski, K. (Eds.). (2020). *Spectral, photon counting computed tomography technology and applications*. CRC Press. ISBN 9781138598126.
8. Torikoshi, M., et al. (2003). Electron density measurement with dual-energy x-ray CT using synchrotron radiation. *Physics in Medicine and Biology, 48*, 673–685.
9. Tsunoo, T., et al. (2004). Measurement of electron density and effective atomic number using dual-energy x-ray CT. *IEEE Nuclear Science Symposium Conference Record, 6*, 3764–3768.
10. MEDIPIX 3, medipix.web.cern.ch /collaboration/ medipix3-collaboration.
11. Hsieh, S., & Sjolin, M. (2018). Digital count summing vs analog charge summing for photon counting detectors. *Medical Physics, 45*, 4085. https://doi.org/10.1002/mp.13098
12. Iniewski, K. (2014). CZT growth, characterization, fabrication and electronics for operation at > 100 Mcps/mm2, Workshop on medical applications of spectroscopic X-ray detectors, CERN, April 2014.
13. Zuber, K., et al. (2015). A comparison of the temporal instabilities found in state-of-the-art CdTe and CZT sensors used in spectral CT measurements with the Medipix3RX detector, IEEE NSS-RSTD-MICS Nov. 2015, San Diego.
14. Ullberg, C., Urech, M., Weber, N., Engman, A., Redz, A., & Henckel, F. (2013, 6 March). Measurements of a dual-energy fast photon counting CdTe detector with integrated charge sharing correction. *Physics of Medical Imaging, 8668*, 86680P. https://doi.org/10.1117/12.2007892
15. Ballabriga, R., et al. (2013). The Medipix3RX: A high resolution, zero dead time pixel detector readout chip allowing spectroscopic imaging. *JINST, 8*, C02016.
16. Fischer, P., & Helmich, A. (2000). A photon-counting pixel chip with energy windowing. *IEEE Transactions on Nuclear Science, 47*, 881–884.
17. Lindner, M., Blanquart, L., Fischer, P., Krüger, H., & Wermes, N. (2001). Medical X-ray imaging with energy windowing. *Nuclear Instruments and Methods in Physics Research Section A: Accelerators, Spectrometers, Detectors and Associated Equipment, 465*, 229–234.
18. Iwanczyk, J. S., et al. (2009). Photon-counting energy-dispersive detector arrays for X-ray imaging. *IEEE Transactions on Nuclear Science, 56*, 59–96.
19. Willemink, M. J., Persson, M., Pourmorteza, A., Pelc, N. J., & Fleischmann, D. (2018). Photon-counting CT: Technical principles and clinical prospects. *Radiology, 289*, 293–312.
20. Richtsmeier, D., Guliyev, E., Iniewski, K., & Bazalova-Carter, M. (2021). Contaminant detection in non-destructive testing using a CZT photon-counting detector. *Journal of Instrumentation, 16*.
21. https://solutioinsilico.com/medical-physics/applications/tasmip-app.php
22. https://physics.nist.gov/PhysRefData/Xcom

23. Methods of calibrating semiconductor radiation detectors using K-edge filters, Elmaddin Guliyev, Georgios Prekas, Michael Rozler, Krzysztof Iniewski, Jean Marcoux, Conny Hansson, 2021/11/19, US, 11169286.
24. Rodesch, P.-A., Richtsmeier, D., Guliyev, E., Iniewski, K., & Bazalova-Carter, M. (2023, March). Comparison of threshold energy calibrations of a photon-counting detector and impact on CT reconstruction. *IEEE Transactions on Radiation and Plasma Medical Sciences, 7*(3) IEEE Paper.
25. Nguyen, J., Rodesch, P.-A., Richtsmeier, D., Iniewski, K., & Bazalova-Carter, M. (2021). Contaminant detection using a CZT photon counting detector with TDI image reconstruction. *Journal of Instrumentation, 16*(11), P11015.

Impact of Blurring Correction Method in Quantitative Image Analysis Using Energy-Resolving Photon Counting Detector (ERPCD)

Hiroaki Hayashi, Daiki Kobayashi, Rina Nishigami, Natsumi Kimoto, Tatsuya Maeda, and Takashi Asahara

1 Introduction

Dr. Röntgen reported the discovery of X-rays in his 1896 paper titled "On a New Kind of Rays" [1], earning him the first Nobel Prize in Physics in 1901 [2]. Dr. Roentgen's paper included radiographs of a compass and of a living person's hand, with the latter receiving particularly significant attention. Although medical science at the time had knowledge of human anatomy, it was not possible to visualize the internal organ structures of a living body. In that era, physicians made diagnoses using sources of information beyond X-rays. However, for us, who commonly utilize X-rays in modern medical diagnoses, it is challenging to fathom the diagnostic methods of that time. In this way, the scientific field that has benefited the most from the discovery of X-rays is medical diagnosis, largely due to advances in technology that can nondestructively image internal organ information.

Figure 1 depicts a conceptual diagram of X-ray diagnosis. X-rays penetrating the object are absorbed by an X-ray imaging detector, subsequently generating an X-ray image by capturing the distribution of absorbed X-ray amounts in a two-dimensional plane. This type of conventional detector is called an energy-integrating detector (EID) [3] because the detector integrates the total energy of X-rays to obtain image density. Due to the varying degrees of penetration of X-rays through objects, X-ray shadowgraphs exhibit diverse image densities, reflecting differences in the amount of penetrating X-rays depending on the materials along the X-ray pathway. Consequently, given the human body's composition of organs with distinct densities and

H. Hayashi (✉) · D. Kobayashi · R. Nishigami · T. Maeda · T. Asahara
Kanazawa University, Kanazawa, Japan
e-mail: hayashi.hiroaki@staff.kanazawa-u.ac.jp

N. Kimoto
JOB CORPORATION, Kanagawa, Japan

K. Iniewski (ed.), *CdTe and CdZnTe Materials*,
https://doi.org/10.1007/978-3-031-64521-1_14

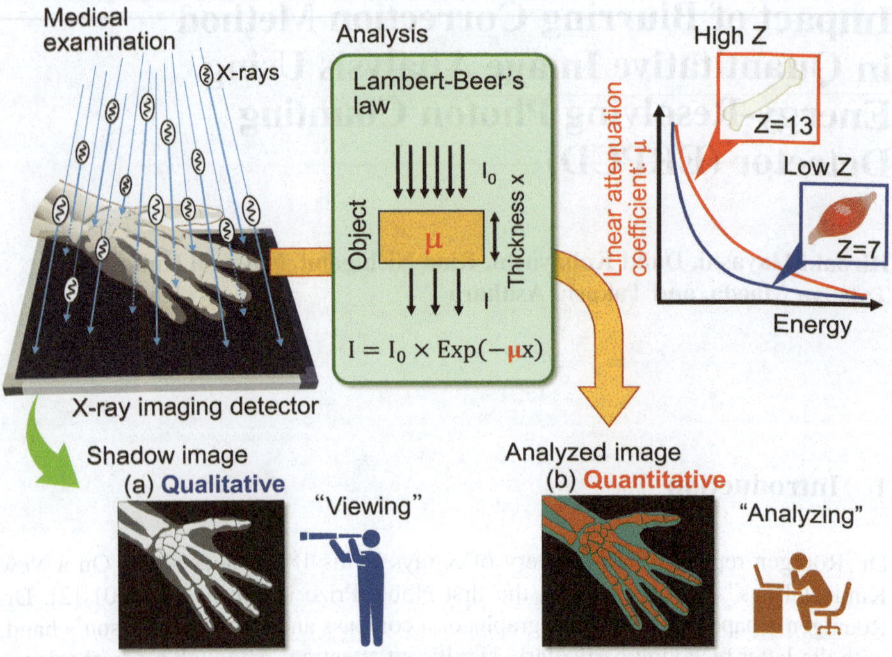

Fig. 1 A conceptual diagram illustrating medical diagnosis utilizing X-rays. Traditionally, diagnoses have relied on the qualitative interpretation of shadow images produced by X-rays. However, by analyzing the attenuation information of X-rays, it is feasible to quantify the physical characteristics of the object. This can be achieved through the utilization of an energy-resolving photon-counting detector (ERPCD)

compositions, these organ disparities are discernible in the projected image. Therefore, medical doctors can diagnose abnormalities by qualitatively viewing X-ray shadow images and comparing them with their anatomical knowledge.

The attenuation law of X-rays is known as Lambert–Beer's law [3] and can be expressed as Exp($-\mu x$), where μ and x are the linear attenuation coefficient [4] and the thickness of the object, respectively. The soft tissues of the human body have effective atomic numbers (Z_{eff}) [5] of 6–7, and the Z_{eff} of bones is approximately 13 [6]. Substances with high Z_{eff} values exhibit high absorption efficiency for low-energy X-rays, so imaging using relatively low-energy X-rays can enhance the difference in the amount of X-ray absorption, leading to the acquisition of high-contrast images [7]. As mentioned above, diagnoses have been made by viewing X-ray images, which are shadow pictures of X-rays. By studying the interaction between X-rays and substances, imaging conditions that provide higher contrast can be derived and used in medical diagnosis. One important idea arises here: if the interaction rate of X-rays can be analyzed based on the laws of physics, it will be feasible to identify the materials within the object and image physical quantities accordingly. An image derived through analysis is called a

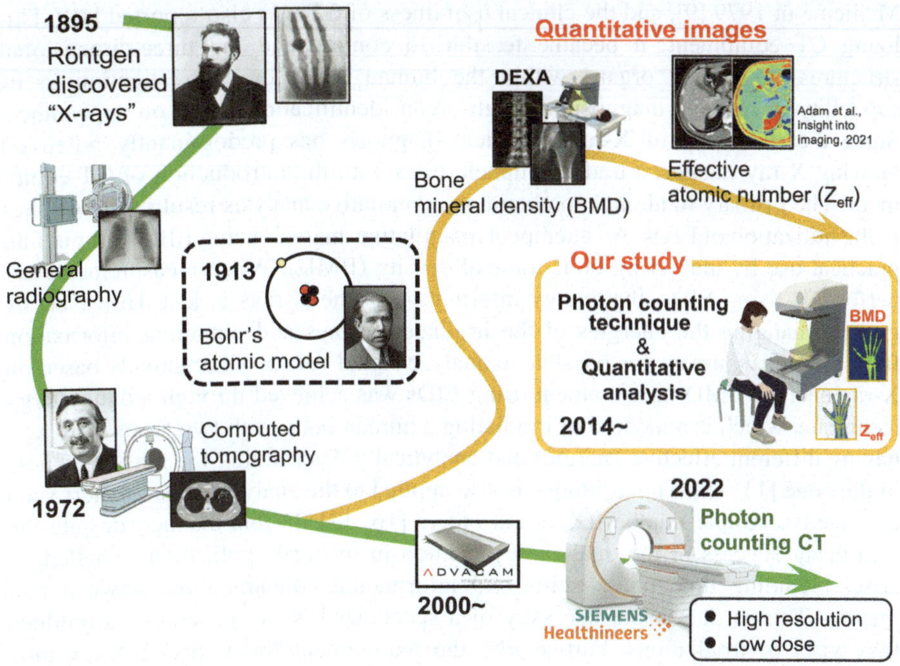

Fig. 2 Relationship between the discovery of X-rays and the development of quantitative analysis imaging technology. Photon-counting-type image detectors were developed around 2000, and clinical CT equipment was released in 2022

quantitative image. To generate such quantitative images, it is necessary to analyze the amount of attenuation of X-rays, and it is essential to develop an energy-resolving photon-counting X-ray detector (ERPCD) that can measure each X-ray photon individually.

Figure 2 shows the relationship between the discovery of X-rays and various related technologies. More than 120 years have passed since the discovery of X-rays, but the method of generating them still follows the method pioneered by Dr. Röntgen. X-rays are generated by irradiating a metal target with accelerated electrons. The interesting point is that medical diagnosis using X-rays was put into practical use before the elucidation of radiophysical science on atomic structures and various interactions. For example, the hydrogen atom was modeled by Dr. Bohr in 1913. From this fact, it can be understood that medical X-ray photographs are analyzed not necessarily based on physics theory but rather on the basis of anatomical knowledge derived from shadow pictures.

Medical X-ray imaging technology advanced significantly with the introduction of computed tomography (CT), which was made possible by innovative ideas, mathematical research, and industrial development. In 1972, the first CT experiment was carried out by Dr. Hounsfield [8], which led to a Nobel Prize in Physiology or

Medicine in 1979 [9], and the clinical usefulness of CT was also reported [10]. Utilizing CT equipment, it became feasible to comprehend the three-dimensional structures of internal organs within the human body, drawing attention to its capability in medical diagnosis through organ identification based on appearance. Since the discovery of X-rays, medical diagnosis has predominantly relied on viewing X-ray images, a tradition upheld even with the introduction of CT equipment. The primary hindrance to obtaining quantitative analysis results with CT lies in the utilization of EIDs. An attempt at quantitative analysis using EIDs was put into practical use by measuring bone mineral density (BMD). When measurements are performed using EIDs, the energy information of the X-rays is lost. However, by greatly changing the energies of the irradiated X-rays and obtaining information from multiple sources, it is possible to analyze signal values quantitatively based on X-ray energy. BMD measurement using EIDs was achieved through a dual-energy technique, which is a method of irradiating a human body with two types of X-rays having different effective energies and analytically determining the mass thickness of the bone [11–15]. The technique is also applied to the analysis of parameters such as effective atomic number (Z_{eff}) and others [16, 17]. In this manner, despite the significant success of the dual-energy method in medical applications, challenges arise, including time phase shifts and information contamination between two images. This is due to the necessity of a specialized X-ray generator to produce two types of irradiations. Particularly, the requirement for a special X-ray tube presents a barrier to the widespread adoption of this method.

To solve this problem, ERPCDs that perform energy discrimination by an X-ray detector have been developed since the 2000s [18–24]. In 2022, Siemens Healthcare brought to market a clinical photon-counting CT machine [24]. Our research group has been developing an ERPCD since 2014 [25–32], which aims to generate quantitative images in general X-ray photography. A report indicated the importance of analyzing the amount of X-ray attenuation by considering the beam hardening effect in the object and the response characteristics of the detector [30]. By appropriately correcting these effects, it was possible to successfully image the Z_{eff} of the subject [29, 30], and it was also shown that bone and soft tissue could be imaged separately [31]. It has also become clear that image correction is necessary when calculating these quantitative images [32]. One of the strengths of ERPCDs is that they can generate quantitative images and obtain various diagnostic information through image analysis. Since quantitative images are obtained as a result of various image-processing technologies, they are very sensitive to error factors such as statistical fluctuations. This is a very different point from diagnosis using ordinary qualitative images. Although objects in images are contoured by identifying differences in pixel values, when image blurring occurs, boundaries become unclear, thus hindering quantitative image analysis. In this chapter, we discuss in detail the importance of image blurring correction.

2 X-Ray Images and the Impact of Image Blurring

2.1 Definition of X-Ray Images

Before discussing the effects of image blurring, we define the X-ray images used in this chapter. We assume an ERPCD system with three energy bins (energy windows). When the X-ray spectrum obtained by the detector is described as $f(E)$, the pixel values for each energy bin are calculated as follows:

$$L = \int_{E=20\ \text{keV}}^{32\ \text{keV}} f(E)dE, \tag{1}$$

$$M = \int_{E=32\ \text{keV}}^{40\ \text{keV}} f(E)dE, \tag{2}$$

$$H = \int_{E=40\ \text{keV}}^{56\ \text{keV}} f(E)dE, \tag{3}$$

where L, M, and H are the pixel values of low-, middle-, and high-energy images, respectively. Our system comprises three energy bins: 20–32 keV for the low-energy range, 32–40 keV for the middle range, and 40–56 keV for the high-energy range. These settings are similar to those used in our previous experiments, in which a tube voltage of 60 kV was used [29–31].

First, we explain the image calculation method used for qualitative evaluation. The corresponding pixel value can be obtained from the integrated value of all radiation energies absorbed by the pixel of interest [20, 21]. Since EIDs are generally equipped with a log amplifier, the pixel value equivalent to the EID image can be calculated as follows:

$$\text{EID image} = \text{Log}\left(E_{\text{eff,Low}}L + E_{\text{eff,Middle}}M + E_{\text{eff,High}}H\right), \tag{4}$$

$$E_{\text{eff,Low}} = \frac{\displaystyle\int_{E=20\ \text{keV}}^{32\ \text{keV}} E \times f_0(E)dE}{\displaystyle\int_{E=20\ \text{keV}}^{32\ \text{keV}} f_0(E)dE}, \tag{5}$$

$$E_{\text{eff,Middle}} = \frac{\displaystyle\int_{E=32\ \text{keV}}^{40\ \text{keV}} E \times f_0(E)dE}{\displaystyle\int_{E=32\ \text{keV}}^{40\ \text{keV}} f_0(E)dE}, \tag{6}$$

$$E_{\text{eff,High}} = \frac{\int_{E=40\text{keV}}^{56\text{keV}} E \times f_0(E)dE}{\int_{E=40\text{keV}}^{56\text{keV}} f_0(E)dE}, \tag{7}$$

where $E_{\text{eff, Low}}$, $E_{\text{eff, Middle}}$, and $E_{\text{eff, High}}$ are the effective energies of the low-, middle-, and high-energy bins, respectively, and $f_0(E)$ represents the incident X-ray spectrum directed toward the object.

Next, to acquire a Z_{eff} image, the following processing was conducted [30], utilizing the information from the M and H datasets. This algorithm calculates the X-ray attenuation rate (μt) for the middle- and high-energy bins, erases thickness information from them, and then converts $(\mu t)_{\text{Middle}}/(\mu t)_{\text{High}}$ to Z_{eff} values. The beam-hardening effect in the object and the influence of the detector response characteristics were properly corrected. The X-ray attenuation rates for the middle- and high-energy bins and their ratio, $\mu^\dagger$, were calculated as follows:

$$(\mu t)_{\text{Middle}} = \ln\left(\frac{\int_{32\text{ keV}}^{40\text{ keV}} f_0(E)dE}{\int_{32\text{ keV}}^{40\text{ keV}} f(E)dE}\right) = \ln\left(\frac{M_0}{M}\right), \tag{8}$$

$$(\mu t)_{\text{High}} = \ln\left(\frac{\int_{40\text{ keV}}^{56\text{ keV}} f_0(E)dE}{\int_{40\text{ keV}}^{56\text{ keV}} f(E)dE}\right) = \ln\left(\frac{H_0}{H}\right), \tag{9}$$

$$\mu^\dagger = \frac{(\mu t)_{\text{High}}}{\sqrt{{(\mu t)_{\text{Middle}}}^2 + {(\mu t)_{\text{High}}}^2}} = \frac{\mu_{\text{High}}}{\sqrt{{\mu_{\text{Middle}}}^2 + {\mu_{\text{High}}}^2}}, \tag{10}$$

where the subscript "$_0$" represents the values corresponding to measurement without an object. Because $\mu^\dagger$ can be described as a function of atomic number Z, its inverse function $Z(\mu^\dagger)$ was determined using reference data of linear attenuation coefficient "μ" (and/or mass attenuation coefficient μ/ρ) [4]. Consequently, Z_{eff} can be determined as:

$$Z_{\text{eff}}\ \text{image} = Z\left(\mu^\dagger\right). \tag{11}$$

The advantage of this conversion method is that μ values can be converted to Z_{eff} values by a simple mathematical transformation. The reference function used for this conversion was determined in advance using the theoretical relationship between $\mu^\dagger$ and atomic numbers.

2.2 Impact of Image Blurring on Z_{eff} Images

X-ray images are affected by image-blurring effects, which are known as point spread functions (PSFs) [33, 34]. The effect of image blurring appears particularly strong at the object edge, so care must be taken when identifying the outline of the object [35, 36]. To improve the visibility of the object's edge, various image processing methods have been proposed. For example, the unsharp masking (UM) method is known as a software-based image processing technique [37–40]. The methods developed in such previous studies are image filters specialized for qualitative analysis and are not necessarily useful for image processing for qualitative analysis. To illustrate this point in this chapter, we present the results of image processing using UM and compare it with the proposed method.

Image blurring occurs even with ERPCDs. Figure 3 is a conceptual diagram summarizing the mechanism by which image blurring occurs in ERPCD. Image blurring is attributed to the following phenomena. As shown by Fig. 3(a), on the right, when the pixel size and the edge of the object do not match, a single pixel contains a mixture of components related to penetrating X-rays and direct X-rays. In this case, it is difficult to accurately represent the object's boundary surface. In particular, images using ERPCDs are usually acquired while moving the detector using a scanogram system, and this effect occurs frequently. X-rays incident on the detector interact with the detector material and generate electric charge clouds (Fig. 3 (b)). However, a physical phenomenon occurs wherein some of the electric charge is shared with adjacent pixels. This phenomenon is known as "cross talk" due to the secondary X-rays that occur succeeding the photoelectric effect and the "charge

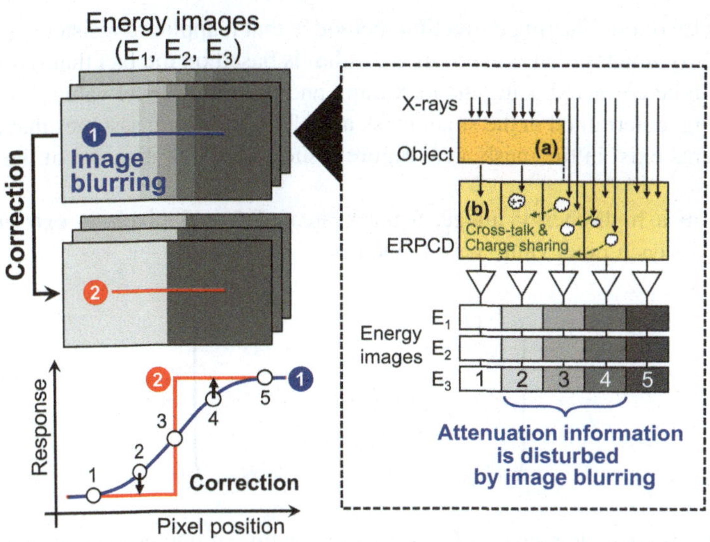

Fig. 3 Conceptual diagram of the mechanism by which image blurring occurs in ERPCDs (right); the phenomena in each energy image and the concept of correction (left)

sharing effect" occurring during the charge collection process facilitated by the electric field within the detector [18, 19, 25, 26]. With our ERPCD, multiple energy images (L: E_1, M: E_2, and H: E_3 in our system) are acquired, as shown on the left side of Fig. 3, and it is assumed that different amounts of image blurring will occur between these energy images. This point of view is very important when considering the problem of blurring in ERPCDs. Namely, to calculate a quantitative image, it is necessary to correct the blurring of all energy images. In the profile of the blurred image on the left in Fig. 3 (No. 1, blue), the pixel values corresponding to pixel numbers 1–5 change gradually, but the profile of the corrected image (No. 2, red) shows the step function.

An imaging technique with ERPCD requires acquisition under low-dose-rate (low-counting-rate) conditions, and attention must be paid to the effect of statistical fluctuations. This is because when the counting rate increases, a pile-up phenomenon occurs in which multiple radiation signals are superimposed, making it impossible to correctly analyze the energy information of each photon. This consideration suggests that only partial blurring correction can be achieved due to the impact of statistical errors. Therefore, there is a need to develop a blurring correction method capable of producing consistent results across multiple energy images affected by statistical fluctuations.

3 Blurring Correction Method

3.1 *Proposed Method*

The novelty of our blurring correction method is that it applies consistent corrections to all energy images. Our correction algorithm is based on the fact that blurred pixel values can be corrected using the maximum and minimum pixel values not affected by blurring, determined in the small mask area. The algorithm assumes that only two components exist in the mask area. Figure 4 shows a block diagram of the analysis algorithm.

First, in a high-energy image, a mask area of 5 × 5 pixels is extracted. The notation for each pixel value was defined as:

$$\begin{pmatrix} H_{1,1} & \cdots & H_{5,1} \\ \vdots & H_{\text{int}} & \vdots \\ H_{5,5} & \cdots & H_{5,5} \end{pmatrix}, \tag{12}$$

where $H_{i,j}$ is each element and H_{int} is the pixel value of interest (center pixel). Then the elements of this matrix were sorted in an ascending order of pixel value, and the sorted elements were renamed as:

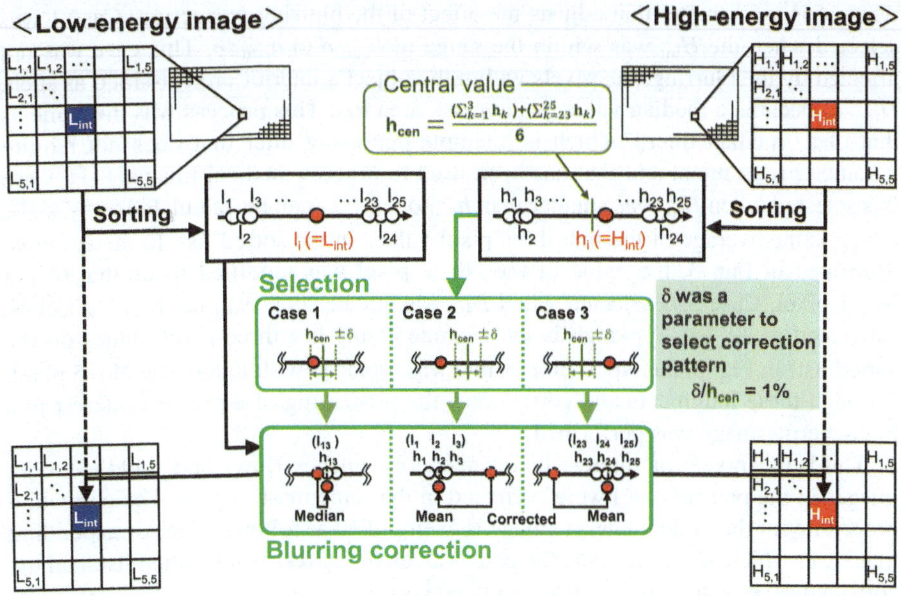

Fig. 4 Our proposed algorithm for blurring the correction of multiple energy images acquired with ERPCD. In the present case, two energy images (middle and high) are corrected, and the pixels near the object edges are identified based only on the high-energy component. This process is performed for a small mask size (5 × 5 pixels)

$$(h_1, h_2, \cdots, h_{13}, \cdots, h_{24}, h_{25}), \tag{13}$$

where h_{13} is the median value in the sorted list. Here, the central criterion value of the mask area h_{cen} was tentatively estimated as:

$$h_{\text{cen}} = \frac{1}{2} \times \left(\frac{h_1 + h_2 + h_3}{3} + \frac{h_{23} + h_{24} + h_{25}}{3} \right). \tag{14}$$

We should note that h_{cen} is just an index used for judgment; it is not the true average value because h_{cen} is an estimated central value using 25% of the information from the mask area.

Secondly, we identified the blurred pixel by comparing H_{int} with h_{cen}. In this process, we considered the following three cases and then derived the corrected value of the pixel of interest, $H_{\text{int, cor}}$, as follows:

$$H_{\text{int,cor}} = \begin{cases} h_{13}, & \text{when } h_{\text{cen}} - \delta \le H_{\text{int}} \le h_{\text{cen}} + \delta \text{ (case 1)} \\ \dfrac{h_1 + h_2 + h_3}{3}, & \text{when } H_{\text{int}} < h_{\text{cen}} - \delta \text{ (case 2)} \\ \dfrac{h_{23} + h_{24} + h_{25}}{3}, & \text{when } h_{\text{cen}} + \delta < H_{\text{int}} \text{ (case 3)} \end{cases}, \tag{15}$$

where δ is a parameter that adjusts the effect of the blurring correction. Case 1 was selected when the H_{int} was within the range of h_{cen}-δ to h_{cen}+δ. This case was not affected by the blurring; the pixels within the object's interior are classified as such. $H_{int,\ cor}$ became a median value of the mask area h_{13}. This process was the same as that of a "median filter," which is a simple denoising filter that does not require parameter adjustment and that has been used to process medical images [41]. Case 2 is selected when H_{int} was smaller than h_{cen}-δ. $H_{int,\ cor}$ can be calculated as $\frac{h_1+h_2+h_3}{3}$, which is the average of the first three pixel values in the sorted list. In the example illustrated in Fig. 3, the value of the No. 2 pixel was modified to be that of the No. 1 pixel. Case 3 is selected when H_{int} was greater than h_{cen}+δ. $H_{int,\ cor}$ can be calculated as $\frac{h_{23}+h_{24}+h_{25}}{3}$, which is the average of the last three pixel values on the sorted list. In Fig. 3, the No. 4 pixel value was replaced with that of the No. 5 pixel. Through these judgments and corrections, the processing of a pixel of interest in a high-energy image was performed.

Third, the pixels of interest in other energy images (low- and middle-energy images in the present case) were corrected in the same mask region. The correction procedure for the middle-energy image is exemplified as follows. The corresponding matrix, in which M_{int} represents the pixel values of interest in the sorted list, and the corrected pixel value $M_{int,\ cor}$ was calculated as:

$$\begin{pmatrix} M_{1,1} & & \cdots & & M_{1,5} \\ & \ddots & & ⋰ & \\ \vdots & & M_{int} & & \vdots \\ & ⋰ & & \ddots & \\ M_{5,1} & & \cdots & & M_{5,5} \end{pmatrix}, \tag{16}$$

$$(m_1, m_2, \cdots, m_{13}, \cdots, m_{24}, m_{25}), \tag{17}$$

$$M_{int,cor} = \begin{cases} m_{13}, & \text{when case 1 was selected in Eq.(15)} \\ \dfrac{m_1+m_2+m_3}{3}, & \text{when case 2 was selected in Eq.(15)} \\ \dfrac{m_{23}+m_{24}+m_{25}}{3}, & \text{when case 3 was selected in Eq.(15)} \end{cases} . \tag{18}$$

It should be noted that this judgment of the middle-energy image was based on the judgment result for the high-energy image, as described in Eq. 15. The corrected values of the pixel of interest, $M_{int,\ cor}$, in cases 1, 2, and 3 can be calculated with m_{13}, $\frac{m_1+m_2+m_3}{3}$, and $\frac{m_{23}+m_{24}+m_{25}}{3}$, respectively. With this algorithm, consistent correction is performed for multiple energy images.

Finally, by performing the above processes pixel by pixel, image blurring can be corrected for the entire image.

3.2 Demonstration of Judgment

Figure 5 exemplifies three judgment cases in the actual X-ray image, in which a comb-shaped phantom composed of aluminum and polymethylmethacrylate (PMMA) with a thickness of 1 cm was used. In the following explanation, the raw count value was divided by h_{cen} for clarity, but in the actual program, the raw count values are used for correction. The parameter δ used in the correction method was tentatively set to $\delta/h_{cen} = 1\%$.

The region of interest (ROI: 5 × 5 pixels) of the black area situated in a uniform PMMA area was judged as case 1. The right graph shows the sorted list. Many pixel values were clustered around h_{cen}. The original H_{int} value was 1.0076, and it was included within the range of h_{cen}-δ to h_{cen}+δ. When our correction method was applied, the corrected $H_{int,\ cor}$ value became 1.0024, which is the median value, h_{13}.

The red ROI area was placed at the aluminum edge and serves as a typical example of case 2. The blurring effect can be observed in the aluminum edge area. The original H_{int} value was 0.9487, which was lower than h_{cen}-δ. When our correction method was applied, $H_{int,\ cor}$ became $\frac{h_1+h_2+h_3}{3} = 0.7324$.

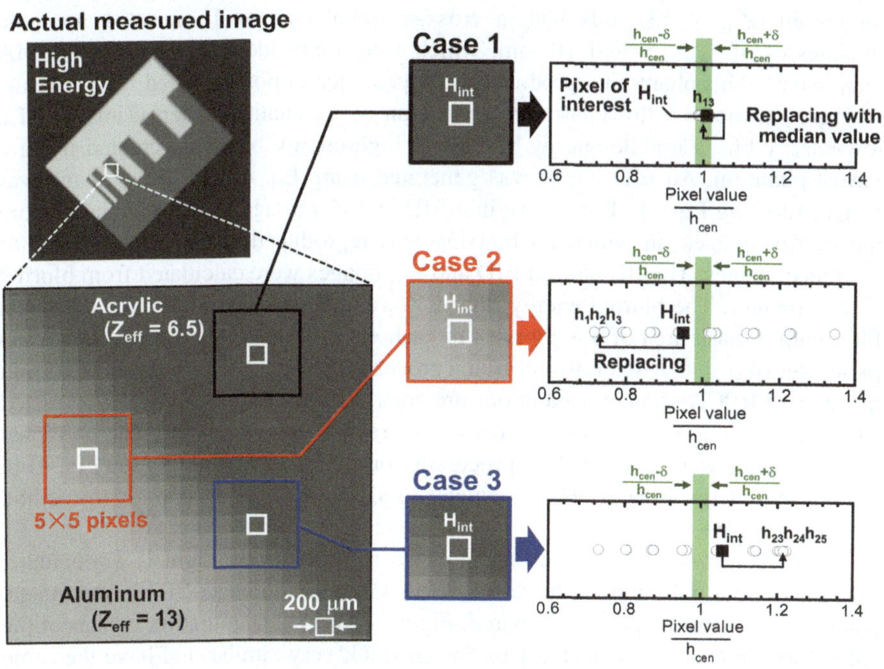

Fig. 5 Example of the judgments of pixels in a comb-shaped phantom consisting of aluminum and PMMA. There are three judgment cases in the close-up view of the image. Case 1 is the PMMA region, case 2 is a boundary region near the aluminum, and case 3 is the opposite boundary region near the PMMA

The blue ROI area was set at the PMMA edge opposite the red ROI area. It serves as a typical example of case 3. The original H_{int} value was 1.0577, which was higher than $h_{cen}+\delta$, and then $H_{int,\ cor}$ became $\frac{h_{23}+h_{24}+h_{25}}{3} = 1.2197$.

4 Verification

4.1 Simulation

The proposed blurring correction method was verified through simulation. A simulation program was used to generate the energy images obtained by the ERPCD, in which a CZT-type imaging detector with a pixel size of 200 μm was assumed. The simulation comprehensively incorporated (a) X-ray attenuations within the object, including the beam hardening effect; (b) the response function of the multipixel-type detector; and (c) statistical fluctuations, all under an incident exposure dose of 15 mGy [42, 43]. The incident X-ray spectrum was calculated using the semi-empirical formula proposed by Tucker et al. [44]. The predetermined conditions were as follows: tube voltage of 60 kV, tungsten target, and a total filtration of 2.5 mm aluminum equivalent thickness. The virtual phantom consisted of four small aluminum ($Z_{eff} = 13$) rods with a cross-sectional size of 1 mm × 1 mm and thicknesses of 1, 3, 5, and 10 mm, which were embedded inside cubic PMMA ($Z_{eff} = 6.5$). This phantom reproduces the appearance of bones buried in soft tissue.

Figure 6 shows a flowchart for verification by simulation. Energy images of a low-energy bin, a middle-energy bin, and a high-energy bin were created using a virtual phantom. An EID image was generated using Eq. 4, and a Z_{eff} image was generated using Eq. 11. The (I) original EID and Z_{eff} images were calculated from the energy images, in which no blurring was reproduced, so they were used as reference images. The (II) blurred EID and Z_{eff} images were calculated from blurred energy images. The blurred energy images were generated by applying a Gaussian filter with a mask area of 3 × 3 pixels to each energy image. A standard deviation parameter of $\sigma = 0.7$ was utilized in this process. This parameter was derived from the actual ERPCD scanner used in our previous studies [30, 31]. The (III) corrected EID and Z_{eff} images were generated from corrected energy images. The corrected energy images were generated by processing our correction method (see Fig. 4) to the blurred energy images. The parameter used in our blurring correction method was $\delta/h_{cen} = 1\%$.

Figure 7 shows a comparison of analyzed results. On the right is a schematic drawing of a virtual sample. The results of the (I) original image, (II) blurred image, and (III) corrected image are compared. Figure 7(a) shows simulation images of the EID. The original and corrected EID images look very similar and have the same image density (pixel value). In a blurred EID image, the edges of the object are not clear. Figure 7(b) shows the distribution of Z_{eff}, in which a numerical histogram was analyzed in the ROI (9 × 9 pixels) for the investigational aluminum samples having a thickness of 10 mm. The results of the ROI analysis can be used to quantitatively

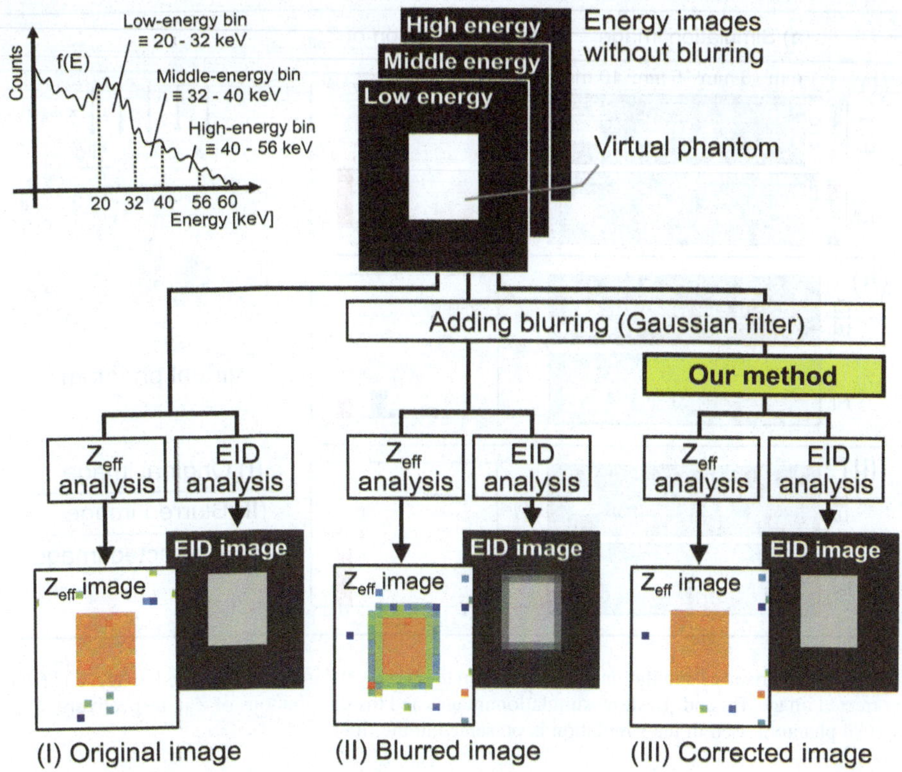

Fig. 6 Conceptual diagram of the verification of the blurring correction method using simulation. EID and Z_{eff} images were generated corresponding to the (I) original image, (II) blurred image, and (III) corrected image

evaluate the effectiveness of our blurring correction. In the (I) original Z_{eff} image, the aluminum area ($Z_{eff} = 13$) was 25 pixels (5 × 5 pixels), and that of PMMA ($Z_{eff} = 6.5$) was 56 pixels around it. In the (II) blurred Z_{eff} image, the aluminum region ($Z_{eff} = 13$) was only 9 pixels, and the blurred pixels showed intermediate Z_{eff} values between those of aluminum and PMMA. This fact indicates that incorrect quantitative values may be calculated at the edges of the object when a quantitative X-ray image is generated using blurred energy images captured by ERPCDs. On the other hand, in the (III) corrected Z_{eff} image, the Z_{eff} distribution within the ROI was the same as that in image I. This underscores the importance of correcting the blurring effect.

4.2 *Experiment Using Our Scanogram Equipment*

In this part of the chapter, we will show an example of a blurring correction method using actually measured images. A slit-scanning-type imaging device was used [30–32]. An X-ray generator (JOB CORPORATION, Japan) using a tungsten target and

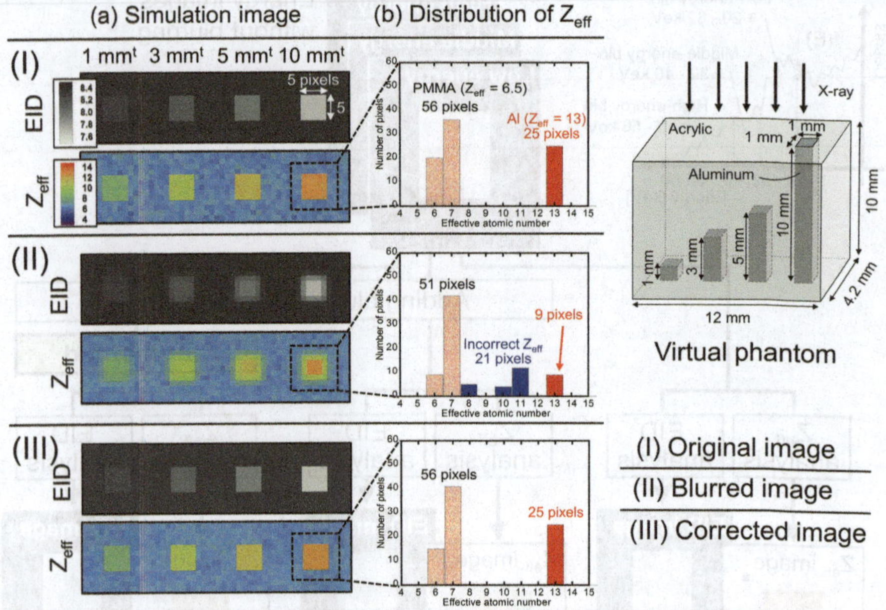

Fig. 7 Comparison of simulation results between the (I) original image, (II) blurred image, and (III) corrected image; (**a**) and (**b**) show simulation images and the distributions of Z_{eff}, respectively. The virtual phantom used in the simulation is presented at the right

total filtration of 2.5 mm aluminum equivalent thickness was used. The tube voltage was 60 kV, and the tube current was 0.5 mA. The experiment was conducted under conditions where the X-ray count rate was lowered to prevent pulse pileup. The image detector consists of a 1.5 mm thick cadmium zinc telluride (CZT) semiconductor detector, with a pixel size of 200 μm. The width of the sensor unit is 20 pixels, and images are obtained by scanning this unit horizontally. The scan speed was 0.125 mm/s. The frame data are reconstructed using the shift and add method. Collimators were placed in front of the X-ray emission port and imaging detector to achieve slit scanning, which prevents the contamination of scattered X-rays. The distance between the source and the detector is 620 mm. The distance between the object and the detector is very large, 370 mm, to prevent the contamination of scattered X-rays. The object was a comb-shaped phantom made of aluminum and PMMA. The outer size was 40 mm × 30 mm. The thickness of the phantom was 10 mm. The phantom had slits of 1, 2, 3, 4, and 5 mm, and these slit widths were precisely manufactured using machining equipment. The comb phantom was placed at a 45 degree angle to the detector movement direction.

Figure 8 shows the analysis results for the comb-shaped phantom. The top row is a photograph of the comb-shaped phantom, the second row is an EID image, the third row is a Z_{eff} image, and the bottom row is a correction amount image. To demonstrate the superiority of our method, the results obtained by the following

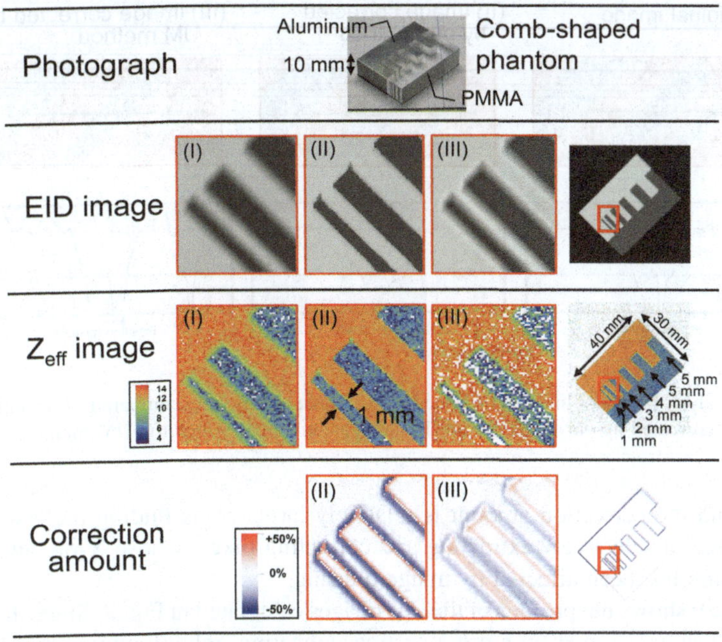

Fig. 8 Analysis results of the comb-shaped phantom obtained in the experiment. The EID image, Z_{eff} image, and correction amount are shown. I, II, and III are the original image, the image corrected by our method, and the corrected image (comparative image) applied through the unsharp masking (UM) method, respectively

three methods were compared. (I) "Original images" were generated from the measured energy images without any blurring correction. (II) "Corrected images" were generated from the energy images corrected by our method. (III) Comparative images were generated from the energy images in which the conventional unsharp masking (UM) method was applied. For the (I) original image in which blurring occurs, the contours between different substances are unclear in the EID image, and incorrect Z_{eff} values are calculated for the corresponding parts. For the (II) corrected image where image blurring was corrected using our method, both the EID and Z_{eff} images have undergone correction. Consequently, accurate quantitative values have been obtained. The white pixels in the Z_{eff} image are pixels for which the effective atomic number could not be calculated; it can be seen that by applying our algorithm, the frequency of occurrence of error pixels is significantly reduced. This suggests that one of the causes of error pixels is that the quality of information in multiple energy images differs. For the (III) comparative image in which object edges were enhanced by applying the UM method, the visibility of the EID image improved slightly, but the image quality of the Z_{eff} image decreased significantly. In particular, the increase in error pixels is a negative result. A comparison of correction amount images between (II) and (III) shows interesting results: our method can extract object

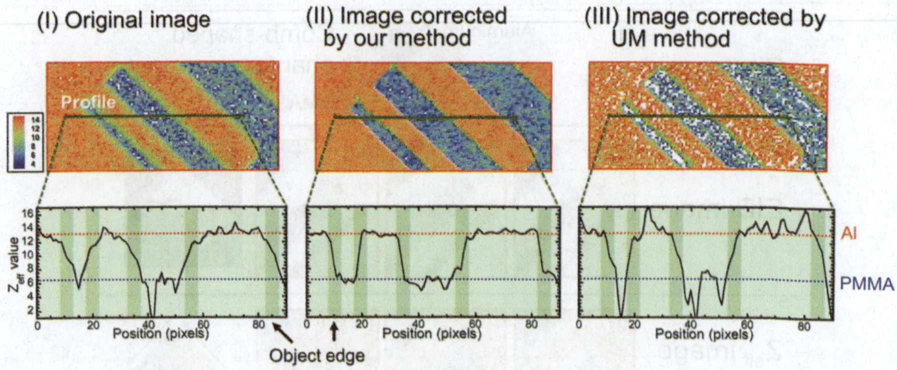

Fig. 9 Analysis results of line profiles (see Fig. 8). Three results were compared: (I) original image, (II) image corrected by our method, and (III) image corrected through the UM method

edges, and the correction amount is relatively larger. This finding underscores the importance of selectively extracting and correcting pixels whose X-ray attenuation information has been affected by image blurring.

Figure 9 shows the profiles of the Z_{eff} images calculated in Fig. 8. Since the profile analysis was performed on a cross-section indicating only aluminum and PMMA, the effective atomic number should be either $Z_{\mathrm{eff}} = 13$ or 6.5. In the (I) original image and the (III) image corrected by the UM method, Z_{eff} values other than aluminum and PMMA were observed. In contrast, (II) the image corrected by our method shows that a profile indicating aluminum or PMMA is obtained.

The unique point of the proposed algorithm in Fig. 4 is that it determines only a single judgment and corrects the two energy images consistently. This idea is especially important when correcting object edges. This difficulty arises because it is very challenging for the algorithm to precisely determine whether pixels near the edge should be modified to higher or lower values. Furthermore, it is also necessary to consider that the spread of blurring differs between energy images. We demonstrate the practicality of our algorithm using Fig. 10. Figure 10(a) is the original image without applying image blurring correction. Figure 10(b) is the proposed method in which image blurring is corrected properly. Because some error pixels remain, further research is required. Figure 10(c) is a very interesting comparative image. We changed our algorithm to perform a "judgment" independently of each energy image, and the Z_{eff} image was calculated after correcting these energy images. Although the image blurring is corrected and inappropriate Z_{eff} value calculations are suppressed, the appearance of error pixels has increased overwhelmingly, and the quality of the quantitative image has decreased significantly. This demonstration shows the importance of applying consistent corrections to each image when calculating another image from multiple images.

Figure 11 shows the results of a demonstration of the importance of blurring correction using a fresh fish sample (also see Fig. 8). The top row is a photograph, the middle row shows EID images, and the bottom row presents Z_{eff} images.

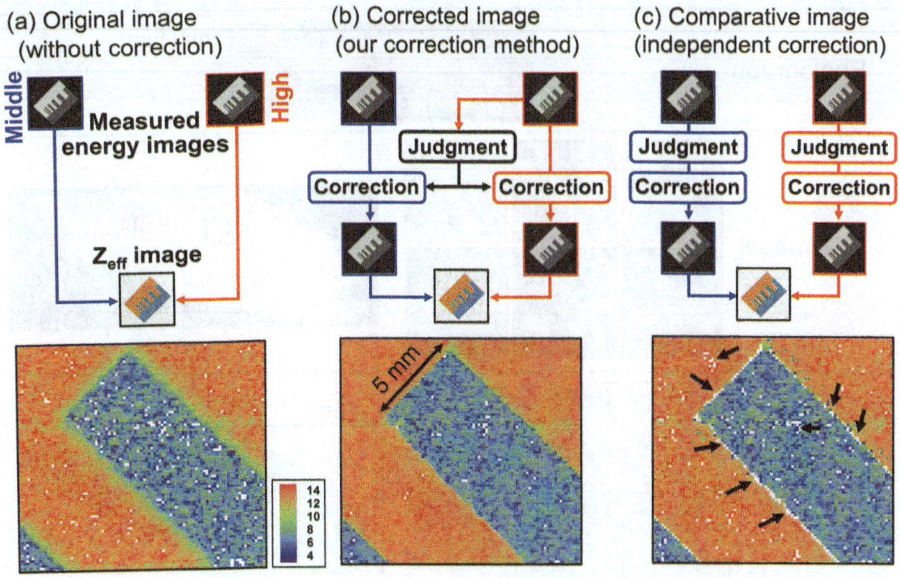

Fig. 10 Comparison of the differences in the algorithms of blurring correction methods when effective atomic number analysis is performed. (**a**) is the original image, where different Z_{eff} values are calculated at the object edges and many error pixels are observed. (**b**) is a blurring-corrected Z_{eff} image calculated through the proposed method. Only one "judgment" was shared between the two images, and proper correction was performed entirely. (**c**) is a comparison image in which independent judgments are set for the two images. The number of error pixels (white pixels) increased near the object edge because the decisions are not uniquely determined around the true edges

Figure 11(a) is a region in which a relatively simple bone, 1 mm thick, is imaged, and Fig. 11(b) is a region of the spine with a complex structure. Image I is the original image, II is the image corrected using the proposed method, and III is the image processed with the UM method. In the EID image, as shown in Fig. 11(a-II), the thin bones can be seen very clearly in the proposed image, which exceeds the visibility obtained with the UM method (see Fig. 11(a-III)). However, for a more complex region of Fig. 11(b), the UM method showed good visibility (Fig. Fig. 11(b-III). As seen in Fig. 11(b-II), it cannot be denied that the proposed method indicated slightly excessive image processing when generating EID images. Regarding Z_{eff} images, image II shows the best results, in which the noise reduction effect is also clearly observed. This is because a median filter is included in our correction algorithm. Images I and III are of low quality and are not suitable for analysis using Z_{eff} images. Although a demonstration using a fish will differ from that using the human body, it is useful to understand that blurring effects can hinder quantitative analysis in living organisms consisting of soft tissue and bone.

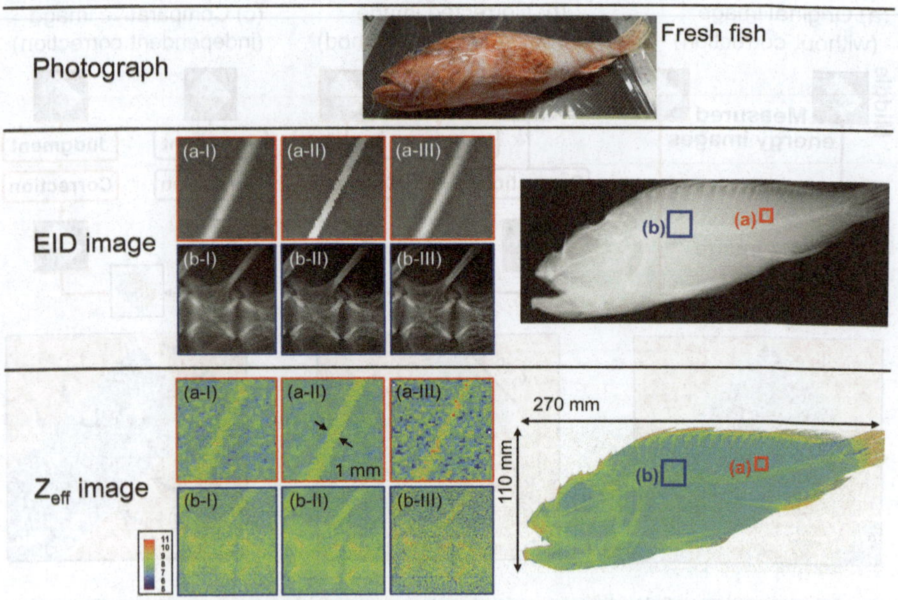

Fig. 11 Results of analyzing a fish. The photograph, EID image, and Z_{eff} image are compared. Images (**a**) and (**b**) are ROIs set in different regions. I, II, and III are the original image, the image with blurring correction using the proposed method, and the image with edge enhancement using the UM method, respectively

4.3 Experiment Using Clinical CT Equipment

To confirm the effectiveness of our correction method, we will show the analysis results using commercial CT equipment. A clinical photon-counting CT (NAEOTOM Alpha, Siemens Healthineers, Germany) can produce a virtual monoenergetic image (VMI) equivalent to a monoenergetic X-ray image. In this demonstration, VMI images of 70 keV and 100 keV were calculated by the vendor's application, and then these images were converted to Z_{eff} images based on Eq. 11. A multienergy phantom was used, within which various component materials were inserted into a water-equivalent cylindrical phantom. Figure 12 shows the results. We can see that image blurring occurs in the original VMI image. In the line profile, the CT values are strongly influenced by the point spread function (PSF), and the boundary surface of the material changes gradually. If the Z_{eff} image is calculated without image blurring correction, incorrect Z_{eff} values will be calculated at the boundaries.

Coincidentally, this blurring phenomenon was very similar to that in the system we described above (see Fig. 3). Therefore, we attempted to correct the VMI image by applying our correction method with the settings described above. The lower part of Fig. 12 shows the VMI image and Z_{eff} image in which image blurring was corrected using the proposed method. As the arrows show, the blurring in the

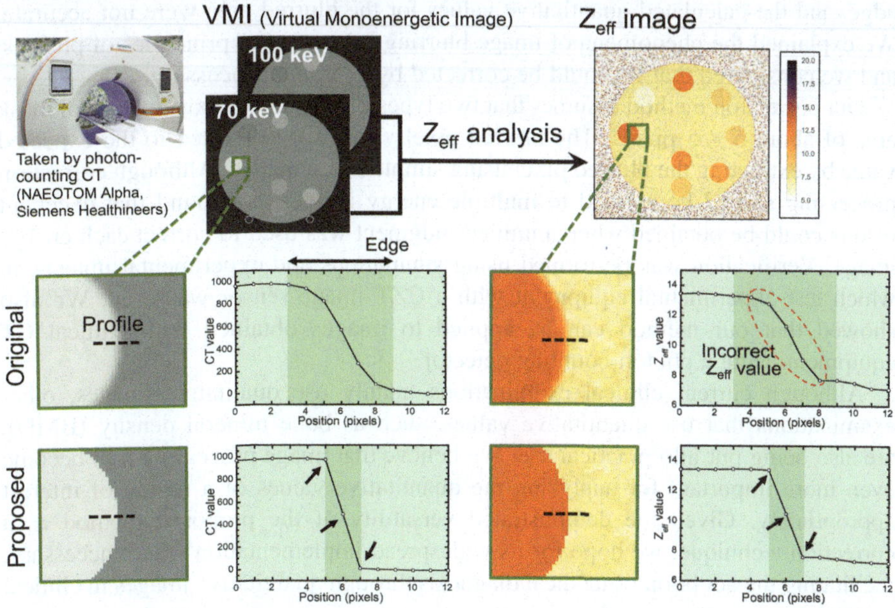

Fig. 12 Demonstration of the applicability of our method using clinical photon-counting CT equipment. Z_{eff} images were generated using virtual monoenergetic images (VMI). Our method was also effective in correcting the blurring of the image obtained with commercial CT equipment

image at the material boundaries is appropriately corrected. This result demonstrates that correction is essential when calculating quantitative values such as Z_{eff}. Thus, our correction method is versatile and can be implemented on commercial devices already in clinical use as well as on future imaging devices.

In actual clinical applications, there are many situations in which the blurring of quantitative images will be expected to significantly affect quantitative values. For example, in a dental X-ray image [45], a tooth of about 1 cm in size is presented in an air environment, which exhibits relatively larger contrast in medical images. This suggests that image blurring significantly affects the observable boundaries of the tooth. In addition, orthopedic surgery targets large dense bones such as the lumbar vertebrae [46], and related medical images show a large contrast between the lumbar vertebrae and the surrounding soft tissues. The bones of the lumbar spine are also the reference bones for calculating BMD [47], and there is a desire to compute quantitative values with greater accuracy.

5 Summary

In this chapter, we focused on the ability of energy-resolving photon-counting detectors (ERPCDs) to visualize quantitative information about objects. During quantitative analysis, the effect of image blurring appeared strongly near the object edge, and the calculated quantitative values for the blurred area were not accurate. We explained the phenomena of image blurring based on the principles of physics, and we considered that it should be corrected by software processing.

Our correction method assumes that two types of substances exist in a small mask area of about 5 × 5 pixels. The blurred pixel value was corrected to the expected value by extracting the blurred pixel using simple mathematics. Although correction processing should be applied to multiple energy images, we found that excellent results could be obtained when a unified judgment was used to correct each energy image. Verification was performed using simulations and experimental images, in which test experimental equipment with a CZT image sensor was used. We also showed that our method can be applied to images obtained with clinical CT equipment using a photon-counting detector.

Although current clinical examinations mainly use qualitative images, other examinations that use quantitative values, such as bone mineral density (BMD), are also being put into practical use. We believe that image processing will become even more important for analyzing the quantitative values of a region of interest appropriately. Given the demonstrated versatility of the proposed method as a correction technique, we hope for its widespread implementation where necessary, facilitating the adoption of medical diagnosis through quantitative images in clinical practice.

References

1. Röentgen, W. C. (1896). On a new kind of rays. *Nature, 53*, 274–276. https://www.nature.com/articles/053274b0
2. The Nobel Prize in Physics. (1901). https://www.nobelprize.org/prizes/physics/1901/summary/
3. Maier, A., Steidl, S., Christlein, V., & Hornegger, J. (2018). *Medical imaging systems* (pp. 1–259) ISBN: 978-3-319-96519-2.
4. Hubbell, J. H., & Seltzer, S. M. (1995). Tables of X-ray mass attenuation coefficients and mass energy-absorption coefficients 1 keV to 20 MeV for elements Z = 1 to 92 and 48 additional substances of dosimetric interest, NIST. https://doi.org/10.6028/NIST.IR.5632.
5. Murty, R. C. (1965). Effective atomic numbers of heterogeneous materials. *Nature, 207*, 398–399. https://www.nature.com/articles/207398a0
6. Yamashita, Y., Kimura, M., Kitahara, M., Hamaguchi, T., Kanno, I., Ohtaka, M., Hashimoto, M., Ara, K., & Onabe, H. (2014). Measurement of effective atomic numbers using energy-resolved computed tomography. *Journal of Nuclear Science and Technology, 51*, 1256–1263. https://doi.org/10.1080/00223131.2014.919881
7. Sandborg, M., Tingberg, A., Ullman, G., Dance, D. R., & Carlsson, G. A. (2006). Comparison of clinical and physical measures of image quality in chest and pelvis computed radiography at different tube voltages. *Medical Physics, 33*, 4169–4175. https://doi.org/10.1118/1.2362871
8. Hounsfield, G. N. (1973). Computerized transverse axial scanning (tomography): Part I. Description of system. *The British Journal of Radiology, 46*, 1016–1022. https://doi.org/10.1259/0007-1285-46-552-1016
9. The Nobel Prize in Physiology or Medicine. (1979). https://www.nobelprize.org/prizes/medicine/1979/summary/

10. Ambrose, J. A. E. (1973). Computerized transverse axial scanning (tomography): Part 2. Clinical application*. *The British Journal of Radiology, 46*, 1023–1047. https://doi.org/10.1259/0007-1285-46-552-1023
11. Cullum, I. D., Ell, P. J., & Ryder, J. P. (2014). X-ray dual-photon absorptiometry: A new method for the measurement of bone density. *The British Journal of Radiology, 62*, 587–592. https://doi.org/10.1259/0007-1285-62-739-587
12. Blake, G. M., McKeeney, D. B., Chhaya, S. C., Ryan, P. J., & Fogelman, I. (1992). Dual energy x-ray absorptiometry: The effects of beam hardening on bone density measurements. *Medical Physics, 19*, 459–465. https://doi.org/10.1118/1.596834
13. Laskey, M. A. (1996). Dual-energy X-ray absorptiometry and body composition. *Nutrition, 12*, 45–51. https://doi.org/10.1016/0899-9007(95)00017-8
14. Theodorou, D. J., Theodorou, S. J., & Sartoris, D. J. (2002). Dual-energy X-ray absorptiometry in diagnosis of osteoporosis: Basic principles, indications, and scan interpretation. *Comprehensive Therapy, 28*, 190–200. https://doi.org/10.1007/s12019-002-0028-6
15. Blake, G. M., & Fogelman, I. (1997). Technical principles of dual energy X-ray absorptiometry. *Seminars in Nuclear Medicine, 27*, 210–228. https://doi.org/10.1016/S0001-2998(97)80025-6
16. Adam, S. Z., Rabinowich, A., Kessner, R., & Blachar, A. (2021). Spectral CT of the abdomen: Where are we now? *Insights Into Imaging, 12*, 138. https://doi.org/10.1186/s13244-021-01082-7
17. Li, K., Li, Y., Qi, Z., Garrett, J. W., Grist, T. M., & Chen, G.-H. (2021). Quantitative lung perfusion blood volume using dual energy CT–based effective atomic number (Z_{eff}) imaging. *Medical Physics, 48*, 6658–6672. https://doi.org/10.1002/mp.15227
18. Taguchi, K., & Iwanczyk, J. S. (2013). Vision 20/20: Single photon counting x-ray detectors in medical imaging. *Medical Physics, 40*, 100901. https://doi.org/10.1118/1.4820371
19. Taguchi, K., Stierstorfer, K., Polster, C., Lee, O., & Kappler, S. (2018). Spatio-energetic cross-talk in photon counting detectors: Numerical detector model (PcTK) and workflow for CT image quality assessment. *Medical Physics, 45*, 1985–1998. https://doi.org/10.1002/mp.12863
20. Willemink, M., J., Persson, M., Pourmorteza, A., Pelc, N. J., & Fleischmann, D. (2018). Photon-counting CT: Technical principles and clinical prospects. *Radiology, 289*, 293–312. https://doi.org/10.1148/radiol.2018172656
21. Leng, S., Bruesewitz, M., Tao, S., Rajendran, K., Halaweish, A. F., Campeau, N. G., Fletcher, J. G., & McCollough, C. H. (2019). Photon-counting detector CT: System design and clinical applications of an emerging technology. *Radiographics, 39*, 729–743. https://doi.org/10.1148/rg.2019180115
22. Danielsson, M., Persson, M., & Sjölin, M. (2021). Photon-counting x-ray detectors for CT. *Physics and Medicine and Biology, 66*, 03TR01. https://doi.org/10.1088/1361-6560/abc5a5
23. McCollough, C. H., Rajendran, K., Leng, S., Lifeng, Y., Fletcher, J. G., Stierstorfer, K., & Flohr, T. G. (2023). The technical development of photon-counting detector CT. *European Radiology, 33*, 5321–5330. https://doi.org/10.1007/s00330-023-09545-9
24. Kreisler, B. (2022). Photon counting detectors: Concept, technical challenges, and clinical outlook. *European Journal of Radiology, 149*, 110229. https://doi.org/10.1016/j.ejrad.2022.110229
25. Hiroaki, H. (2017). Chapter 4. In S. Reza (Ed.), *Response function of two-dimensional CdTe detector, Semiconductor radiation detectors, technology, and applications* (1st ed., pp. 85–108). CRC Press. ISBN:9781138710344.
26. Hiroaki, H., Natsumi, K., Takashi, A., Takumi, A., Cheonghae, L., & Akitoshi, K. (2021). Chapter 1–5. In *Photon counting detectors for X-ray imaging: Physics and applications* (pp. 1–119). Springer. ISBN: 978-3-030-62679-2.
27. Natsumi, K., Hiroaki, H., Cheonghae, L., Tatsuya, M., & Akitoshi, K. (2022). Chapter 4. In K. Iniewski (Ed.), *Algorithm for generating effective atomic number, soft-tissue, and bone images based on analysis using an energy-resolving photon-counting detector, advanced X-ray radiation detection: Medical imaging and industrial applications* (pp. 75–106). Springer. ISBN: 978-3-030-92988-6.

28. Natsumi, K., Hiroaki, H., Cheonghae, L., Tatsuya, M., Kobayashi, D., Nishigami, R., & Katsumata, A. (2023). Chapter 11. In S. Hsieh & K. Iniewski (Eds.), *Quantitative analysis methodology of X-ray attenuation for medical diagnostic imaging: Algorithm to derive effective atomic number and soft tissue and bone images, photon counting computed tomography: Clinical applications, image reconstruction and material discrimination* (pp. 219–245). Springer. ISBN: 978-3-031-26061-2.
29. Natsumi, K., Hayashi, H., Takashi, A., Yoshiki, M., Kanazawa, Y., Yamakawa, T., Yamamoto, S., Yamasaki, M., & Okada, M. (2017). Precise material identification method based on a photon counting technique with correction of the beam hardening effect in X-ray spectra. *Applied Radiation and Isotopes, 124*, 16–26. https://doi.org/10.1016/j.apradiso.2017.01.049
30. Kimoto, N., Hayashi, H., Asakawa, T., Lee, C., Asahara, T., Maeda, T., Goto, S., Kanazawa, Y., Katsumata, A., Yamamoto, S., & Okada, M. (2021). Effective atomic number image determination with an energy-resolving photon-counting detector using polychromatic X-ray attenuation by correcting for the beam hardening effect and detector response. *Applied Radiation and Isotopes, 170*, 109617. (14 pages. https://doi.org/10.1016/j.apradiso.2021.109617
31. Natsumi, K., Hiroaki, H., Cheonghae, L., Tatsuya, M., Miku, A., Yuki, K., Akitoshi, K., Shuichiro, Y., & Masahiro, O. (2021). A novel algorithm for extracting soft-tissue and bone images measured using a photon-counting type X-ray imaging detector with the help of effective atomic number analysis. *Applied Radiation and Isotopes, 176*, 109822. (12 pages). https://doi.org/10.1016/j.apradiso.2021.109822
32. Kobayashi, D., Hayashi, H., Nishigami, R., Maeda, T., Asahara, T., Kanazawa, Y., Katsumata, A., Kimoto, N., & Yamamoto, S. (2024). A blurring correction method suitable to analyze quantitative X-ray images derived from energy-resolving photon counting detector. *Physics in Medicine and Biology, 69*, 075023. https://doi.org/10.1088/1361-6560/ad3119
33. Moy, J. P. (2000). Signal-to-noise ratio and spatial resolution in x-ray electronic imagers: Is the MTF a relevant parameter? *Medical Physics, 27*, 0094–2405. https://doi.org/10.1118/1.598859
34. Ohkubo, M., Wada, S., Kunii, M., Matsumoto, T., & Nishizawa, K. (2008). Imaging of small spherical structures in CT: Simulation study using measured point spread function. *Medical and Biological Engineering and Computing, 46*, 273–282. https://doi.org/10.1007/s11517-007-0283-x
35. Tromans, C. E., Cocker, M. R., & Brady, S. M. (2012). A model of primary and scattered photon fluence for mammographic x-ray image quantification. *Physics in Medicine and Biology, 57*, 6541–6570. https://doi.org/10.1088/0031-9155/57/20/6541
36. Kueh, A., Warnett, J. M., Gibbons, G. J., Brettschneider, J., Nichols, T. E., Williams, M. A., & Kendall, W. S. (2016). Modelling the penumbra in computed tomography 1. *Journal of X-Ray Science and Technology, 24*, 583–597. https://doi.org/10.3233/XST-160576
37. Ramponi, G. (1998). A cubic unsharp masking technique for contrast enhancement. *Signal Processing, 67*, 211–222. https://doi.org/10.1016/S0165-1684(98)00038-3
38. Brettle, D., & Carmichael, F. (2011). The impact of digital image processing artefacts mimicking pathological features associated with restorations. *British Dental Journal, 211*, 167–170. https://doi.org/10.1038/sj.bdj.2011.676
39. Trentacoste, M., Mantiuk, R., Heidrich, W., & Dufrot, F. (2012). Unsharp masking, countershading and halos: Enhancements or artifacts? *Computer Graphics Forum, 31*, 555. https://doi.org/10.1111/j.1467-8659.2012.03056.x
40. Clark, J. L., Wadhwani, C. P., Abramovitch, K., Rice, D. D., & Kattadiyil, W. T. (2018). Effect of image sharpening on radiographic image quality. *The Journal of Prosthetic Dentistry, 120*, 927–933. https://doi.org/10.1016/j.prosdent.2018.03.034
41. Mohd, S. V., Sagheer, S. N., & George. (2020). A review on medical image denoising algorithms. *Biomedical Signal Processing and Control, 61*, 102036. https://doi.org/10.1016/j.bspc.2020.102036
42. Nishigami, R., Hayashi, H., Kobayashi, D., Maeda, T., Asahara, T., Kanazawa, Y., Kimoto, N., & Yamamoto, S. (2024). Applicability of high tube voltage imaging to achieve accurate quantitative images when applying photon counting detectors to general radiography. *Proceedings of SPIE.*

43. Nishigami, R., Hayashi, H., Kobayashi, D., Lee, C., Maeda, T., Asahara, T., Kanazawa, Y., Kimoto, N., & Yamamoto, S. (2023). Simulation study on functional images to optimize radiographic condition using energy resolving photon counting detector. *Proceedings Of IEEE*. https://doi.org/10.1109/NSSMICRTSD49126.2023.10338161
44. Tucker, D. M., Barnes, G. T., & Chakraborty, D. P. (1991). Semiempirical model for generating tungsten target x-ray spectra. *Medical Physics, 18*, 211–218. https://doi.org/10.1118/1.596709
45. Katsumata, A., Kohinata, K., Esaki, Y., & Kawai, M. (2022). Variance of radiographical alveolar bone mineral density by the anatomical morphology of mandibular bone. *Heliyon, 8*, e11507. https://doi.org/10.1016/j.heliyon.2022.e11507
46. Yamashita, K., Tamaki, Y., Nakajima, D., Omichi, Y., Takahashi, Y., Takai, M., Goto, T., Hayashi, H., Higashino, K., Tsuruo, Y., & Sairyo, K. (2023). A cadaveric simulation study of radiation exposure to the surgical team during fluoroscopic spinal surgery: How much are we exposed? *Spine Surgery and Related Research, 7*, 341–349. https://doi.org/10.22603/ssrr.2022-0184
47. Raisz, L. G., & Seeman, E. (2001). Causes of age-related bone loss and bone fragility: An alternative view. *Journal of Bone and Mineral Research, 16*, 1948–1952. https://doi.org/10.1359/jbmr.2001.16.11.1948

Index

K. Iniewski (ed.), *CdTe and CdZnTe Materials*,
https://doi.org/10.1007/978-3-031-64521-1

SPRINGER NATURE

GPSR Compliance

The European Union's (EU) General Product Safety Regulation (GPSR) is a set of rules that requires consumer products to be safe and our obligations to ensure this.

If you have any concerns about our products, you can contact us on ProductSafety@springernature.com

In case Publisher is established outside the EU, the EU authorized representative is:

Springer Nature Customer Service Center GmbH
Europaplatz 3
69115 Heidelberg, Germany

The manufacturer's authorised representative in the EU is Springer Nature Customer Service Centre GmbH, Europaplatz 3, 69115 Heidelberg, Germany. If you have any concerns regarding our products, please contact ProductSafety@springernature.com

Printed and bound by CPI Group (UK) Ltd, Croydon, CR0 4YY
28/09/2026
02224081-0010